Pathways to African Food Security

This book examines Africa's current food system and future challenges for food security over the next 25 years.

Africa is on the rise and by 2050, the continent will be home to a quarter of the world's population. The analysis presented in this book clearly shows that the African food system needs to transform at a much faster pace to ensure that the people it serves are food secure. This book begins with four contrasting case studies that focus on country-specific challenges in Egypt, Ethiopia, Senegal and Zimbabwe. This is followed by 15 thematic chapters organised in three sections on challenges, threats and opportunities. Individual chapters address a wide range of topics including climate change, water security, farm sizes, crop yields, conservation trade-offs, food prices, trade, conflict and structural change. The book concludes by discussing key pathways to improve Africa's food system and food security for the decades ahead.

This book is essential reading for students, scholars and practitioners who work on global food security, sustainable food systems, food, health and nutrition and African development.

Michiel de Haas is Assistant Professor of Economic and Environmental History at Wageningen University and Research, the Netherlands.

Ken E. Giller is Emeritus Professor of Plant Production Systems at Wageningen University and Research, the Netherlands.

Earthscan Food and Agriculture

Globalisation and Livelihood Transformations in the Indonesian Seaweed Industry
Edited by Zannie Langford

Principles of Sustainable Aquaculture
Promoting Social, Economic and Environmental Resilience, 2nd edition
Stuart W. Bunting

The Spatial Organization of Urban Agriculture in the Global South
Food Security and Sustainable Cities
Ada Górna

Digital Communication for Agricultural and Rural Development
Participatory Practices in a Post-COVID Age
Edited by Ataharul Chowdhury and Gordon Gow

Regenerative Farming and Sustainable Diets
Human, Animal and Planetary Health
Edited by Joyce D'Silva and Carol McKenna

Sustainable Cacao Cultivation in Latin America
Edited by Luz Cecilia García, Naga Raju Maddela, Freddy Zambrano Gavilanes and Carolina Aguilar Duarte

Food Policy in the United States
An Introduction, 3rd edition
Parke Wilde

Agricultural and Climate Change Adaptation Law in Africa
Reflections from Kenya, Nigeria and South Africa
Habib Sani Usman

Pathways to African Food Security
Challenges, Threats and Opportunities towards 2050
Edited by Michiel de Haas and Ken E. Giller

For more information about this series, please visit: https://www.routledge.com/Earthscan-Food-and-Agriculture/book-series/ECEFA

Pathways to African Food Security

Challenges, Threats and Opportunities towards 2050

Edited by
Michiel de Haas and Ken E. Giller

Routledge
Taylor & Francis Group
LONDON AND NEW YORK

earthscan
from Routledge

Cover image: Ken E. Giller

First published 2025
by Routledge
4 Park Square, Milton Park, Abingdon, Oxon, OX14 4RN

and by Routledge
605 Third Avenue, New York, NY 10158

Routledge is an imprint of the Taylor & Francis Group, an informa business

We thank the Dutch Research Council/Nederlandse Organisatie voor Wetenschappelijk Onderzoek (NWO) for funding the Open Access publication.

British Library Cataloguing-in-Publication Data
A catalogue record for this book is available from the British Library

ISBN: 978-1-032-64967-2 (hbk)
ISBN: 978-1-032-64965-8 (pbk)
ISBN: 978-1-032-64969-6 (ebk)

DOI: 10.4324/9781032649696

Typeset in Times New Roman
by codeMantra

Contents

List of figures ix
List of tables xiii
List of contributors xv
Preface xix

Introduction 1

1 **Food systems and food security in Africa: looking back and projecting forwards** 3
MICHIEL DE HAAS, KATRIEN DESCHEEMAEKER, PETER EBANYAT, EWOUT FRANKEMA, KEN E. GILLER, IBRAHIMA HATHIE, AND LEO VAN WISSEN

PART 1
Diverse pathways on the country level: past, present and future 21

2 **Senegal's journey to feed its growing population by 2050: pathways to accelerated food systems transformation** 23
IBRAHIMA HATHIE

3 **Feeding a fast-growing population by 2050 through accelerated agricultural transformation: potential avenues for Ethiopia** 35
TESFAYE SHIFERAW SIDA, TEMESGEN DESALEGN, GASHAW T. ABATE, AND TILAHUN AMEDE

4 **Past, present and future of food security in Egypt** 49
RACHA RAMADAN AND ANTOINE CASTET

5 **Zimbabwe's agriculture and food security: past, present and future (1960–2050)** 59
MARK NYANDORO AND JENS A. ANDERSSON

PART 2
Challenges 73

6 **Africa's demographic challenge** 75
LEO VAN WISSEN

7 **From yield and nutrient gaps to assessing future food self-sufficiency in Africa** 89
ANTONIUS G.T. SCHUT AND MARTIN K. VAN ITTERSUM

8 **Global market and food security in Africa: more trade or more domestic production?** 100
SIEMEN VAN BERKUM AND BART DE STEENHUIJSEN PITERS

9 **Why is food expensive in sub-Saharan African cities?** 111
EWOUT FRANKEMA

10 **Addressing malnutrition in all its forms in Africa requires a radical paradigm shift** 124
ALIDA MELSE-BOONSTRA, INGE D. BROUWER, FOLAKE SAMUEL, AND MOFU MUSONDA

PART 3
Threats 137

11 **Political instability and food security: the long view towards 2050** 139
HAN VAN DIJK AND NAKAR SYNTYCHE DJINDIL

12 **Climate change: impacts and adaptation in smallholder farming systems in sub-Saharan Africa** 150
KATRIEN DESCHEEMAEKER, PATRICIA MASIKATI, AND DILYS SEFAKOR MACCARTHY

13 **Reconciling food security and forest conservation in Africa: challenges and opportunities** 160
MARIEKE SASSEN

14 **Food security for Africa's growing population: a labour market perspective** 175
JANNEKE PIETERS

15 **How small is beautiful? Farm size and economic development in Africa** 187
KEN E. GILLER AND JENS A. ANDERSSON

PART 4
Opportunities 199

16 Transformative trees: forests for water and food security 201
DOUGLAS SHEIL

17 A water angle on Africa's 2050 food security challenge 213
GERT JAN VELDWISCH, JONATHAN DENISON, GABRIELLA IZZI, JEAN KAMWAMBA-MTETHIWA, HANS KOMAKECH, BANCY MATI, AND PIETER WAALEWIJN

18 Harnessing biotechnology to release the potential of the bioeconomy for Africa 223
ENOCH KIKULWE AND JUSTUS WESSELER

19 The multifunctional role of livestock in East African food systems: the case for dairy 232
SIMON J. OOSTING, TODD A. CRANE, AN NOTENBAERT, ADEMOLA BRAIMOH, ASAAH NDAMBI, AUGUSTINE AYANTUNDE, ESTHER KIHORO, CORINA VAN MIDDELAAR, AND JAN VAN DER LEE

20 Extending the smallholder value-creation frontier: a business perspective on food security 243
PAUL T.M. INGENBLEEK, OLAWALE ROTIMI OPEYEMI, DOROTHY KANORIO MURUGU, COBUS OBERHOLSTER, AND LUCAS E. URBANO

21 Agricultural development in the poorest countries: insights from the poultry sector in Sierra Leone 251
ERWIN BULTE, NICCOLÒ MERIGGI, MICHAEL ROZELLE, AND MAARTEN VOORS

Conclusion 265

22 Growing towards a food-secure Africa in 2050: reflections and pathways 267
MICHIEL DE HAAS, EWOUT FRANKEMA AND KEN E. GILLER

Index *281*

Figures

1.1 Correlation between log GDP per capita and food security at country scale 6
1.2 Agricultural employment and food security in the early 2020s 7
1.3 Cereal yields and food security in the early 2020s 9
1.4 Change in cereal yield and the area cropped with cereals, 1965–2020 10
1.5 Cereal yield and agricultural employment share in 48 African countries, Bangladesh and Vietnam in 2021, compared to averages for Africa (solid lines) and low- and middle-income countries (LMICs) (dotted lines) 13
1.6 Cereal yield and agricultural employment changes in 48 African countries, Bangladesh and Vietnam, 1991–2021, compared to averages for Africa (solid lines) and low- and middle-income countries (LMICs) (dotted lines) 14
2.1 Development of the rural and total population (1960–2050) 25
2.2 (a) Share of cereal crops per decade (1961–2020); (b) cereal production per decade (1961–2020) 28
2.3 (a) Evolution of cereal yields in Senegal (1961–2021); (b) ranking among African countries in terms of cereal yields 28
2.4 Food import dependency and food net import dependency ratios for Senegal (1961–2021) 29
3.1 Relationship between population growth and change in cereal production between 1961 and 2022 (a) and cereal production per capita (b) 39
3.2 Historical and projected cereal production and deficit in Ethiopia (1961–2050). BAU stands for business-as-usual 40
3.3 Land size per farm and average cereal productivity for non-clustered typical smallholder farms across four crops and from four states in Ethiopia 41
3.4 Land size per farm and average cereal productivity for clustered farms organized under the Agricultural Commercialization Clusters (ACC) implemented by the Ethiopian government across four crops and in four states in Ethiopia 42
4.1 Agriculture area in Egypt (% of total land area), 1961–2022 50
4.2 Agricultural production of the main crops products (tons) 52
4.3 The quantity of wheat imports since 1961 52

5.1 Import dependency ratio = imports/(production + imports - exports) and net import dependency ratio = (imports - exports)/(production + imports - exports) for (colonial) Zimbabwe, 1961–2020 61
5.2 Maize production (× 1,000 tonnes) of commercial (blue-dotted line) and smallholder (red-dotted line) farming sectors and five-year moving averages, 1970–2000 63
5.3 Estimated percentage share of different farming sub-sectors in national maize output, 2010–2017 67
6.1 Total population estimates and medium variant projections by World Regions 1950–2050 76
6.2 Relative (left axis) and absolute (right axis) population growth of Asia and Africa compared 77
6.3 Total, urban, and rural relative population growth (1990–2020) of African countries 82
6.4 Population growth and urban and rural contributions to population growth (1990–2020) of African countries 83
6.5 Total, urban, and rural relative population growth (2020–2050) of African countries 85
7.1 Cereal self-sufficiency ratios of ten selected countries in 2050 in relation to current farmers' yields (Ya) for scenarios where current yield trends are extrapolated and where yield gaps are close to 50 or 80% of water-limited yield (Y_W) under rainfed conditions 92
8.1 Wheat price developments 2002–2023 (nominal terms) 101
8.2 Daily caloric contribution and net import dependency (%) of the five most important staple foods in 19 African countries 103
9.1 Relative food prices versus per capita GDP, 2017 112
9.2 Monthly wholesale maize prices in Nairobi (Kenya), Lagos (Nigeria) and Chicago (USA), in US$/kg, 2000–2023 115
9.3 Real-world market price index for an unweighted basket of maize, rice and wheat, 1846–2021 (1900 = 100) 117
10.1 Essential elements for accelerating the systemic transformation of food systems 133
11.1 Adults height by cohort and village of individuals born between 1960 and 1985 145
13.1 Forest-food linkages: forest for crop and grazing land, non-timber forest products (e.g. mushrooms, fish, medicine), water regulation, fuelwood, and tools 161
13.2 Projected habitat loss for all species as a result of projected changes in land use between 2005 and 2050 in Africa under a "strong regional integration but reactive governance" scenario with climate change as a given, which is considered a plausible scenario for the region 166
13.3 The forest transition curve: an example for moist forest areas in Africa 169
14.1 Employment in agriculture, in low- and middle-income countries in sub-Saharan Africa 175
14.2 Self-reported intended destination of output from family farming 176
14.3 Share of total employment in agriculture and manufacturing, 2000 and 2015 178
14.4 Share of wage and salaried workers in total employment, 2000 and 2015 179
15.1 Density distributions of farm size across different agro-ecological zones for selected countries in sub-Saharan Africa 189

15.2 Remittance inflows, percentage of GDP and percentage of the population dependent on (inter)national remittances 194
16.1 Modelled distribution of closed forest (1900–2000) based on a modelling approach using various reference data 202
16.2 Schematic relationship between tree cover and groundwater recharge 204
16.3 Proportion of rainfall recycled from land by where it ends up (a) and where it comes from (b) and the inferred role of forest (c). (a, b) Annual average continental precipitation recycling ratio in which darker colours denote that more than 50% of this rainfall moisture is derived from land 206
16.4 The estimated sources by volume contributed to the mean annual precipitation across major African watersheds (1981–2016). The size of each pie chart represents the magnitude of the mean annual precipitation 207
18.1 Food, feed, fibre, fuel, and further materials (4+1Fs) in the bioeconomy 224
21.1 Percentage of the population that is "not able to afford a healthy diet" 251
21.2 Protein supply (2000–2019) for selected countries in West Africa and Europe 252
21.3 Chicken production and trade 253
21.4 Changes in consumer food price index for selected countries in West Africa 253
21.5 Static and dynamic analysis of trade measures 258
22.1 Focus areas for food systems change 268

Tables

6.1	Fertility and life expectancy assumptions and resulting projected population size for Africa of UN DESA 2022 population projections	79
6.2	Population projections for sub-Saharan Africa by UN DESA, WCD, and IHME for 2050 and 2100	80
6.3	Demographic characteristics of the four paths: total, urban, and rural population growth (1990–2020) and total fertility rate (TFR) and life expectancy (2020)	81
6.4	Population density (persons per km^2) by continent, 1950, 2020, and 2050	84
8.1	Consumption share (CS) and net import dependence (NID) of the top five staples	103
9.1	Staple food prices in sub-Saharan Africa and South and Southeast Asia compared, 2011 and 2017 (in current US$)	114
10.1	Prevalence of key nutritional outcomes in example countries categorized by their situation along four paths of agricultural and economic change	128
10.2	Food system typologies based on a composite of four indicators	130
18.1	Examples of CRISPR/Cas9-based crops for Africa	226
19.1	Human populations living in farming systems and cattle populations (data from years 2017–2021)	235
19.2	Food system objectives	236
19.3	Roles of (dairy) cattle in farming systems and of dairy farming systems in food systems	237
19.4	Performance of farming systems with (dairy) livestock on food system objectives based on an evaluation done by the authors	238

Contributors

Gashaw T. Abate, International Food Policy Research Institute (IFPRI), Washington DC, USA.

Tilahun Amede, AGRA, Nairobi, Kenya.

Jens A. Andersson, Plant Production Systems, Wageningen University & Research, The Netherlands.

Augustine Ayantunde, Animal Breeding & Genomics, Wageningen University & Research, The Netherlands.

Ademola Braimoh, The World Bank, Washington DC, USA.

Inge D. Brouwer, Division of Human Nutrition and Health, Wageningen University & Research, The Netherlands.

Erwin Bulte, Development Economics, Wageningen University & Research, The Netherlands.

Antoine Castet, UMR Dévelopement & Sociétés, Sorbonne Institute of Development Studies (IEDES), Paris 1 Panthéon-Sorbonne University, Paris, France.

Todd Crane, International Livestock Research Institute, Nairobi, Kenya.

Michiel de Haas, Economic and Environmental History, Wageningen University & Research, The Netherlands.

Bart de Steenhuijsen Piters, Wageningen Economic Research, Wageningen University & Research, The Netherlands.

Jonathan Denison, World Bank, Washington DC, USA.

Temesgen Desalegn, Ethiopian Institute of Agricultural Research (EIAR), Addis Ababa, Ethiopia.

Katrien Descheemaeker, Plant Production Systems, Wageningen University & Research, The Netherlands.

Nakar Syntyche Djindil, Independent researcher, Ottawa, Canada.

Peter Ebanyat, Department of Agricultural Production, Makerere University, Kampala, Uganda.

Ewout Frankema, Economic and Environmental History, Wageningen University & Research, The Netherlands.

Ken E. Giller, Plant Production Systems, Wageningen University & Research, The Netherlands.

Ibrahima Hathie, Initiative Prospective Agricole et Rural (IPAR), Senegal.

Paul T.M. Ingenbleek, Marketing and Consumer Behaviour, Wageningen University & Research, The Netherlands.

Gabriella Izzi, Lilongwe University of Agriculture and Natural Resources (LUANAR), Department of Land and Water Resources, Lilongwe, Malawi.

Jean Kamwamba-Mtethiwa, World Bank, Washington DC, USA.

Esther Kihoro, International Livestock Research Institute, Nairobi, Kenya.

Enoch Kikulwe, Alliance of Bioversity International and International Center for Tropical Agriculture (CIAT), Nairobi, Kenya.

Hans Komakech, Nelson Mandela African Institution of Science and Technology (NM-AIST), Water and Environmental Science and Engineering Department and Water Infrastructure and Sustainable Energy Futures (WISE-Futures) Centre of Excellence, Arusha, Tanzania.

Dilys Sefakor MacCarthy, School of Agriculture, University of Ghana, Accra, Ghana.

Patricia Masikati, CIFOR-ICRAF, Zambia.

Bancy Mati, Department of Soil, Water and Environmental Engineering, Jomo Kenyatta University of Agriculture and Technology (JKUAT), Nairobi, Kenya.

Alida Melse-Boonstra, Division of Human Nutrition and Health, Wageningen University & Research, The Netherlands.

Niccolò Meriggi, Development Economics, Wageningen University & Research, The Netherlands, Centre for the Study of the African Economies, University of Oxford, United Kingdom, International Growth Centre, London School of Economics and Political Science, United Kingdom.

Dorothy Kanorio Murugu, Jomo Kenyatta University of Agriculture and Technology (JKUAT), Kenya.

Mofu Musonda, Independent consultant, Zambia.

Asaah Ndambi, Animal Breeding & Genomics, Wageningen University & Research, The Netherlands.

An Notenbaert, International Livestock Research Institute, Nairobi, Kenya.

Mark Nyandoro, Economic History, University of Zimbabwe and School of Social Sciences, North-West University, South Africa.

Cobus Oberholster, Livestock, Properties, Auctioneers, BKB, South Africa.

Simon Oosting, Animal Production Systems, Wageningen University & Research, The Netherlands.

Olawale Rotimi Opeyemi, JR Farms, Nigeria.

Janneke Pieters, Development Economics, Wageningen University & Research, The Netherlands.

Racha Ramadan, Faculty of Economics and Political Science, Cairo University and Agence Universitaire de la Francophonie, Cairo, Egypt.

Michael Rozelle, Development Economics, Wageningen University & Research, The Netherlands.

Folake Samuel, Department of Nutrition, University of Ibadan, Nigeria.

Marieke Sassen, Plant Production Systems and Wageningen Environmental Research, Wageningen University & Research, The Netherlands.

Antonius G.T. Schut, Plant Production Systems, Wageningen University & Research, The Netherlands.

Douglas Sheil, Forest Ecology and Forest Management, Wageningen University & Research, The Netherlands.

Tesfaye Shiferaw Sida, International Maize and Wheat Improvement Center (CIMMYT), Addis Ababa, Ethiopia.

Lucas Urbano, Regenerative Agriculture and Sustainable Sourcing, Unilever, The Netherlands.

Siemen van Berkum, Wageningen Economic Research, Wageningen University & Research, The Netherlands.

Jan van der Lee, Animal Breeding & Genomics, Wageningen University & Research, The Netherlands.

Han van Dijk, Sociology of Development and Change, Wageningen University & Research, The Netherlands.

Martin K. van Ittersum, Plant Production Systems, Wageningen University & Research, The Netherlands.

Corina van Middelaar, Animal Production Systems, Wageningen University & Research, The Netherlands.

Leo van Wissen, Netherlands Interdisciplinary Demographic Institute (NIDI), and University of Groningen, The Netherlands.

Gert Jan Veldwisch, Water Resources Management, Wageningen University & Research, The Netherlands.

Maarten Voors, Development Economics, Wageningen University & Research, The Netherlands.

Pieter Waalewijn, World Bank, Washington DC, USA.

Justus Wesseler, Agriculture Economics and Rural Policy, Wageningen University & Research, The Netherlands.

Preface

Africa is on the rise. By 2050, the continent will be home to a quarter of the world's population. Whereas discussions of future food security take place in the context of populations that are shrinking in China and many wealthier economies, and plateauing in South Asia, Africa's population will continue to grow rapidly in the foreseeable future. This demographic ascendancy, coupled with urbanisation and changes in income and consumption patterns, demands a transformation of agri-food systems. Thus, the question how Africa's growing population will achieve food security in 2050 and beyond should be firmly on the table.

The idea for this book arose from a series of discussions reaching out across disciplinary divides. When we initiated this book, Michiel had just finished an edited volume on African migration history, and had embarked on a research project about long-run African agricultural trade and prices, funded by the Dutch Research Council (NWO). This prompted questions about the profound ongoing shift from an abundance to scarcity of land and resources and its consequences for food security. Ken was working on a generous grant from the NWO-CGIAR Strategic Expert Programme, which freed up his time to explore questions of equity in current African smallholder farming systems. When sharing experiences and ideas, we continually confronted each other with questions about how our analysis of the past and current trends would pan out in the future. We realised that a dispassionate, open and interdisciplinary conversation about the challenge of achieving future food security in Africa is crucial. Thus, we set out to broaden the discussion by involving the co-authors of the volume's introduction and experts on a wide range of relevant topics based at our own university and other institutions across Africa and Europe.

Our invitation to contribute was met with enthusiasm. Initial ideas for essays were presented and debated at a workshop held in Wageningen in November 2022 with support from the Wageningen School of Social Sciences. From these discussions, it became clear that food systems and food security cannot be studied, let alone governed, in isolation but are embedded in broader processes of ecological, economic and demographic change. Recognising this context and working through the many complexities and contradictions that it entails might be frustrating for policy makers working on food systems change – let alone transformation – but it is vital for pursuing and ultimately achieving durable improvements. Meanwhile, the opposite is also true: economic growth will not automatically translate into a sustainable improvement of nutrition security, and the many challenges and trade-offs require deliberate and effective governance and decisive political leadership. We challenged authors to be free-thinking in their visions concerning Africa's food security challenge towards 2050.

While individual country experiences vary, and sub-national variation within countries is perhaps even greater, we believe that thinking about Africa's food security challenge at a continental scale can be productive for three reasons. The first reason is pragmatic. The reality is that policy makers, donors and the general public across the globe already frequently refer to Africa as if it were a region with a high degree of coherence and unity. Relevant institutions, projects and policies, with regard to trade, agriculture or migration, target (sub-Saharan) Africa. Taking a continental scope means that we can link to such discourses, while providing more disaggregated and case-based perspectives where needed. Our selection of four country chapters highlights this diversity. Other chapters also provide a multitude of contrasts and country examples. Second, a continental approach encourages comparative thinking, reveals the diversity of experiences among countries, regions and households and has the potential to map patterns and challenges at an appropriate spatial level. Many of the book's chapters take such a comparative approach, explicitly or implicitly. Third, the continent's diversity may prove crucial towards finding solutions, as countries can learn from each other's experiences and develop synergies, for example, through intra-African migration, economic specialisation and (food) trade. Some of this potential for synergy is woven into the book's chapters and emerges from its totality and the closing chapter.

All chapters have been critically read by at least two reviewers and by both of us as editors, in some cases over several rounds. This process has been rewarding in terms of our learning and has led to the identification of inconsistencies among arguments presented in different chapters that have deepened our understanding. We are grateful for the authors' commitment to the project and their willingness to contribute some of their core ideas and insights. The whole book was subjected to independent peer review by the publisher, fulfilling a condition for a grant to fund the open-source, online publication, for which we thank NWO. We also thank our counterparts at Routledge for their professional guidance through the publication process. Finally, for contributing to the project and participating in our workshop, we thank Abigail Fagbohun, Betelhem Negede, Clare Nakimpi, Evans Osabuohien, Gerrit-Jan van Uffelen, Gonne Beekman, Stephen Adaawen, Tanik Joshipura and Wouter van der Weijden.

As we were finalising the complete manuscript, the 2024 revision of the World Population Prospects was released (UN-DESA, 2024). For consistency, we have chosen to maintain reference to the 2022 projections as the differences in assumptions and outcomes for the African continent are relatively minor, with slightly less projected growth according to the latest projections (about 20 million less in 2050 and 100 million less in 2100).

We have learned a huge amount over the past two years and express our sincere gratitude to all of the contributors and reviewers for their intellectual stimulation and their patience with our repeated questions and comments. We hope that this volume will catalyse further discussion, debate and action towards a food secure Africa by 2050.

Michiel de Haas & Ken E. Giller
Wageningen
02/09/2024

Reference

UN-DESA. (2024). *World Population Prospects 2024 Revision*.

Introduction

1 Food systems and food security in Africa: Looking back and projecting forwards

Michiel de Haas, Katrien Descheemaeker, Peter Ebanyat, Ewout Frankema, Ken E. Giller, Ibrahima Hathie, and Leo van Wissen

1 A growing challenge

Achieving food and nutrition security for Africa's growing population is a pressing global challenge and a crucial pillar of prosperity, well-being and development. While political awareness of this challenge, in Africa and beyond, has grown in recent years, a resolution remains nowhere in sight. African agri-food systems are fragile and strained by growing domestic demand, large inequities, intensifying resource constraints and numerous governance and supply-chain challenges (Giller et al., 2021a). Universal access to healthy and sustainable diets remains far from achieved (FAO, IFAD, UNICEF, WFP & WHO, 2023).

Despite recurrent alarmism about an "African food crisis" (Berry, 1984; Jaeger, 1992; Devereux, 2009; Sasson, 2012; Bjornlund et al., 2022), in the aggregate and in the long term, the supply of basic nutrients has kept pace with the increase in food demand from rapidly growing and urbanizing populations. Since 1960, the *per capita* supply of calories, proteins and fats in Africa grew steadily, albeit much slower than in other world regions (FAO, 2024). Over the past decades, in most African countries, child mortality has declined, and life expectancy has increased substantially (UN DESA, 2022).

However, in the past decade, trajectories towards sufficient and nutritionally adequate diets have stagnated and food insecurity has affected growing numbers of people. Since about 2010, after several decades of steady improvement, the number and share of undernourished people in the continent as a whole has again increased, from 159 million (15.1%) to 282 million (19.7%) in 2022 (FAO, IFAD, UNICEF, WFP & WHO, 2023, p. 10). In the latter year, 342 million Africans experienced "severe food insecurity", and an additional 527 million "moderate food insecurity". Jointly, these two categories comprise 61% of Africa's total population, up from 45% in 2015 (FAO, IFAD, UNICEF, WFP, and WHO, 2023, p. 20).

Economic growth, which was promising in many African countries until 2008, has also slowed, deflating hopes for a rapid and job-generating take-off (Frankema & Van Waijenburg, 2018). This in turn limits the potential for investing in the transformation of food systems, despite numerous policy declarations and commitments by African governments. Moreover, recent years have seen a wave of civil strife in the Sahel, the Sudan, Central Africa and the Horn of Africa, which undermines effective governance, and compounds food security challenges, both today and for the generations growing up in conditions of scarcity and insecurity.

Meanwhile, the continent is still undergoing a demographic shift of a magnitude and speed that is unprecedented in human history (Paice, 2021). In 1960, 9% of the world's population lived in Africa, compared to 17% in 2020 (UN DESA, 2022). Over this period, Africa's population grew sixfold, about double the rate of global population growth. A fifth of global population growth during this period took place in Africa. According to the "medium variant" predictions

DOI: 10.4324/9781032649696-2

of the United Nations, Africa's share of the global population will have risen to 25% in 2050, and 38% in 2100, although the latter prediction is highly uncertain. In the three decades until 2050, Africa's population is projected to grow by 83%, and by 2100 it is expected to have tripled compared to 2020. Between 2020 and 2050, 1.88 billion people will be added to the world's population, of which 1.12 billion in Africa (UN DESA, 2022). By 2050, Nigeria is expected to be the world's third most populous country (currently sixth), with 377 million inhabitants, while the Democratic Republic of the Congo (DRC) (217 million) and Ethiopia (215 million) will also enter the top ten (UN DESA, 2022).

Urbanization has also been rapid and is projected to continue apace in the coming decades. Africa's urban population share increased from 19% in 1960 to 44% in 2020, corresponding to an 11-fold absolute expansion of Africa's urban population. In East Africa, which had the lowest initial urban population share, the increase was 20-fold (UN DESA, 2018).[1] Most future population growth will take place in urban areas (Van Wissen, this volume). It is projected that by 2050 59% of all Africans will be living in cities (UN DESA, 2018). This is driven by rural-urban migration, but even more by endogenous urban growth (high per capita fertility) resulting from cities' young populations, and by the conversion of former towns and rural areas into new cities (Menashe-Oren & Bocquier, 2021; Meier zu Selhausen, 2022). The challenges this expansion poses for urban food systems are daunting (Battersby, 2017; Blekking et al., 2020).

The scale of population growth in Africa and a recent slowing of fertility decline (Bongaarts, 2017; Schoumaker, 2019) have incited widespread anxieties about mass exodus, starvation and environmental disaster. Such fears are compounded by the adverse effects of land degradation and climate change, the reality of persistent poverty and resurgent conflict in large parts of the continent (Hall et al., 2017). Africa's demographic transition also unfolds at a time when concerns about the breach of planetary boundaries are urgent (Steffen et al., 2015).

Certainly, the repercussions of Africa's ongoing demographic transition are more multifaceted and uncertain than as portrayed in alarmist projections. Population growth might even deliver "demographic dividends" (Paice, 2021). Yet it is beyond doubt that Africa's demographic ascendancy brings major challenges for achieving sustainable reductions of poverty and food insecurity. The reasons for this are at least threefold. First, as Africa's share of the world population grows, this also implies that the share of the world population growing up under food-insecure conditions increases. Second, not only for basic numerical reasons (more mouths to nourish), but also because of a hoped-for reduction of the incidence of poverty (resulting in expanding and diversifying diets), most of the additional global demand for food in 2050 will have to reach African consumers. Third, population growth itself brings new challenges of resource pressure, land fragmentation and urban and youth unemployment, each of which places further stresses on food systems and food security. Moreover, these resource pressures rise in a context of high insecurity about the long-run effects of climate change and biodiversity loss on Africa's food production potential.

This introductory chapter unpacks the challenge of transforming Africa's agri-food systems to achieve food security for a rapidly growing population. We first discuss how food systems and security are embedded in complex and slow-moving economic and political systems. We then move to argue that both major domestic agricultural gains and food imports are necessary for providing sufficient and adequate food supply. Next, we plot African countries along two axes that drive food security outcomes: intensification (proxied by cereal yields) and structural change (proxied by the agricultural workforce share). We show that gains on both dimensions over the past decades have been slow. We argue that a change of pace is needed, but that the challenges faced in both agriculture and the broader economic context are large, and jeopardized by a range of threats. Understanding these challenges and threats and identifying potential opportunities motivate and structure the remainder of this edited volume, culminating in our concluding chapter, which synthesizes the key findings and insights of this collaborative effort.

2 Food security and food systems

To nourish a rapidly growing number of African consumers in the face of increasing pressure on resources in the coming 30 years, agri-food systems must transform at a much faster pace than has been the case over the past 60 years (Headey & Jayne, 2014; Van Ittersum et al., 2016; Suri & Udry, 2022). The demographic dimensions of this challenge, involving the joint processes of population growth and urbanization, are covered in a chapter by Van Wissen (this volume) This challenge also involves many interlinked and contested interfaces: the development and application of food and agricultural technologies; improved coordination among national, regional and global institutions and policy arenas; the steering of complex decisions about private and public investment; the upgrading of domestic and international supply chains for both agricultural inputs and food; responsivity towards shifting lifestyles and consumer preferences; and adaption to resource scarcity, and climate change and uncertainty. Moreover, African agri-food systems are not self-contained, but are embedded in broader local and global economic, social and political trajectories, as the 2019–2022 Covid-19 pandemic clearly demonstrated (Arndt et al., 2020; Nechifor et al., 2021). All this motivates the interdisciplinary approach of this book.

2.1 What is food and nutrition (in)security?

A growing number of dimensions are recognized to be crucial for food security. People do not just require a set amount of calories, proteins or even nutrients (*availability* and *access*) at the right time (*stability*) and into the future (*sustainability*), but also have preferences (*agency*) and capabilities (*utilization*) which shape the types of food they eat, where and how (Burchi & De Muro, 2016; Clapp et al., 2022). As societies urbanize and economies become more interdependent and complex, consumption patterns change. These changes may affect different social groups, households and individual household members differently. They may not only foster access to healthy and more sustainable diets but also increase the consumption of unhealthy and processed foods (Ambikapathi et al., 2022). Thus providing food and nutrition security for all requires, yet goes far beyond, access to sufficient calories. Healthy diets and a consideration of unequal access are imperative to counter the multiple burdens of malnutrition which already affect many African nations: undernutrition and hidden hunger (deficiencies of micronutrients) on the one hand and overweight and obesity, and their associated diet-related, non-communicable diseases, on the other hand (Reardon et al., 2021). In their chapter, Melse-Boonstra et al. (this volume) explore the nutritional dimensions of Africa's food security challenge further.

Measuring food security is needed to make comparisons across time and space. However, measurement is also complex, given the breadth and multidimensionality of the concept. A host of global food security indicators exist, each with their own priorities, emphasis and underlying variables (Poudel & Gopinath, 2021; Manikas, Ali & Sundarakani, 2023). Several different measures are used throughout this volume, but in the remainder of this introduction we illustrate our main points using two very different indicators. First, we use the Global Food Security Index (GFSI): a broad, composite indicator based on a large number of variables that proxy for food affordability, availability, quality, safety and sustainability (EIU, 2023; Izraelov & Silber, 2019). Second, we use Food and Agriculture Organization (FAO)'s food insecurity measure based on its monitoring instrument, the Food Insecurity Experience Scale (FIES).

2.2 *Can Africa's agri-food systems be "transformed"?*

Food security is underpinned by well-functioning agri-food systems, which contribute to providing a range of other services, including income security and poverty alleviation, climate change mitigation, nature conservation and gender equality. These diverse outcomes are not always aligned, as Oosting et al. (this volume) show for the case of the dairy sector in East Africa, and Sassen (this volume) for the case of forest conservation across the continent. In their chapter, Kikulwe and Wesseler (this volume) make a case that more widespread adoption of biotechnology might help align a wide range of desired outcomes. Agri-food systems encompass the consumers of food, the producers of the agricultural outputs and a wide range of value chain actors linking farm to fork. Suppliers of inputs, finance, knowledge, infrastructure and institutions are also involved in the systems (Reardon et al., 2019; Barrett et al., 2022). In their chapter, Ingenbleek et al. (this volume) provide a business perspective on agri-food systems and argue for the importance of expanding business opportunities even in remote regions.

Food systems also interact with broader human (political, social, economic) and ecological systems, on multiple levels. Van Dijk and Syntyche (this volume) evaluate how one particularly destructive man-made system, that of protracted conflict, disrupts food systems and undermines food security. On a very different note, Sheil (this volume) argues that food production relies crucially on continental moisture systems, in which forests play a vital role. The degree to which food systems are embedded in economic systems is clearly illustrated by the close association of various distinct indicators of food security with income per head of the population (Figure 1.1). How food systems function is also closely related to a host of other factors, including trade freedom, political commitment to climate adaptation and access to agricultural inputs (EIU, 2023). Given these many interactions and interdependencies, food systems transformation is a complex and multifaceted process.

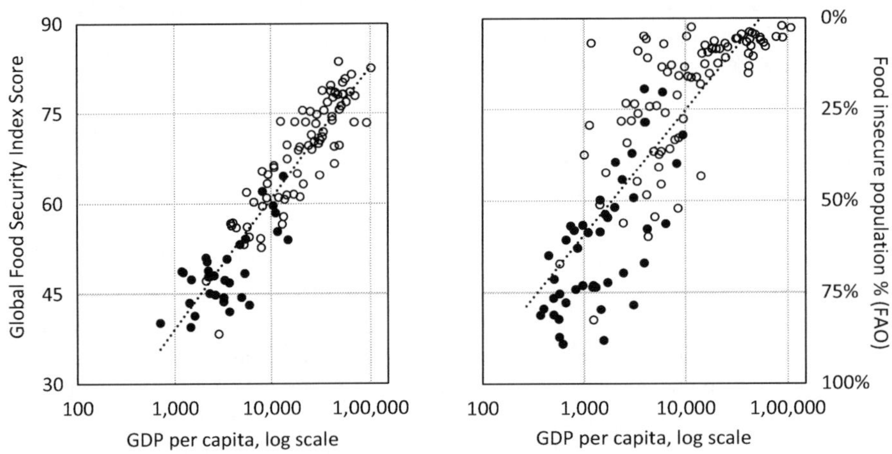

Figure 1.1 Correlation between log GDP per capita and food security at country scale. African countries are shown in solid black, others as open circles.

Sources: Economist Intelligence Unit, 2023, World Bank Development Indicators, 2023; FAO, 2024.

Notes: All countries for which data was available are included. In the left panel, food security is measured using the latest version of the Global Food Security Index (GFSI) (2022), a multidimensional indicator which contains 25 variables covering affordability, availability, quality and safety, and sustainability and adaptation dimensions. GDP per capita (2021) is from the World Bank, using international US dollars at purchasing power parity (2017 constant prices). In the right panel, food security is measured using the FAO (Food and Agriculture Organization) indicator "prevalence of moderate or severe food insecurity in the total population (percent)" for 2020–2022. For this panel, GDP per capita (2021) is from FAOStat and expressed in 2015 constant prices.

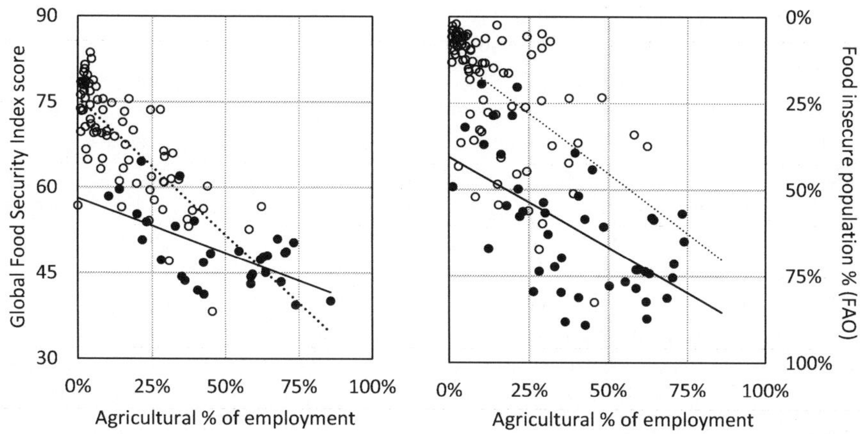

Figure 1.2 Agricultural employment and food security in the early 2020s. African countries are shown as black dots (with a black linear trend line), all other countries as white dots (with a dashed linear trend line).

Sources: The agricultural percentage of employment is based on International Labour Organization (ILO)-modelled estimates for 2021 (World Bank Development Indicators, version 18/12/2023). For the food security measures used, see Figure 1.1.

While food security vitally hinges on agricultural production, there is a strong inverse relationship – both between countries and across time – between the share of the population employed in agriculture and the degree of food security (Giller et al., 2021b). Figure 1.2 shows that while African countries follow this pattern, they also have relatively poor food security outcomes compared to other countries at similar stages of structural change. This applies even to countries that have already achieved low agricultural labour shares, such as Egypt, Senegal and Nigeria. Why is this the case? Food systems face not only distinct challenges but also new opportunities when a large share of the population is non-rural and non-food producing (Battersby, 2017; Barrett et al., 2017; Marshall et al., 2021). Rapid growth of a non-agricultural and urban workforce means lengthening supply chains, and a disentangling of domestic food production and consumption (Barrett et al., 2022). If, in such a situation, a country wants to avoid rising import dependency, farmers inevitably need to grow more (and more diverse) food which must reach the appropriate consumers in a palatable, nutritious, safe and affordable state. This means that feeding a growing share of urban consumers requires not only augmenting agricultural output, but also better-functioning supply chains and storage facilities to connect producers and consumers spatially and temporally (i.e. seasonally) with low transaction costs and limited food waste. Addressing the puzzle why food prices are high in African cities, Frankema (this volume) explores inefficiencies across the entire value chain. Pieters (this volume) explores the particular challenges African nations face in their process of structural transformation, especially when it comes to the provision of non-agricultural jobs.

3 Can domestic agricultural production meet growing demand?

The "raw material" for the food we eat is, and will at least in the coming 30 years remain, largely agricultural in origin (Conceição et al., 2016; Fader et al., 2013; Van Ittersum et al., 2016). Most African nations produce the bulk of agricultural output that ultimately ends up on their citizens' plates in their own rural areas, partly for producers' self-consumption and partly to be marketed through domestic value chains. Such domestic production can be sufficient or fall short of domestic demand.

Is substantial African agricultural output growth necessary to feed growing populations? One alternative would be reduction of food loss and waste along the agri-food supply chain, which can augment total food availability and enhance the food security of producers and consumers (Stathers et al., 2020). However, while the potential direct gains of reducing food loss and waste for total food supply are substantial, they are also uncertain and limited. FAO long estimated cereal waste in Africa to be in the range of 20%–30%, and even higher for other crops, but these are likely overestimates (Reardon et al., 2019). But even if we accept the upper-bound estimates, reducing waste will far from suffice to provide for the projected increased demand as a result of population growth and, especially, urbanization. That said, the indirect effects of reducing loss and waste may be substantial, increasing the profitability of agriculture and food processing and thus generating incentives for farmers and agri-food businesses to invest, intensify and step up production.

Several authors in this volume make the case for stimulating a more diverse, nutritious and resource-efficient crop mix, with a greater focus on local varieties. Again, however, the potential benefits of such a shift for overall food security are uncertain and limited. An example of a promising resource-efficient crop would be cassava, which is more drought-resistant and has substantially higher yields per unit of land than cereals. However, such a switch can only work on a large scale if consumers' diets also change in this direction. This is not happening: the share of cereals versus roots and tubers in Africa has hardly changed over the long run. It is worth recounting Asia's experience in recent decades: as consumers' purchasing power increased, the share of roots and tubers in total consumption has *declined* relative to the consumption of cereals (FAO, 2024; Reardon et al., 2019). A potential shift to benefit nutrition would be from maize, rice and wheat back to traditional coarse grains such as millet and sorghum. Again, however, shifts in both consumption and production have so far pushed in the opposite direction.

Substantial overall agricultural output growth will be an inevitable, primary driver towards achieving food security for Africa's growing population. How can this be achieved? Augmenting agricultural production extensively, through the expansion of cropland, requires little in the way of systemic change and is comparatively easy to achieve from a policy and technological perspective. However, the potential of cropland expansion – without concomitant lifting of labour bottlenecks through substantial capital investment – for the improvement of rural livelihoods, and thus rural access to affordable healthy diets, is limited. Extensive agricultural growth (i.e. agricultural area expansion) is particularly problematic in land-scarce contexts, as it competes with other land uses, and comes with high environmental costs. Thus, in countries facing rapid population growth, growing land pressure and land fragmentation, cropland expansion becomes increasingly untenable (Chamberlin, Jayne & Headey, 2014). But also in cases of low population densities, cropland expansion may have environmental implications, such as the loss of ecosystem services of forests.

Augmenting land productivity through sustainable intensification is viewed by many as the most important direction of change towards greater future food self- sufficiency in Africa (Pretty, Williams & Toulmin, 2012; Conceição et al., 2016; Headey & Jayne, 2014; Binswanger-Mkhize & Savastano, 2017). The increasing use of synthetic fertilizers is essential not only to restore the strongly depleted resource base, but also to prevent further soil degradation (Falconnier et al., 2023). Notably, intensification efforts tend to be uneven across crops, with more calorie-dense staples like maize and rice receiving most attention from agricultural input suppliers, research institutions and policy makers. This means that intensification, although it increases the supply of calories, does not necessarily translate into healthier, more nutritious diets.

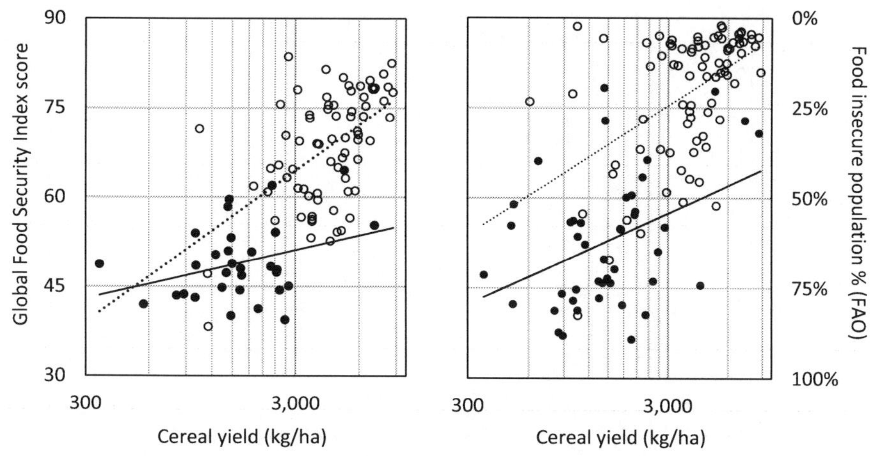

Figure 1.3 Cereal yields and food security in the early 2020s.

Sources: Cereal yields are for 2021 (World Bank Development Indicators, version 18/12/2023). For the food security measures used, see Figure 1.1.

Notes: African countries are shown as black dots (with a black linear trend line), all other countries as white dots (with a dashed linear trend line).

In Figure 1.3, we show how cereal yields correlate with food security. While indicative for broad trends and level differences, several caveats of national-level cereal yield data should be borne in mind: these represent a national average; both actual and potential yields vary substantially according to agro-ecological conditions; yields of non-cereal agricultural output may follow a different pattern; and there are known measurement and reporting gaps and errors. Despite these caveats, the two variables associate positively on a global scale. However, in Africa the relationship is less clear. This suggests that even countries with more productive agricultural sectors have so far not been able to translate these gains into enhanced food security.

Over the past 60 years, gains in land productivity have been key for growing food supply: African cereal yields per hectare more than doubled, increasing by 112% between 1961 and 2020 albeit from a very low base. However, the expansion of land was equally essential. Overall cropland increased by 75%, and cereal area by 118% (FAO, 2024, also Figure 1.2). In a global context, Africa stands out for its slow yield gains (despite also coming from a low starting point) and rapid cropland expansion, as global cereal yields expanded by a factor of 203% while cropland area increased only by 16%, and cereal area by a mere 13% (FAO, 2024; World Bank, 2023, also Giller et al., 2021b). Indeed, many have emphasized that African agricultural output growth has largely been of an extensive nature for most of the 20th and early 21st century (Hinderink & Sterkenburg, 1987; Platteau & Hayami, 1998; Austin, 2008, Giller et al., 2021b). Yet these analyses at a continental scale conceal important differences among African regions (Figure 1.4). Cereal yields in North Africa increased fourfold and those in Southern Africa increased by a factor of 2.5, although the cropland area increased relatively little. By contrast, the cereal yield increases in other regions in 2020 were less than double those in 1961, whilst the area of cropland increased strongly – by a factor of 2.5 in Western Africa, 3.0 in Eastern Africa and 3.6 in Central Africa. In fact, the cereal yield increases in Southern Africa were also much smaller when South Africa is excluded. Essentially, over the past six decades, cropland expansion was by far more important in ensuring that food production kept pace with population growth than yield improvement for sub-Saharan Africa as a whole.

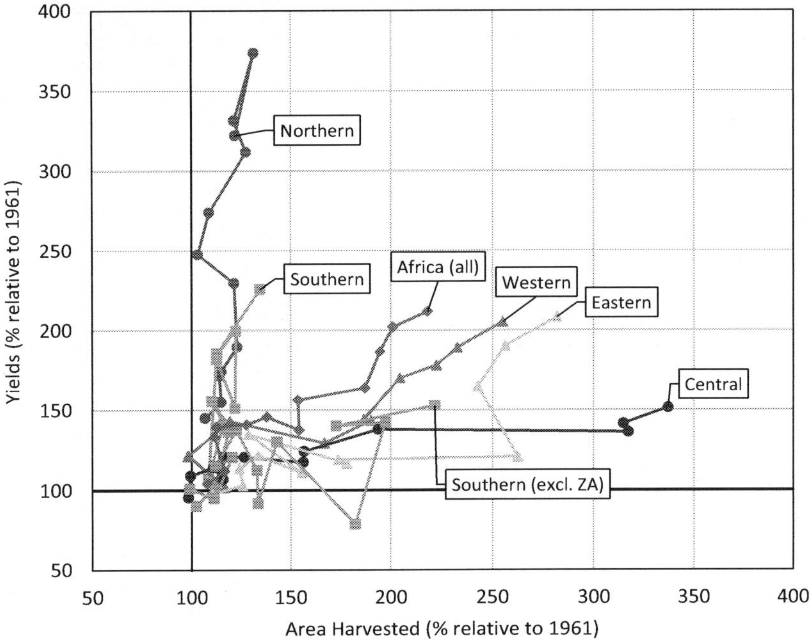

Figure 1.4 Change in cereal yield and the area cropped with cereals, 1965–2020 (%, 1961 = 100, five-year intervals).

Notes: All figures are expressed in relative changes since 1961 (set to 100%). For a listing of countries included in each region see https://en.wikipedia.org/wiki/Regions_of_the_African_Union.

Is there potential for a shift in the direction of African agricultural growth in the near future? Schut and Van Ittersum (this volume) argue that yield gains are crucial to sustainably improve food and nutrition security. However, intensification of food crops requires rural households to prioritize investment in agriculture above other livelihood activities. In reality, a substantial share of agricultural producers themselves are net-purchasers of food in many parts of Africa (Clapp, 2017; Reardon et al., 2019; Giller et al., 2021a). Giller and Andersson (this volume) argue that ever smaller farm sizes in many African regions work against intensification and reduce its benefits for poverty alleviation and food security (also see Giller, 2020). As Descheemaeker et al. (this volume) show in their chapter, climate change threatens farmers' ability to earn a sufficient income from agriculture and contribute to food security, let alone increase their output. Veldwisch et al. (this volume) discuss the potential to expand irrigation for sustainable intensification.

Intensification further requires substantial public investment and political stability, conditions that have mostly not been in place over the past decades. Since the late-colonial period, agriculture typically received relatively modest impetus from governments and was sometimes used as an outright "cash cow" to finance (urban) development projects (Lipton, 1977; Bates, 1981; Barrett et al., 2017). Meanwhile, the implementation of structural adjustment policy in the 1980s and 1990s resulted in the dismantling of agricultural support structures and lower producer prices for farmers' crops (Bryceson, 2002; Clapp, 1997; Havnevik et al., 2007). Although patterns vary widely among countries, at a continental level the use of fertilizers stagnated (at very low levels) for a period of about 30 years after 1980 (Suri & Udry, 2022). This reduction in fertilizer use undermines sustainable food production and security in Africa. It has triggered further mining of nutrients leading to soil degradation and making intensification all the more challenging (Wiig et al., 2001).

Nonetheless, in recent years, there are some signals that intensification is happening on an aggregate scale. While still low today, cereal yields have increased across all regions of Africa (Figure 1.4), against a backdrop of declining cropland per capita, ranging between 0.71 hectares per capita (Southern Africa) and 0.47 (Eastern Africa) in 1961, and between 0.17 (Eastern Africa) and 0.27 (Western Africa) in 2021 (FAOStat). In recent decades, the use of fertilizer per hectare has expanded significantly again across Africa, although recent disruptions in global fertilizer supply chains since the Russian invasion of Ukraine have interrupted this trend (Malpass, 2022). Arguably, these relationships signify responses to "Boserupian pressures": growing land pressures and demand from an increasing non-agricultural population share. However, it is too early to tell if these trends will be sustained (Headey & Jayne, 2014; Jayne, Chamberlin & Headey, 2014).

A promising sign is that after the 2003 Maputo Declaration and the 2014 Malabo Declaration on Accelerated Agricultural Growth and Transformation for Shared Prosperity and Improved Livelihoods, African states have, as one of the seven Malabo Commitments, aimed to step up investment in agriculture to at least 10% of their budgets (African Union, 2021). Many African nations have developed concrete plans for food systems transformation, a theme that also features on the agenda of the African Union (Ighobor, 2023). Some studies have reported positive effects of increased agricultural investment on food security (Kamenya et al., 2022), but evidence so far remains limited.

4 What about food imports?

A deficit of domestic food output over consumption (or misalignment of the types of food that are demanded and supplied) can be overcome through food imports. In the wake of disruptive structural adjustment programmes of the 1980s and 1990s, and export bans and resultant staple food price hikes following global shocks, dependency on food imports has gained a rather odious reputation among many scholars and policy makers in Africa. However, the case against food imports is far from decided. The future of food prices is uncertain (Koning et al., 2008), but for most of the past half century, global food prices have declined relative to other prices, and even despite recent increases, the real prices (i.e. expressed in constant US$) of key staples such as maize, wheat and rice currently remain about half their average during the 1960s and 1970s, as shown by Frankema (this volume, Figure 9.3). For most African countries, food imports (from within but especially beyond Africa) have therefore become an attractive strategy to address domestic food demand in cases where developing domestic value chains and sustainable intensification are difficult and expensive, or run into resource constraints (Beltran-Peña, Rosa & D'Odorico, 2020; Fader et al., 2013; Janssens et al., 2020; Joshipura, 2024). The reality is that, currently, about one fifth of Africa's caloric demand is obtained from outside the continent, with a heavy focus on cereals (in particular rice and wheat), oils and livestock products. The current situation is an outcome of a gradual upward trend of food import dependency from the mid-20th century onwards, when less than 5% of all calories were imported (Luan et al., 2013; Joshipura, forthcoming). Imports are widely consumed, catering for diversifying (urban) consumption portfolios, often of wealthier income groups (Rakotoarisoa, Iafrate & Paschali, 2011; Liverpool-Tasie et al., 2023).

African countries have typically financed their food imports through the export of cash crops and, increasingly, sub-soil minerals and crude oil. Indeed, food import dependency correlates positively with national income, and is larger on average in coastal nations (Luan et al., 2013).

International food aid has, in times of crisis, also augmented supply. With growing populations, continued urbanization and adverse effects of climate change in some regions, it is likely that intra-African food trade and food imports from outside the region will continue and even increase further in decades to come (Egbendewe et al., 2017; Fader et al., 2013; Janssens et al., 2020; Kinnunen et al., 2020). Moreover, it is hard to envision, with some major exceptions such as Egypt and Ethiopia, that sizeable quantities of wheat, the demand for which is likely to stay strong in the coming decades, can be efficiently produced domestically. Van Berkum and de Steenhuijsen Piters (this volume) explore to what extent and if so how African countries can and should reduce their reliance on food imports.

Whether the net effect of food imports on food insecurity is positive or negative is and will remain highly context-dependent (Clapp, 2017). Non-food specialization (for example in export crops like cocoa or coffee) can increase international purchasing power, thus boosting food access. An advantage of imports is that they can mitigate pressure on scarce resources and contribute to easing land and resource pressures in Africa, and thus indirectly even contribute to nature conservation. A trend towards increasing imports would have downsides too. First, countries with low or volatile export revenue may struggle to import enough food to supplement domestic supply reliably, and are vulnerable to global shocks in supply, as exemplified by price spikes in 2008 and the more recent shortages triggered by the Russian invasion of Ukraine. Thus overreliance on imports can hurt food security, and especially its accessibility dimension. Second, reliance on food imports may undermine investment in the agricultural (food crop) sector and domestic agri-food value chains and create situations where domestic farmers are unable to compete. Bulte et al. (this volume) explore how the poultry sector in Sierra Leone is affected by imports and evaluate the potential for and benefits of import substitution.

5 Four paths of change

In Figure 1.5, we plot 48 African countries for which data is available for 2021 on the two axes described above. We also include Vietnam and Bangladesh, two Asian economies that have made substantial progress in both agricultural productivity and structural change in recent decades (Liu et al., 2020; Mujeri & Mujeri, 2021). For reference, we include the (unweighted) average for Africa as a whole, and the (weighted) average of all low- and middle-income countries (LMICs). The exercise generates four quadrants. The upper-right quadrant represents a situation of "agriculture-inclusive" structural change (Q2), where a comparatively low share of the population in agriculture occurs in conjunction with high productivity in the agricultural sector. The lower-right quadrant, instead, represents a situation of "agriculture-exclusive" structural change (Q3), where a low agricultural employment share does not coincide with high agricultural productivity, a situation which we might expect, for example, in food import-dependent, oil-exporting economies. The upper-left quadrant represents a situation of agricultural growth (Q1), where agricultural productivity is comparatively high, but a large share of the population remains in agriculture. Finally, the lower-left quadrant includes countries that have low yields and a high agricultural employment share, suggesting an undesirable situation of agricultural stagnation (Q4).

Overall, this exercise confirms that most African countries remain in early phases of structural change and have agricultural sectors with low productivity. Of all mainland African countries, only Egypt and South Africa have cereal yields comparable to Vietnam and Bangladesh, and higher than the average for all LMICs. Only about a third of all African countries have a smaller share of their workforce in agriculture than the average for all LMICs, and most of these are island nations or located in north or southern Africa. The figure clearly illustrates that there is large scope for "catch-up" growth, involving yield gap closure, and for non-agricultural

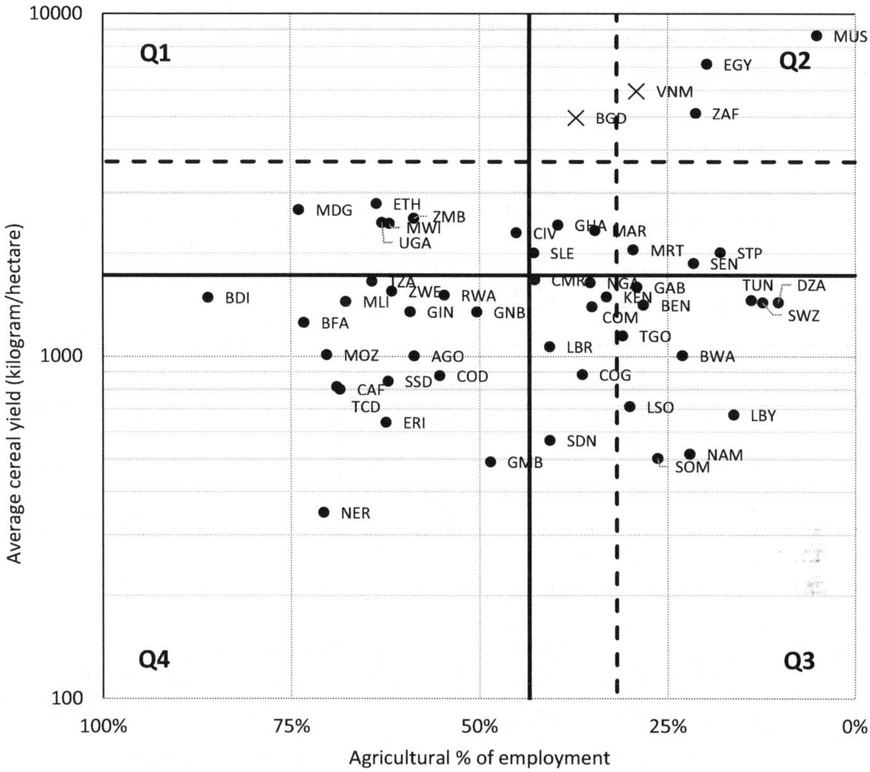

Figure 1.5 Cereal yield and agricultural employment share in 48 African countries, Bangladesh and Vietnam in 2021, compared to averages for Africa (solid lines) and low- and middle-income countries (LMICs) (dotted lines).

Sources: See Figures 1.2 and 1.3.

Notes: The four quadrants (divided by solid lines) are constructed by taking the (unweighted) average of all African countries for both axes. The figure also shows what the quadrants would have looked like had the average for "LMICs" worldwide been taken to divide the quadrants (dotted lines).

employment growth. Such catch-up would contribute substantially to a structural context in which food system transformation is feasible and food security enhanced. But, as noted, improvements in food security will not occur automatically, and rapid change also brings major challenges for food systems that require effective governance on many levels.

To what extent have African nations been catching up regarding agricultural productivity and structural change? In Figure 1.6, we replicate the exercise in Figure 1.5, but focus on changes over the three decades from 1991 to 2021. With a few exceptions, all African countries have experienced a decline in their agricultural labour share (although it is important to note that the absolute number of workers in agriculture in many cases continued to increase due to fast population growth). Most have managed to increase their cereal yields per hectare. But as should be expected given their low yields in 2021 (Figure 1.5), most African countries have performed comparatively poorly in terms of raising cereal yields over the past decades. Others, however, such as Botswana, Côte d'Ivoire, Ethiopia, Senegal and South Africa have made considerable headway (as well as Angola and Mozambique, although this rather signifies recovery from devastating civil conflicts at the start of the period): each more than doubled their yields per hectare.

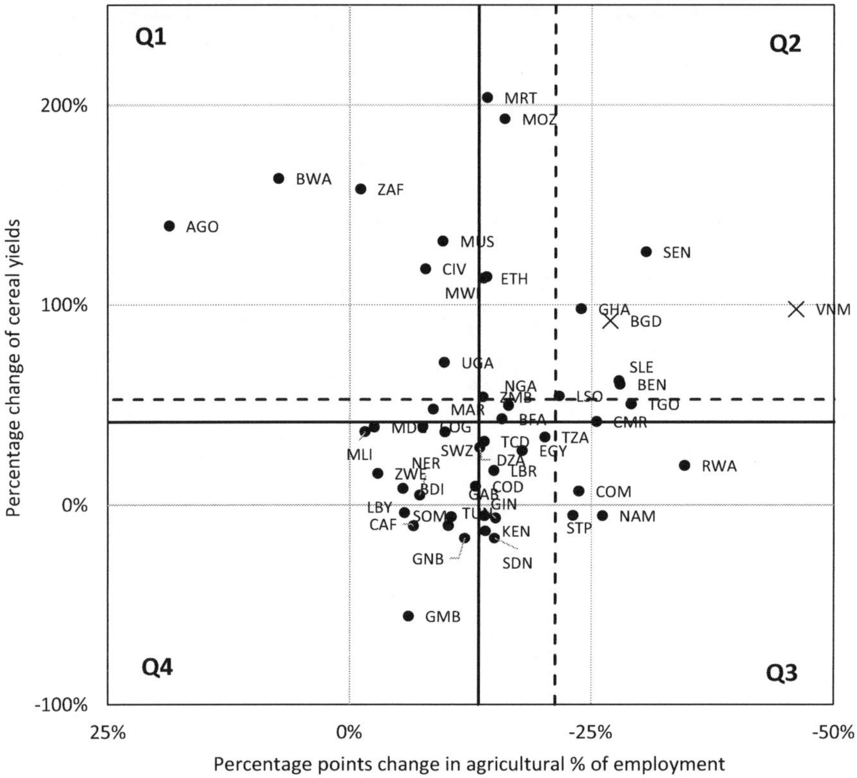

Figure 1.6 Cereal yield and agricultural employment changes in 48 African countries, Bangladesh and Vietnam, 1991–2021, compared to averages for Africa (solid lines) and low- and middle-income countries (LMICs) (dotted lines).

Sources: See Figures 1.2 and 1.3.

Notes: The four quadrants (divided by solid lines) are created by taking the (unweighted) average of all African countries for both axes as the dividing line. The figure also shows what the quadrants would have looked like had the average for "LMICs across the world" been taken to divide the quadrants (dotted lines).

Figures 1.5 and 1.6 provide an overall picture of Africa's lagging past performance, comparatively poor present-day outcomes and large overall potential to catch up, in terms of both agricultural intensification and structural change. Moreover, Figures 1.1–1.3 illustrate that African countries have struggled to translate economic and agricultural growth into improvements in food security. Further, the exercise highlights the large diversity in country experiences across Africa, which is only further compounded if issues such as food import dependency, population growth patterns and resource scarcities are taken into consideration, as done in several of the volume's chapters.

For the first section of the volume, four countries have been selected, with different positions in the four quadrant framework. To each, a chapter has been devoted, led by scholars based in the country, and with close knowledge of and involvement in domestic policy debates on issues of agricultural change, food security and food systems transformation. Ramadan and Castet discuss the case of Egypt, which has experienced a long period of agricultural productivity growth and structural change (past experience on the path of inclusive structural change). Sida et al. look into the case of Ethiopia, which in recent decades has embarked on remarkable agricultural growth, but remains in the early phases of structural change (the path of agricultural growth). Hathie evaluates the case of Senegal, which has seen comparatively fast gains in agricultural productivity as well as structural change (the path of inclusive structural change). Nyandoro and Andersson cover Zimbabwe, which has faced decades of stagnation and even regression in its agricultural sector as well as the economy at large (the path of agricultural stagnation).

6 A change of pace is needed

The use of the four paths to structure our thinking has revealed substantial variety across Africa over the past three decades, which will likely persist in the coming 25 years as well. There are large and relevant differences in the ways in which food systems function across African nations, and in the structural conditions that shape such systems. This has repercussions, for example, for how we think about food imports, which may be an affordable solution in some countries facing resource constraints or undergoing fast structural change, whereas they are a clear signal of malfunctioning food systems and agricultural stagnation in others. In countries where structural change happens without agricultural intensification, large differences in performance exist within national borders and even within the agricultural sector. This means not only that rural food insecurity is perpetuated, but also that poverty and food insecurity can easily spill over into urban areas when (internal) migration pressures increase. Situations of agricultural intensification with limited structural change may be a viable route for densely populated countries with high rural population shares, but under conditions of fast population growth and land scarcity, it may also present a dead end, in which yield gains are undermined by ever-shrinking farm sizes. Clearly, raising yields through sustainable intensification or pushing for agriculture-led growth is no universal panacea for food security *per se* (Dercon & Gollin, 2014; Giller, 2020). Relatedly, further increasing food imports might undermine agricultural vibrancy in some countries, but cannot be excluded as a major route to achieve greater food security in others, given a context of fast population and resource constraints (Fader et al., 2013).

While highlighting diversity, Figures 1.5 and 1.6 also offer an overarching perspective. If we compare the performance of African countries to the average performance of all LMICs across the world, it is notable that agricultural sectors in Africa remain important for employment but with only low productivity, even in cases where the outflow of labour from agriculture has already progressed relatively far. This raises concerns for both rural and urban food security as it suggests the countries are trapped in poverty. Less than a handful of countries achieve both structural change and agricultural intensification, against the global LMIC reference point. Thus, even in countries with more intensive agricultural development or structural change, it is questionable if the pace of change is fast enough to absorb growing demand, let alone improve nutrition security, as it will unfold in the coming decades. A change of pace is undoubtedly necessary, and perhaps population growth will provide the required impetus.

While there is reason for concern and past trends are hardly encouraging, trajectories can shift, and the future may be brighter. Along which paths will African countries and regions develop in the coming 30 years, and to what extent will this translate into great food and nutrition security? These are complex questions, not least because of the large degree of variation within Africa, but also because of the complexity of food systems themselves.

To open up new perspectives and synergies between them, contributors to this volume were invited to reflect on long-term trends and, where helpful, to refer to the four paths introduced here. We solicited input from experts in a wide range of sciences working on soils, crops, forests, water, land use, climate change and nutrition, as well as social scientists working on conflict, institutions, markets, consumption, labour, pricing and governance. We asked them to reflect on challenges, threats and opportunities in these different domains. This has resulted in the 21 chapters that make up the three main sections of this volume, followed by a concluding synthesis.

Note

1 The urbanization data linked to the latest round of population projections (UN DESA, 2022) has yet to be released. Therefore, we use the data from the previous round of projections (UN DESA, 2018) whenever urbanization is discussed.

References

African Union. (2021). *AU Business Plan to Implement The CAADP-Malabo* Dec*laration 2022–2025 (3rd CAADP Biennia Review Report)*. Addis Abeba: African Union.

Ambikapathi, R., Schneider, K.R., Davis, B., Herrero, M., Winters, P., & Fanzo, J.C. (2022). Global food systems transitions have enabled affordable diets but had less favourable outcomes for nutrition, environmental health, inclusion and equity. *Nature Food*, *3*(9), 764–779.

Arndt, C., Davies, R., Gabriel, S., Harris, L., Makrelov, K., Robinson, S., ... & Anderson, L. (2020). Covid-19 lockdowns, income distribution, and food security: An analysis for South Africa. *Global Food Security*, *26*, 100410.

Austin, G. (2008). Resources, techniques, and strategies south of the Sahara: revising the factor endowments perspective on African economic development, 1500–2000. The Economic *History Review, 61*(3), 587–624.

Barrett, C.B., Christiaensen, L., Sheahan, M., & Shimeles, A. (2017). On the structural transformation of rural Africa. *Journal of African Economies*, *26*(suppl_1), i11–i35.

Barrett, C.B., Reardon, T., Swinnen, J., & Zilberman, D. (2022). Agri-food value chain revolutions in low-and middle-income countries. *Journal of Economic Literature*, *60*(4), 1316–1377.

Bates, R.H. (1981). *Markets and states in tropical Africa: The political basis of agricultural policies.* University of California Press.

Battersby, J. (2017). MDGs to SDGs–new goals, same gaps: The continued absence of urban food security in the post-2015 global development agenda. *African Geographical Review*, *36*(1), 115–129.

Beltran-Peña, A., Rosa, L., & D'Odorico, P. (2020). Global food self-sufficiency in the 21st century under sustainable intensification of agriculture. *Environmental Research Letters*, *15*(9), 095004.

Berry, S.S. (1984). The food crisis and agrarian change in Africa: A review essay. *African Studies Review*, *27*(2), 59–112.

Binswanger-Mkhize, H.P., & Savastano, S. (2017). Agricultural intensification: The status in six African countries. *Food Policy*, *67*, 26–40.

Bjornlund, V., Bjornlund, H., & van Rooyen, A. (2022). Why food insecurity persists in sub-Saharan Africa: A review of existing evidence. *Food Security*, *14*(4), 845–864.

Blekking, J., Waldman, K., Tuholske, C., & Evans, T. (2020). Formal/informal employment and urban food security in Sub-Saharan Africa. *Applied Geography*, *114*, 102131.

Bongaarts, J. (2017). Africa's unique fertility transition. *Population and Development Review*, *43*, 39–58.

Bryceson, D.F. (2002). The scramble in Africa: Reorienting rural livelihoods. *World Development*, *30*(5), 725–739.

Burchi, F., & De Muro, P. (2016). From food availability to nutritional capabilities: Advancing food security analysis. *Food Policy*, *60*, 10–19.

Chamberlin, J., Jayne, T.S., & Headey, D. (2014). Scarcity amidst abundance? Reassessing the potential for cropland expansion in Africa. *Food Policy*, *48*, 51–65.

Clapp, J. (1997). *Adjustment and agriculture in Africa. Farmers, the state and the World Bank in Guinea.* St. Martin's Press.

Clapp, J. (2017). Food self-sufficiency: Making sense of it, and when it makes sense. *Food Policy*, *66*, 88–96.

Clapp, J., Moseley, W.G., Burlingame, B., & Termine, P. (2022). The case for a six-dimensional food security framework. *Food Policy*, *106*, 102164.

Conceição, P., Levine, S., Lipton, M., & Warren-Rodríguez, A. (2016). Toward a food secure future: Ensuring food security for sustainable human development in Sub-Saharan Africa. *Food Policy*, *60*, 1–9.

Dercon, S., & Gollin, D. (2014). Agriculture in African development: Theories and strategies. *Annual Review of Resource Economics*, *6*(1), 471–492.

Devereux, S. (2009). Why does famine persist in Africa? *Food Security*, *1*, 25–35.

Egbendewe, A.Y., Lokonon, B.O.K., Atewemba, C., & Coulibaly, N. (2017). Can intra-regional food trade increase food availability in the context of global climatic change in West Africa? *Climatic Change*, *145*, 101–116.

EIU (The Economist Intelligence Unit). (2015). *Global food security index 2015. An annual measure of the state of global food security*. London, The Economist Intelligence Unit Limited.

Fader, M., Gerten, D., Krause, M., Lucht, W., & Cramer, W. (2013). Spatial decoupling of agricultural production and consumption: Quantifying dependences of countries on food imports due to domestic land and water constraints. *Environmental Research Letters*, *8*(1), 014046.

Falconnier, G.N., Cardinael, R., Corbeels, M., Baudron, F., Chivenge, P., Couëdel, A., … & Giller, K.E. (2023). The input reduction principle of agroecology is wrong when it comes to mineral fertilizer use in sub-Saharan Africa. *Outlook on Agriculture*, *52*(3), 311–326.

FAO (Food and Agriculture Organization of the United Nations). (2024). *FAOSTAT statistical database*. FAO, Rome.

FAO, IFAD, UNICEF, WFP, & WHO (2023). *The state of food security and nutrition in the world. Urbanization, agrifood systems transformation and healthy diets across the rural-urban continuum*. FAO, Rome.

Frankema, E., & Van Waijenburg, M. (2018). Africa rising? A historical perspective. *African Affairs*, *117*(469), 543–568.

Giller, K.E. (2020). The food security conundrum of sub-Saharan Africa. *Global Food Security*, *26*, 100431.

Giller, K.E., Delaune, T., Silva, J.V., van Wijk, M., Hammond, J., Descheemaeker, K., … & Andersson, J.A. (2021a). Small farms and development in sub-Saharan Africa: Farming for food, for income or for lack of better options? *Food Security*, *13*(6), 1431–1454.

Giller, K.E., Delaune, T., Silva, J.V., Descheemaeker, K., van de Ven, G., Schut, A.G., … & van Ittersum, M.K. (2021b). The future of farming: Who will produce our food? *Food Security*, *13*(5), 1073–1099.

Hall, C., Dawson, T.P., Macdiarmid, J.I., Matthews, R.B., & Smith, P. (2017). The impact of population growth and climate change on food security in Africa: Looking ahead to 2050. *International Journal of Agricultural Sustainability*, *15*(2), 124–135.

Havnevik, K., Bryceson, D., Birgegård, L.E., Matondi, P., & Beyene, A. (2007). *African agriculture and the World Bank: Development or impoverishment?* Nordiska Afrikainstitutet.

Headey, D.D., & Jayne, T.S. (2014). Adaptation to land constraints: Is Africa different? *Food Policy*, *48*, 18–33.

Hinderink, J.J., & Sterkenburg, J.J. (1987). *Agricultural commercialization and government policy in Africa*. Routledge.

Ighobor, K. (2023). Transforming Africa's food systems: The challenges and opportunities. A conversation with Ibrahim Mayaki, African Union Special Envoy. *Africa Renewal* (June). https://www.un.org/africarenewal/magazine/june-2023/transforming-africa%E2%80%99s-food-systems-challenges-and-opportunities

Izraelov, M., & Silber, J. (2019). An assessment of the global food security index. *Food Security*, *11*(5), 1135–1152.

Jaeger, W.K. (1992). The causes of Africa's food crisis. *World Development*, *20*(11), 1631–1645.

Janssens, C., Havlík, P., Krisztin, T., Baker, J., Frank, S., Hasegawa, T., … & Maertens, M. (2020). Global hunger and climate change adaptation through international trade. *Nature Climate Change*, *10*(9), 829–835.

Jayne, T.S., Chamberlin, J., & Headey, D.D. (2014). Land pressures, the evolution of farming systems, and development strategies in Africa: A synthesis. *Food Policy*, *48*, 1–17.

Joshipura, T. (2024). Feeding African cities: Hinterland suitability and urban growth in twentieth-century Sub-Saharan Africa. *Economic History of Developing Regions*, online first view. https://doi.org/10.1080/20780389.2024.2376549

Joshipura, T. (forthcoming). How Sub-Saharan Africa became food import dependent – A long-run perspective (Unpublished doctoral dissertation). Wageningen University.

Kamenya, M.A., Hendriks, S.L., Gandidzanwa, C., Ulimwengu, J., & Odjo, S. (2022). Public agriculture investment and food security in ECOWAS. *Food Policy*, *113*, 102349.

Kinnunen, P., Guillaume, J.H., Taka, M., D'odorico, P., Siebert, S., Puma, M.J., … & Kummu, M. (2020). Local food crop production can fulfil demand for less than one-third of the population. *Nature Food*, *1*(4), 229–237.

Koning, N.B.J., Van Ittersum, M.K., Becx, G.A., Van Boekel, M.A.J.S., Brandenburg, W.A., Van Den Broek, J.A., … & Smies, M. (2008). Long-term global availability of food: Continued abundance or new scarcity? *NJAS: Wageningen Journal of Life Sciences*, *55*(3), 229–292.

Lipton, M. (1977). *Why poor people stay poor: A study of urban bias in world development*. Temple Smith; Australian National University Press.

Liu, Y., Barrett, C.B., Pham, T., & Violette, W. (2020). The intertemporal evolution of agriculture and labor over a rapid structural transformation: Lessons from Vietnam. *Food Policy*, *94*, 101913.

Liverpool-Tasie, L.S.O., Reardon, T., Parkhi, C.M., & Dolislager, M. (2023). Nigerians in poverty consume little wheat and wheat self-sufficiency programmes will not protect them from price shocks related to the Russia–Ukraine conflict. *Nature Food*, *4(4)*, 288–293. https://doi.org/10.1038/s43016-023-00722-z

Luan, Y., Cui, X., & Ferrat, M. (2013). Historical trends of food self-sufficiency in Africa. *Food Security*, *5*, 393–405.

Malpass, D. (2022). A transformed fertilizer market is needed in response to the food crisis in Africa. Voices: Perspectives on development (World Bank Blogs), 21 December 2022.

Manikas, I., Ali, B.M., & Sundarakani, B. (2023). A systematic literature review of indicators measuring food security. *Agriculture & Food Security*, *12*(1), 10.

Marshall, Q., Fanzo, J., Barrett, C.B., Jones, A.D., Herforth, A., & McLaren, R. (2021). Building a global food systems typology: A new tool for reducing complexity in food systems analysis. *Frontiers in Sustainable Food Systems*, *5*, 432.

Meier zu Selhausen, F. (2022). Urban migration East and West Africa since 1950. In M. de Haas and E. Frankema (Eds.), *Migration in Africa: Shifting patterns of mobility from the 19th to the 21st century* (pp. 281–307). Routledge.

Menashe-Oren, A., & Bocquier, P. (2021). Urbanization is no longer driven by migration in low-and middle-income countries (1985–2015). *Population and Development Review*, *47*(3), 639–663.

Mujeri, M.K., & Mujeri, N. (2021). *Structural transformation of Bangladesh economy: A South Asian perspective*. Springer.

Nechifor, V., Ramos, M.P., Ferrari, E., Laichena, J., Kihiu, E., Omanyo, D., … & Kiriga, B. (2021). Food security and welfare changes under COVID-19 in Sub-Saharan Africa: Impacts and responses in Kenya. *Global Food Security*, *28*, 100514.

Paice, E. (2021). *Youthquake: Why African demography should matter to the world*. Head of Zeus.

Platteau, J.-P., & Hayami, Y. (1998). Resource endowments and agricultural development: Africa versus Asia. In Y. Hayami and M. Aoki (Eds.), *The institutional foundations of East Asian economic development* (pp. 357–412). Palgrave McMillan.

Poudel, D., & Gopinath, M. (2021). Exploring the disparity in global food security indicators. *Global Food Security*, *29*, 100549.

Pretty, J.N., Williams, S., & Toulmin, C. (2012). *Sustainable intensification: Increasing productivity in African food and agricultural systems*. Routledge.

Rakotoarisoa, M., Iafrate, M., & Paschali, M. (2011). *Why has Africa become a net food importer*. FAO.

Reardon, T., Echeverria, R., Berdegué, J., Minten, B., Liverpool-Tasie, S., Tschirley, D., & Zilberman, D. (2019). Rapid transformation of food systems in developing regions: Highlighting the role of agricultural research & innovations. *Agricultural Systems*, *172*, 47–59.

Reardon, T., Tschirley, D., Liverpool-Tasie, L.S.O., Awokuse, T., Fanzo, J., Minten, B., … & Popkin, B.M. (2021). The processed food revolution in African food systems and the double burden of malnutrition. *Global Food Security*, *28*, 100466.

Sasson, A. (2012). Food security for Africa: An urgent global challenge. *Agriculture & Food Security*, *1*, 1–16.

Schoumaker, B. (2019). Stalls in fertility transitions in sub-Saharan Africa: Revisiting the evidence. *Studies in Family Planning*, *50*(3), 257–278.

Stathers, T., Holcroft, D., Kitinoja, L., Mvumi, B.M., English, A., Omotilewa, O., … & Torero, M. (2020). A scoping review of interventions for crop postharvest loss reduction in sub-Saharan Africa and South Asia. *Nature Sustainability*, *3*(10), 821–835.

Steffen, W., Richardson, K., Rockström, J., Cornell, S.E., Fetzer, I., Bennett, E.M., … & Sörlin, S. (2015). Planetary boundaries: Guiding human development on a changing planet. *Science*, *347*(6223), 1259855.

Suri, T., & Udry, C. (2022). Agricultural technology in Africa. *Journal of Economic Perspectives*, *36*(1), 33–56.

UN DESA (United Nations Population Division Department of Economic and Social Affairs). (2018). *World urbanization prospects 2018*. United Nations.

UN DESA (United Nations Population Division Department of Economic and Social Affairs). (2022). *World population prospects 2022*. United Nations.

Van Ittersum, M.K., Van Bussel, L.G., Wolf, J., Grassini, P., Van Wart, J., Guilpart, N., … & Cassman, K.G. (2016). Can sub-Saharan Africa feed itself? *Proceedings of the National Academy of Sciences*, *113*(52), 14964–14969.

Wiig, H., Aune, J.B., Glomsrød, S., & Iversen, V. (2001). Structural adjustment and soil degradation in Tanzania. A CGE model approach with endogenous soil productivity. *Agricultural Economics*, *24*, 263–287.

World Bank. (2023). *World development indicators*.

Part 1

Diverse pathways on the country level

Past, present and future

2 Senegal's journey to feed its growing population by 2050: pathways to accelerated food systems transformation

Ibrahima Hathie

1 Introduction

For decades, food security has been a priority for Senegal. How to feed the steadily growing and urbanizing population through domestic agricultural production has become a key challenge for the government. The country faces rapid population growth and high urbanization, which are weighing on food demand.

Senegal[1] is located in the Sahel, a semi-arid region of West Africa subject to recurrent climatic challenges, severe land degradation, and structural deficits in food production. Agriculture in Senegal is mainly rain-fed with many smallholder farmers dependent on it for their livelihoods. Cereals and legumes dominate the farming systems along with livestock, which plays a significant role in the functioning of the systems. Fishing is important for food security, providing essential animal proteins, income, and jobs to a large segment of the population living along the Senegalese coast. Six agro-ecological regions structure the spatial distribution of crop and livestock production: the Senegal River valley in the north, home to irrigated crops, the sylvo-pastoral region in the central north, the groundnut basin in the centre, the Niayes zone with its microclimate favourable to horticulture, rainy Casamance, and eastern Senegal.

Senegal displays a strong dietary dependence on food imports with broken rice and wheat as two emblematic examples. The country is also vulnerable to world food prices. In the 2008 food crisis, the pass-through of world to local prices was close to unity for heavily imported staples such as rice, eroding the purchasing power of urban households and rural people sourcing locally. Okou et al. (2022) point out that external factors drive food price inflation, but that domestic factors can mitigate these vulnerabilities. Domestically, food price inflation is lower for staple foods in countries with more local production and for products with a lower consumption share. Consequently, strengthening local production helps reduce vulnerability to external shocks, particularly imported inflation. Many authors agree that Senegal's dependence on food imports is more a result of neglecting domestic food production and over-specializing in a single cash crop, namely groundnuts (John, 2015; Resnick, 2014; Ba et al., 2009).

In recent years, the country has made significant progress, particularly in increasing agricultural production and improving farmers' incomes. Agriculture has once again become a priority in the Emerging Senegal Plan (2014–2035). Despite these improvements, agriculture has remained unable to satisfy the country's food needs, which are covered largely by imports. The food price crisis of 2008 prompted the government to launch the Great Agricultural Offensive for Food and Abundance, with a flagship programme for rice self-sufficiency. Recently, in a context of multifaceted crises (COVID-19, conflicts, climate change), Senegal adopted a strategy aiming to achieve food sovereignty in response to the risks of disruptions to world market supplies and the subsequent high prices of imported products. Will this strategy sustainably

DOI: 10.4324/9781032649696-4

resolve Senegal's food dependency on imports and benefit food security? Alternatively, would it be better to develop a thriving export agriculture of high-value products and import staples? So far, recent trends question the country's capacity to feed itself through its agriculture by 2050. Will profound structural changes be possible and under what conditions?

This chapter traces the trajectory of Senegalese food systems by first highlighting current and future challenges. To place ongoing trends in historical perspective, Section 3 describes how the country fed itself between 1961 and 2020. Section 4 explores possible pathways enabling Senegalese agriculture to cover food needs by 2050.

2 Current and future challenges

Population growth, rapid urbanization, and a young population are shaping the future of Senegalese agriculture as these factors underpin profound changes in food demand and contribute to the structuring of food markets. Senegalese agriculture should address the resulting changes in food demand while addressing the challenges of climate change and soil degradation. If policymakers fail to act appropriately, their inertia could compromise the future ability of the Senegalese agriculture to produce sufficient, high-quality food.

2.1 Rapid population growth and strong urbanization

Like most African countries, Senegal's population is growing rapidly, from 3.3 million in 1960, 7.1 million in 1988 to 17.8 million in 2023, and is estimated to reach 32.6 million by 2050 (UN DESA, 2022). The demographic structure has also changed with a significant increase in the working population. Urbanization is also on the rise. The urban population share, which was 23% in 1960, reached 48% in 2020. By 2050, the urban population is projected to represent 65% of the total population (UN DESA, 2018). This high degree of urbanization has implications for food security and nutrition. Urban dwellers depend on the rural world or imports from abroad for their food. Urbanization induces changes in diets, reinforcing the dependence of food demand on products sourced from abroad such as broken rice, wheat, and milk powder (Figure 2.1).

2.2 Climate change and variability

Agriculture plays a key role in Senegal's economy, contributing 17% of GDP, yet employs more than 60% of the active population, an indication of low productivity. Rain-fed agriculture dominates the sector, exposing it to climate change and variability. Like the Sahel, Senegal experienced a rise in temperatures over the last three decades, along with high spatial and temporal irregularity (Noblet et al., 2018). Rainfall displayed negative trends with increased variability. The 1980–1990 decade was particularly dry in all agro-climatic zones. In the mid-1990s, a trend of wetter conditions emerged with more intense, but fewer precipitation events (Nicholson, Funk & Fink, 2018; Trisos et al., 2022). Recent trends indicate a slight increase in annual rainfall with increased variability (Noblet et al., 2018).

Intergovernmental Panel on Climate Change (IPCC) climate projections (Trisos et al., 2022) for both the near (2040–2060) and distant (2080–2100) future predict an intensification of the average warming, a variability of precipitation, and a greater frequency and intensification of extreme phenomena (see also Descheemaeker et al., this volume). Simulations targeting Senegal corroborate these IPCC projections. They show a temperature increase of between 1.5°C–2.5°C and between 2.5°C–6°C respectively for the RCP4.5 and RCP8.5 scenarios. Rainfall simulations highlight seasonal changes and extreme precipitation events. False start, early end of the rainy

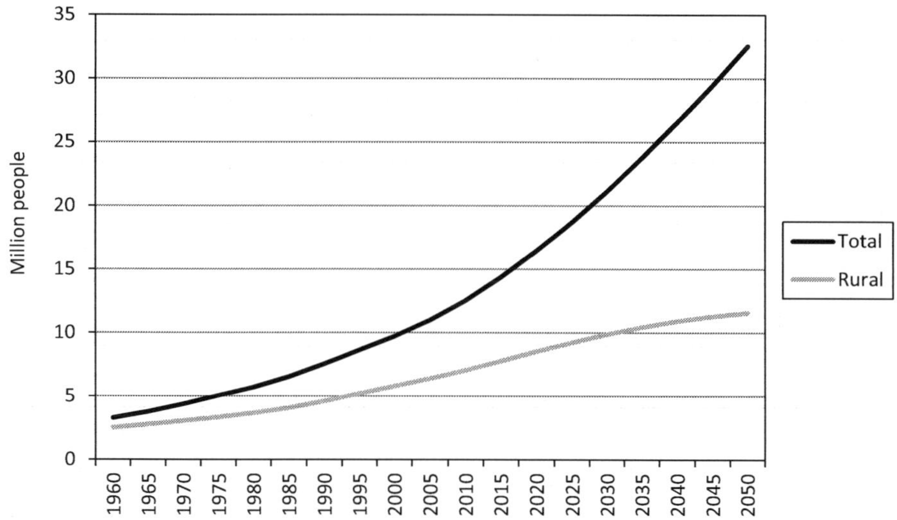

Figure 2.1 Development of the rural and total population (1960–2050).
Source: UN DESA PD (2018, 2022).

season, higher frequency of daily showers, and increase in the number of hot nights and days constitute real threats to the country's food security (Noblet et al., 2018; MacCarthy et al., 2021).

2.3 Soil degradation

Senegal faces a major challenge of land degradation, as nearly 65% of the agricultural land is degraded (World Bank, 2008). Between 1982 and 2006, 20% of the grassland area experienced degradation while the sparse vegetation diminished by 36%, exposing the soil to water and wind erosion. Annual bush fires affect the grassland while charcoal burning also contributes to deforestation (Sow et al., 2016).

Anthropogenic soil erosion, the second largest form of land degradation, affects 7.15% of the total land area. Soil nutrient mining is a major cause of land degradation in the groundnut basin and Casamance, especially when combined with a low fertilizer application rate (World Bank, 2008).

Senegal experiences severe salinity problems due to the country's downstream position of several large rivers and deltas and its largely flat topography leading to poor drainage. Salinity affects about 645,000 ha of agricultural land (Sow et al., 2016). Poorly drained irrigated areas lead to waterlogging and salinity. Niang et al. (2018) note progressive soil degradation in production systems, particularly of the Senegal River Delta where drainage is lacking due to inadequate private and public developments resulting in transformation of cropland into saline soils at the *Sine* and *Saloum* estuaries. Land degradation is a real threat to food security as it greatly reduces the productive base. If inadequately addressed, it will contribute to a long-term decline in the agricultural productivity.

2.4 Changes in human diets

The colonial system had a strong influence on the diets of the Senegalese people. France, Senegal's colonizer, developed groundnut production to supply the metropolis with oil and imported cheap broken rice from Indochina to feed the cities. During this period, the rural population consumed mainly local cereals (millet, sorghum, and maize), supplemented by leafy sauces and dairy products. Post-independence policies followed the same path, continuing to favour groundnut production for export and rice import from Asia, which is cheaper and easier to prepare than millet.

In the last few decades, Senegal has experienced profound changes in its diets. Rapid population growth, strong urbanization, rising income levels, and changing lifestyles are the main drivers of the changing food patterns. The consumption of rice and wheat-based bread, once limited to large cities, has conquered rural areas, where rice has become a staple for lunch and wheat-based bread for breakfast. Overall, rice consumption per capita has increased from 60 kg in 1990 to 78 kg in 2017 while millet per capita consumption declined from 78 kg in 1990 to 30 kg in 2017 (Marivoet et al., 2021; IPAR, 2017). Urban diets have diversified with increased consumption of vegetables and animal proteins (fish and poultry mainly). These changes in food demand are likely to persist and deepen, posing a challenge to the country's food systems to adequately and competitively feed the Senegalese population.

3 How has Senegal fed itself from 1960 to the present day?

Senegal's food security is a legacy of colonial and post-independence choices focused on developing a groundnut economy for exports, relying on rice imports to feed the urban population, and neglecting cereal production mainly directed to self-subsistence.

3.1 Three periods that structure the development of agriculture and food

Agriculture and food in Senegal have undergone major structural changes since 1960. Three main periods can be recognized. The 1960–1979 period constitutes the "welfare state" era marked by direct government intervention. Public monopolies whose purpose was to provide services to farmers organized in cooperatives, supplied groundnut seeds, fertilizer, and equipment on credit, and marketed groundnut. Various climatic, economic, and financial shocks[2] undermined this system and led to the structural adjustment period (1980–1999).The subsequent liberalization of the economy prescribed by the Bretton Woods institutions led to a restructuring of food chains and defined a new role for government: regulate the various economic players, while fostering an environment conducive to private investment. Consequently, public authorities dismantled the credit, marketing, and farmer support systems. This government disengagement led to the decline of the groundnut economy and the emergence of a diversified agricultural export portfolio (green beans, cherry tomatoes, mangoes).

The third phase (2000–2019) coincides with political changes and displays a more interventionist agricultural policy. During 2000–2007, the government focused on special operations aiming to increase the production of maize, cowpea, and cassava. In response to the 2008 food crisis, the government launched the Great Agricultural Offensive for Food and Abundance, along with the National Rice Self-Sufficiency Programme. The Emerging Senegal Plan (2014–2035) and its agricultural programme (Programme to speed up the pace of Senegalese agriculture) confirm the priority of rice self-sufficiency along with groundnut and onion. Horticultural production has taken off during these last two decades, which has also seen the emergence of institutional innovations (financing, technologies, contract farming, and regulation) and promising value chains (rice, onions, potatoes).

3.2 *The restructuring of agricultural markets in Senegal*

Structural adjustment programmes in the 1990s restructured agricultural markets, with varying impacts on groundnuts, rice, and horticultural products.

Groundnut. During 1980–1984, the government dismantled the former state-controlled system (marketing board, input supply system, and credit) and launched, in 1984, the privatization of the oil mills. Consequently, farmers faced increased difficulties in accessing seeds, fertilizers, and equipment, and had to cope with uncertain market conditions. The volume of groundnuts processed for export fell, hurting farmers' incomes (Hathie & Lopez, 2002). In the 2000s, fluctuating production levels due to irregular rainfall, weak sector governance, and unfavourable world market conditions further depressed groundnut production and exports.

Chaotic governance of the groundnut sector combined with climatic shocks resulted in poor performance of groundnut production in recent decades. The consequences are dramatic for many rural households whose income and livelihoods depend on groundnut production. The groundnut industry employs two-thirds of the rural population and makes a major contribution to food security in rural areas, as more than one-third of total production is self-consumed. Its role in feeding livestock and its agronomic contribution in fixing nitrogen in the poor soils of the groundnut basin make groundnuts a strategic crop for the country.

Irrigated rice. From 1984 to 1996, major changes took place in the rice sector, including the liberalization of inputs access, the privatization of the credit system, the growth of private sector irrigation, the credit crisis, the Franc of the Financial Community of Africa devaluation, the end of the state monopoly, and the liberalization of rice imports. These reforms have accelerated the emergence of a local private sector across the entire value chain. In the 2000s, diversified interventions by projects and programmes generated remarkable performances in the valley (higher yields and production, improved quality, greater competitiveness).

High-value horticulture. Thanks to favourable agro-climatic conditions in the western region of the Niayes and targeted policies, horticulture developed in the region with high-value export products (French bean and mangoes) despite greater international requirements with high norms and standards (Maertens & Swinnen, 2009). However, these initiatives are spatially localized and involve relatively few farmers compared to those involved in groundnut and rice.

3.3 *Evolution of production and productivity*

Agricultural production grew modestly and unevenly during the 1980–1990 period due to its high exposure to climate variability and dependence on rain-fed production systems (Ba et al., 2009; Hathie & Ba, 2018). During the 2000–2016 period, crop production was the main driver of agricultural output and growth, regularly accounting for more than half of overall agricultural GDP (World Bank, 2018).

Cereal production follows an irregular trajectory with an upward trend. From 1.02 million tons in 2000, cereal production reached 1.77 million tons in 2010 and 2.76 million tons in 2019 (FAOSTAT, 2023). Millet, sorghum, maize, and rice make up the bulk of cereal production. Throughout 1961–2000, millet accounted for at least three-fifths of cereal production, while rice and maize averaged 16% and 8%, respectively. From the 2000s onwards, because of policy choices, rice and maize production became more important at the expense of millet. In the last decade, rice (40%) supplanted millet (34%) in the contribution to the cereal domestic supply (Figure 2.2a and 2.2b). Under its new food sovereignty programme, the government aims to

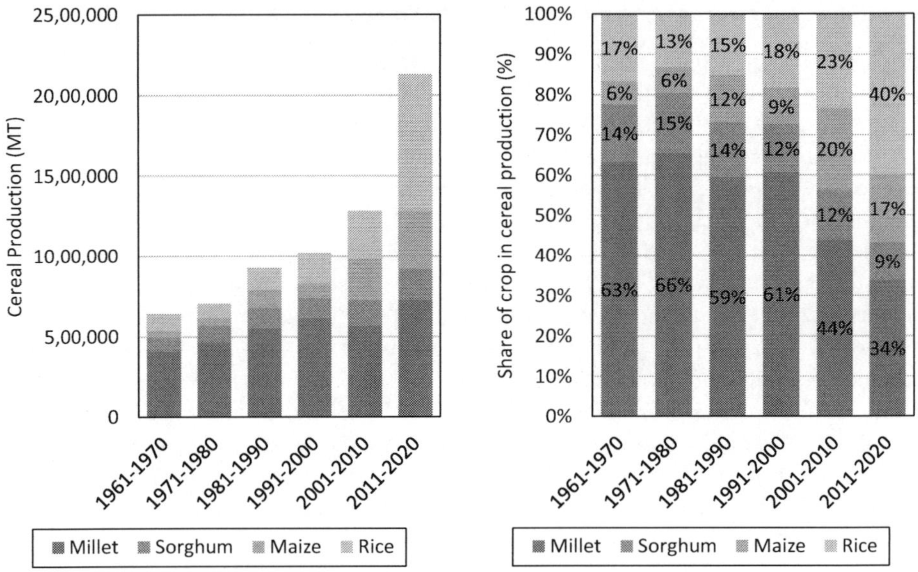

Figure 2.2 (a) Share of cereal crops per decade (1961–2020); (b) cereal production per decade (1961–2020).

Source: FAOSTAT (2023).

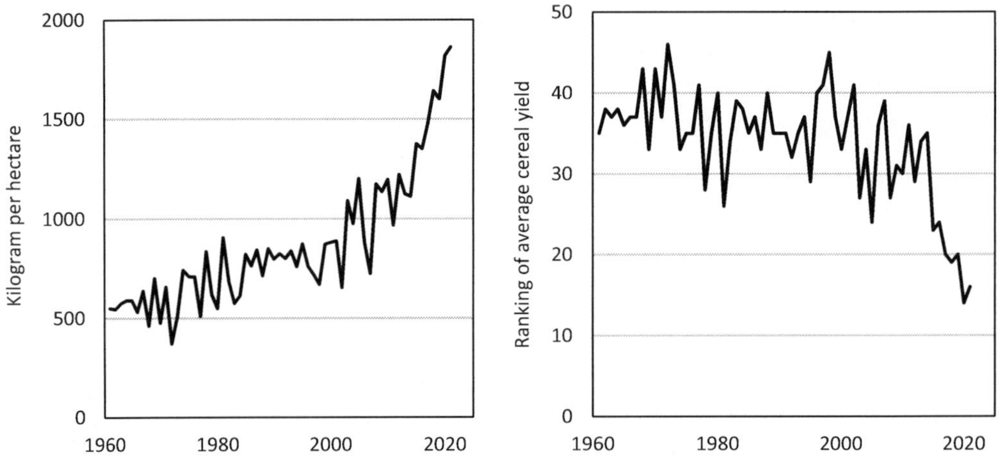

Figure 2.3 (a) Evolution of cereal yields in Senegal (1961–2021); (b) ranking among African countries in terms of cereal yields.

Source: WDI (2023).

produce half of its wheat needs by 2027, an unrealistic projection, not least given that wheat is not well-suited to the agro-ecological conditions that prevail in Senegal (Figure 2.2).

Cereal yields were relatively low in 1961–2000, displayed a slight increase from 2003, and took off in 2015 (Figure 2.3a). Compared with other African countries, Senegal was among the worst performers, ranking between 25th and 45th place during the 1961–2000 period. Over the last decade, Senegal has done much better, moving into the top 20 African countries

(Figure 2.3b), with a ranking of 14th in 2020. Rice's performance partly explains this upturn, as yields of most other crops remained stagnant (World Bank, 2018; Hathie et al., 2017) (Figure 2.3).

3.4 High dependence on food imports

Food imports weigh heavily on Senegal's trade balance and exceeded € 1.22 billion in 2021 or 28% of the country's total exports. Despite progress in production, cereal imports increased steadily during the last decade and reached 2.36 million tons in 2021 for a value of € 740 million (ANSD, 2021). The share of calories imported corroborates these trends and displays a high level of food imports and a steady increase in the 1961–2021 period. While the share stood at 21% on average in the first two decades (1961–1979), it climbed to 31% on average during the 1980s–1990s. In the 2000s, the share of calories imported jumped to reach 42% on average (Joshipura, forthcoming). The net import dependency ratio is much lower early in the period because groundnuts are a food crop and large exports thus push down the total net import dependency ratio (Figure 2.4).

Senegal's dependence on food imports is structural. Diets depend on foodstuffs insufficiently produced in the country. For example, wheat-based bread has become a staple in urban and rural diets while Senegal imports its entire needs in wheat. Net dependence on rice imports[3] is 56%, down from 70% in 2010 following the implementation of the rice self-sufficiency strategy. Senegal is heavily dependent on palm oil imports, with a net import dependency rate of 91% (see also van Berkum and de Steenhuijsen Piters, this volume). Thanks to import regulation, net dependence on onion imports (27%) is relatively low, with a boost to local production, which now covers needs for nine months.

The polycrisis revealed the unsustainability of Senegal's production and consumption patterns. Urban preferences for imported rice, milk powder, and wheat-based bread weigh heavily on the trade balance and increase the country's vulnerability (Resnick, 2014).

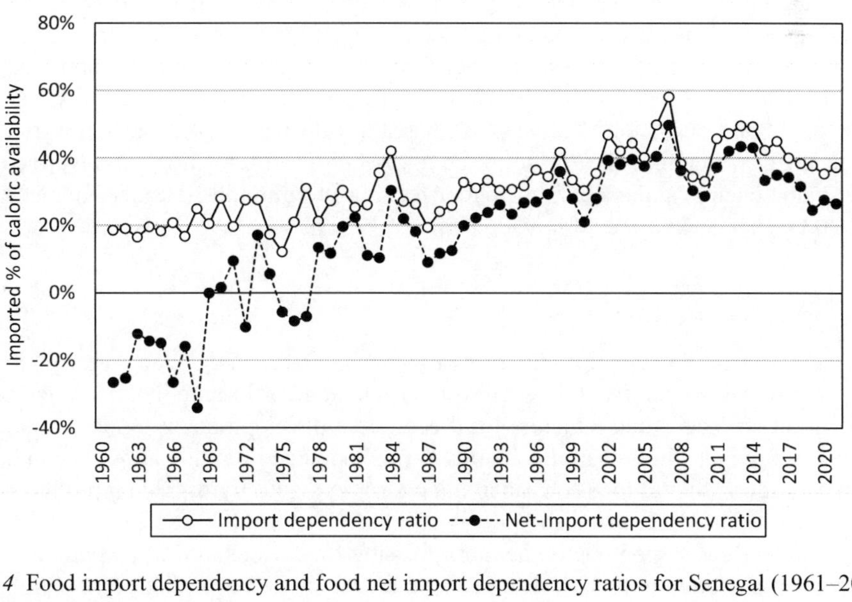

Figure 2.4 Food import dependency and food net import dependency ratios for Senegal (1961–2021).
Source: Joshipura (2024).

4 Possible pathways for Senegalese agriculture

The current situation shows the inability of agriculture to cover the country's food needs. Prospects are not very reassuring due to key variables such as population growth, urbanization, climate change, soil degradation, and diets dependent on imported foodstuff, which constitute serious threats to the coverage of future food needs. Faced with these challenges, are there alternatives to feed Senegal by 2050 and reduce its dependence on imports?

Senegal can pursue at least three avenues. First, to accept a situation of heavy dependence on food imports and find ways of covering these expenses through export revenues. This option, advocated by the Bretton Woods institutions, proved its limits and vulnerability in the face of international shocks. Today, with Senegal's oil and gas discovery, this option is becoming attractive, although it has several additional drawbacks: risk of neglecting agriculture, country exposure to fluctuations of world oil prices, and insufficient consideration of long-term effects of global decarbonization uncertainties. A second option for the government is to move towards food sovereignty based on an import-substitution strategy. This approach has technical and methodological limitations. The third alternative consists in transforming domestic food systems to cover the essential food needs of the population. This option is more difficult, given the constraints mentioned above, but it is possible and desirable. Below, we propose a number of avenues for its implementation.

4.1 Develop sustainable and resilient food systems

Until now, Senegalese food security policies have focused on low-cost food provision regardless of its origin. However, recent events have demonstrated the vulnerability of countries that are highly dependent on world supply. In the context of export bans, soaring oil prices, and rising dollar and freight costs, the strong transmission of world prices of essential products to domestic prices is aggravating food insecurity and threatening the stability of these countries.

Developing sustainable and resilient food systems requires investment in food system infrastructure, support for risk reduction, and incentives for private investment (UNDP, 2022b). Recent progress in Senegal's rice value chain is attributable to this holistic approach. Interventions in good agricultural practices, contract farming, integrated financing, and upgrading of processing plants have increased yields, improved paddy and white rice quality, and met the demands of urban consumers (Hathie, 2016). Scaling up many localized innovations would be beneficial. These include the "drive for an agro-ecological transition", innovative contracts in cereal value chains, and scaling up school feeding programmes (Bezner Kerr et al., 2021; Gálvez, 2022; Verguet et al., 2020).

4.2 Rely on an adequate scale

Senegal cannot ensure food sovereignty in autarky. It should rather develop a strategy focused on a three-dimensional scale. First, it should develop local markets through short circuits. Innovative territorial food projects could structure local economic development, provide consumers with fresh, quality products, and guarantee family farms' adequate remuneration. Second, national policy interventions should focus on satisfying consumer needs by supplying products from the domestic market. The government should provide incentives and other support to strengthen this domestic market, thus responding to the most pressing food needs of the population.

The Economic Community of West African States region is the third relevant scale where the country can source food to fill its local deficit. The interlocking of local, national, and regional markets, such as proposed through the African Continental Free Trade Area (AfCFTA), would reduce the dependence by diversifying the sources of food exchanges and by developing these respective markets at the same time.

4.3 Industrialize the agricultural sector

Links between the agricultural sector and the agri-food industry are weak, with poorly structured agricultural value chains and minimal product processing, limiting the possibilities for urban consumers to feed themselves locally. The agri-food industry imports most of its inputs and domestic supply barely meets urban demand, reinforcing the weight of food imports.

Underdeveloped value chains, with insufficient linkages between different segments, are a major constraint. For example, production of fruits and vegetables is increasing steadily, with considerable post-harvest losses and little processing. Senegal imports directly most of its dairy products or processes them from imported milk powder. Indeed, the country imports many processed products it could produce locally and exports raw materials that, if processed, could be sold at much higher profit margins (World Bank, 2018).

Appropriate agro-processing policies, including incentivizing sustainable practices, enhancing linkages between agriculture and industry, fostering innovation, and ensuring policy integration, would boost productivity, create new jobs, and enable family farms to obtain higher and more stable incomes. Premises already exist in the Senegal River Valley with the development of the rice value chain (Hathie, 2016). Scaling up with other food value chains in relevant territories is possible with adequate interventions addressing issues such as storage, packaging, financing, and infrastructure.

4.4 Diversify diets

The dependence of current diets on foreign sources is a big obstacle to transforming food systems. However, recent history teaches us that diets are dynamic. Their evolution depends on economic, social, cultural, religious, and geopolitical factors. Therefore, future policies must take care of consumer imaginaries, connect messages to culture, health, and ethical values, and ensure availability and access of these products (Cliffer et al., 2019). For example, although a majority of Senegalese would consume more millet-based products, the lack of processing is a powerful barrier in urban areas.[4] In the suburbs, local cereals constitute most of the street food. In hotels, customers rush for dishes based on local cereals.

The millet example shows the importance of a holistic intervention to hope for sustainable change (Ingram et al., 2023; OECD, 2021). This would involve the following. First, increasing millet productivity using adequate inputs and formal contracts with the processing industry that ensures decent incomes for producers. Second, processing millet industrially to make quality flour available to businesses and households. Third, diversifying culinary and other manufactured foods.

In recent years, cities witnessed a sharp rise in the consumption of animal-based products (especially poultry). Similarly, there is enormous potential for fruits and vegetables, whose impact on nutritional quality is undeniable (see Melse-Boonstra et al., this volume).

5 Conclusion

Senegal is heavily dependent on food imports, which places the country in a vulnerable situation in a context of recurring shocks. Food dependency is the result of historical factors that have shaped food production and consumption patterns. Agricultural policy failure is another reason for this dependency.

Population growth, urbanization, climate change, soil degradation, and diets' dependency of foreign sources are all factors that make the distant prospects of covering the food needs of populations through domestic production more than uncertain. This raises the need to make a breakthrough by promoting the materialization of transitions that, in the long term, would favour a transformation of sustainable food systems coupled with the adoption of healthy and nutritious diets.

For these changes to materialize, it is imperative to focus interventions on the development of sustainable and resilient food systems. The relevant market for the development of these food systems includes local, national, and regional markets. Scaling up innovations in the production segment will only be sustainable if there is a clear strategy of processing agricultural products within the main value chains. The downstream part of the value chains is of paramount importance. Diversifying diets (cereals, vegetables, and animal products) and linking them to farmers' choices would strengthen value chains and help to meet the food demands of a growing and increasingly urban population.

Notes

1 Senegal's Human Development Index was 0.511 in 2021, placing the country in the "low human development" category and 170th among 191 countries and territories (UNDP, 2022a).
2 The drought of 1968–1974 and the two oil shocks of 1973 (Yom Kippur War) and 1979 (the Iranian revolution).
3 Net import dependence is calculated based on the formula: NID = (100*(import-export))/(prod+import-export).
4 The time and labour opportunity cost of preparing millet-based dishes is high in urban areas. Further processing, packaging, and distributing these cereal-based meals would increase demand.

References

Agence Nationale de la Statistique et de la démographie (ANSD), (2021). Note d'analyse du commerce extérieur. Various years: 2009–2014–2019–2021.

Ba, C.O., Diagana, B., Dièye, P.N., Hathie, I., & Niang, M. (2009). *Changements structurels des economies rurales dans la mondialisation. Programme RuralStruc-Phase II.* World Bank.

Bezner Kerr, R., Madsen, S., Stüber, M., Liebert, J., Enloe, S., Borghino, N., … & Wezel, A. (2021). Can agroecology improve food security and nutrition? A review. *Global Food Security*, *29*, 100540.

Cliffer, I., Masters, W.A., Trevino, J.A., Webb, P., & Ghosh, S. (2019, October 29). *Food systems and nutrition: Emerging evidence and research opportunities.* usaid.gov. https://pdf.usaid.gov/pdf_docs/PA00W5W9.pdf

FAOSTAT (2023). FAO database. https://www.fao.org/faostat/en/#home

Gálvez, E. (2022). *Scaling up inclusive innovation in agrifood chains in Asia and the Pacific.* Bangkok, FAO.

Hathie, I. (2016, December). *Food sovereignty brief n°23- promoting rice self-sufficiency in West Africa: Achievements, limits and issues for debate - inter-réseaux.* Inter-reseaux.org. https://www.inter-reseaux.org/en/publication/food-sovereignty-brief-n23-promoting-rice-self-sufficiency-in-west-africa-achievements-limits-and-issues-for-debate/

Hathie, I., & Ba, C.O. (2018). Family farming confronted by drought and liberalization in Senegal. In P.M. Bosc, J.M. Sourisseau, P. Bonnal, P. Gasselin, E. Valette, & J.F. Bélières (Eds.), *Diversity of family farming around the world.* Springer.

Hathie, I., & Lopez, R.A. (2002). The impact of market reforms on the Senegalese peanut economy. *Journal of International Development*, *14*(5), 543–554.

Hathie, I., Seydi, B., Samaké, L., & Sakho-Jimbira, S. (2017, October). *Ending rural hunger - brookings.* Brookings.edu. https://www.brookings.edu/articles/ending-rural-hunger-the-case-of-senegal/

Ingram et al., (2023). Further concepts and approaches for enhancing food system resilience. *Nature Food, 4*, 440–441. https://doi.org/10.1038/s43016-023-00762-5

IPAR. (2017). Étude de la consommation des céréales de base au Sénégal. Initiative Prospective Agricole et Rurale, Feed the Future Senegal Naatal Mbay.

John, A. (2015). The origins of Senegal's dependency on rice imports. *Rice Today, 14*(3), 28–30.

Joshipura, T. (forthcoming). *How Sub-Saharan Africa became food import dependent – A long-run perspective* (Unpublished doctoral dissertation). Wageningen University.

MacCarthy, D.S., Hathie, I., Freduah, B.S., Ly, M., Adam, M., Ly, A., ... & Valdivia, R.O. (2021). Potential impacts of agricultural intensification and climate change on the livelihoods of farmers in Nioro, Senegal, West Africa. In C. Rosenzweig, C.Z. Mutter, & E.M. Contreras (Eds.), *Handbook of climate change and agrosystems: Climate change and farming system planning in Africa and South Asia: AgMIP stakeholder-driven research, Part 2.* World Scientific Publishing Company. https://doi.org/10.1142/9781786348814_fmatter

Maertens, M., & Swinnen, J.F.M. (2009). Trade, standards, and poverty: Evidence from Senegal. *World Development, 37*(1), 161–178.

Marivoet, W., Ulimwengu, J.M., Sall, L.M., Gueye, A., Savadogo, K., & Dia, K. (2021). Hidden hunger: Understanding dietary adequacy in urban and rural food consumption in Senegal. IFPRI Discussion Paper 02036. July.

Niang, S., Faye, A.T., & Diop, T. (2018). Dégradation chimique des sols dans les systèmes de production du Sénégal: Analyse à partir des environnements du haut Saloum, du delta du fleuve Sénégal et du Gandiolais. *Revue de géographie du Laboratoire Leïdi «DTD », 17*, 246–266.

Nicholson, S.E., Funk, C., & Fink, A.H. (2018). Rainfall over the African continent from the 19th through the 21st century. *Global and Planetary Change, 165*, 114–127. https://doi.org/10.1016/j.gloplacha.2017.12.014

Noblet, M., Faye, A., Camara, I., Seck, A., Sadio, M., & Bah, A. (2018). Etat des lieux des connaissances scientifiques sur les changements climatiques pour les secteurs des ressources en eau, de l'agriculture et de la zone côtière. Report produced under the project "Projet d'Appui Scientifique aux processus de Plans Nationaux d'Adaptation dans les pays francophones les moins avancés d'Afrique subsaharienne". Climate Analytics GmbH, Berlin. 76 pages.

OECD. (2021). *Making better policies for food systems.* OECD Publishing. https://doi.org/10.1787/ddfba4de-en.

Okou, C., Spray, J., & Unsal, D.F. (2022). Staple food prices in Sub-Saharan Africa: An empirical assessment, IMF Working Papers, WP/22/135.

Resnick, D. (2014). The political economy of food price policy in Senegal", In P. Pinstrup-Andersen (Ed.), *Food price policy in an era of market instability: A political economy analysis.* Oxford University Press. Downloaded from https://academic.oup.com/book/42636/chapter/358110195 by guest on 30 March 2023.

Sow, S., Nkonya, E., Meyer, S., & Kato, E. (2016). Cost, drivers and action against land degradation in Senegal. In E. Nkonya, A. Mirzabaev & J. von Braun (Eds.), *Economics of land degradation and improvement – A global assessment for sustainable development.* Springer. https://doi.org/10.1007/978-3-319-19168-3_19

Trisos, C.H., Adelekan, I.O., Totin, E., Ayanlade, A., Efitre, J., Gemeda, A., ... & Zakieldeen, S. (2022). Africa. In H.-O. Pörtner, D.C. Roberts, M. Tignor, E.S. Poloczanska, K. Mintenbeck, A. Alegría, ... & B. Rama (Eds.), *Climate change 2022: Impacts, adaptation and vulnerability. contribution of working group II to the sixth assessment report of the intergovernmental panel on climate change.* Cambridge University Press, Cambridge, UK and New York, NY, USA, pp. 1285–1455. https://doi.org/10.1017/9781009325844.011

United Nations Development Programme. (2022a). Uncertain times, unsettled lives: Shaping our future in a transforming world. The 2021/2022 Human Development Report. Retrieved at https://hdr.undp.org/system/files/documents/global-report-document/hdr2021-22pdf_1.pdf

United Nations Development Programme. (2022b). Towards food security and sovereignty in Africa. UNDP-RBA-Food Security-Working Paper-Sept 2022

United Nations, Department of Economic and Social Affairs, Population Division. (2018). *World urbanization prospects: The 2018 revision, online edition.*

United Nations, Department of Economic and Social Affairs, Population Division. (2022). *World population prospects 2022, online edition.*

Verguet, S., Limasalle, P., Chakrabarti, A., Husain, A., Burbano, C., Drake, L., & Bundy, D.A.P. (2020). The broader economic value of school feeding programs in low- and middle-income countries: Estimating the multi-sectoral returns to public health, human capital, social protection, and the local economy. *Frontiers in Public Health*, *8*, 587046. https://doi.org/10.3389/fpubh.2020.587046. PMID: 33344398; PMCID: PMC7744483.

World Bank. (2008). Senegal—country environmental analysis. Report No. 48804-SN. Sustainable Development Department Africa Region. Washington

World Bank. (2018). Senegal economic update: Recent growth drivers in Senegal, and the role of agriculture in developing a resilient and inclusive economy. Avril.

3 Feeding a fast-growing population by 2050 through accelerated agricultural transformation: potential avenues for Ethiopia

Tesfaye Shiferaw Sida, Temesgen Desalegn, Gashaw T. Abate, and Tilahun Amede

1 Introduction

In 2020, Africa's population reached 1.3 billion, and projections show it will rise to 2.5 billion by 2050 (UN-DESA, 2022). Meeting the escalating food demand in the continent while improving nutrition and safeguarding the environment and production resources poses significant challenges. Ethiopia offers a compelling case study to explore potential avenues for feeding a fast-growing population through accelerated agricultural transformation. Ethiopia has grappled with a challenging history marked by food insecurity, culminating in the famine of the 1980s. This severe famine, which unfolded against the backdrop of political and economic instability, brought Ethiopia's food crisis to a climax. This tragic historical chapter prompted an immediate need for humanitarian aid as well as a broader examination of systemic issues contributing to food insecurity. In the decades following the 1980s famine, Ethiopia has made notable strides in addressing food insecurity through various initiatives and international collaborations. The country's resilience and commitment to positive change serve as a testament to its ongoing efforts to overcome the shadows cast by historical challenges and build a more secure and prosperous future (Diriba, 2020).

As of 2023, Ethiopia's estimated population stands at 120 million. Projections indicate an increase to 135 million by 2033 (UN-DESA, 2022). By the middle of the century, the country is expected to become one of the world's top ten most populous countries, with a staggering 213 million inhabitants (CSA, 2020; UN-DESA, 2022). For the timeframe from 1993 to 2021, government statistics show that annual cereal production exhibited a consistent growth rate of 11.9%, increasing from 6 million to 210 million tons per year (FAO, 2023). Despite this remarkable growth, over a quarter of the population still lives below the poverty line, food insecurity persists, 12–16 million people face chronic hunger, and malnutrition is widespread (FAO et al., 2022). According to the Food and Agriculture Organization (FAO) estimates, approximately 20% of the country's population was experiencing acute food insecurity in 2021 (FAO, 2021). Furthermore, the Global Hunger Index (GHI)[1] report (GHI, 2022) assigns Ethiopia a score of 27.6, indicating a serious degree of hunger.

This reeling mismatch presents a substantial strain on the country's agricultural sector, which is already hampered by limited arable land, fragmented farms, suboptimal productivity, and environmental degradation (Yigezu Wendimu, 2021). While Ethiopia has achieved noteworthy successes in increasing agricultural productivity in recent decades (Diriba, 2020), the formidable task of matching this progress with its rapid population expansion has never been easy (Moreland and Smith, 2012). The deleterious effects of climate change and water scarcity further compound the country's food security concerns (Onoja et al., 2019). Increasing urban

DOI: 10.4324/9781032649696-5

population poses an additional challenge to food security in Ethiopia. By 2050, projections indicate that about 83.3 million inhabitants, more than one-third (39%) of the country's projected population, will live in urban centres (UN-DESA, 2022).

Ethiopia's historic success in improving agriculture production and productivity can be attributed to several key strategies and initiatives (FAO, 2018). Yet, considerable room for improvement in crop productivity remains, as evidenced by the average maize yield of less than 4 tons per hectare, far less than half of what could be achieved (Assefa et al., 2020; Belachew et al., 2022; Van Dijk et al., 2020). Additionally, the country imports approximately US$1 billion worth of wheat annually. This is despite Ethiopia being one of the few African countries with great potential to produce rainfed wheat (Senbeta and Worku, 2023). The Ethiopian government has set a bold target for agricultural development, aiming for a 6.2% annual increase in agricultural output from 2020 baseline over the subsequent decade to sustain past achievements (MOF, 2020). Considering Ethiopia's commendable history of progress in agriculture, achieving this goal appears feasible. However, even if Ethiopia achieves this output target, it will still fall short of attaining complete food and nutrition security (FAOSTAT, 2023). To ensure food security for all its citizens by 2030, Ethiopia's agriculture must grow 13.3% annually, accounting for current deficiencies and the increased demand resulting from additional population (authors' calculation from FAOSTAT, 2023). Achieving such a substantial transformation requires unparalleled excellence and dedication. In this chapter, we explore pathways for accelerated agricultural transformation that can meet the increasing food demands while minimizing the reliance on unsustainable, energy-intensive, and resource-depleting production practices. We suggest moving beyond the traditional agriculture-led structural change pathway to achieve food systems transformation. In addition to structural change, where labour rapidly moves out of agriculture, and intensive agricultural growth, where labour gradually transitions from agriculture to more labour-productive sectors, we propose transformative strategies tailored for Ethiopia and Africa at large.

2 Agricultural development in Ethiopia: exemplary initiatives and lessons

Ethiopia has implemented several strategies and initiatives aimed at transforming its agricultural sector and achieving food security. Noteworthy development strategies and programmes such as the Agricultural Development-Led Industrialization (ADLI) strategy, the Agricultural Growth Program (AGP), and, more recently, the Ten Commodities in Ten Years (10-in-10) and Sustainable and Resilient Food System Transformation (SRFST) programmes have played a crucial role in the country's agricultural progress (UNCTAD, 2022). In this section, we review the impact of these initiatives, highlighting their successes, challenges, and lessons learned.

2.1 Agricultural Development Led Industrialization (ADLI) strategy

The Ethiopian government initiated the ADLI in the early 1990s with the aim of using agriculture as a driver for industrialization and overall economic growth (MOFED, 2010). The strategy drew upon theories from the 1960s that emphasized the development of smallholder agriculture to create demand for industrial goods and inputs (Mellor and Dorosh, 2010). The fundamental assumption underlying the ADLI is that enhancing agricultural productivity in rural areas generates surpluses that can be utilized by urban industries. Consequently, these industries can absorb additional labour, enabling the rural agricultural land to be utilized for even more intensive production. In addition, agricultural productivity growth can lead to higher rural incomes which generate demand for manufactured products. This approach was deemed appropriate because agriculture represents the largest sector of the economy, employing the majority of the population and contributing substantially to the country's Gross Domestic Product (GDP) and export earnings.

While it is difficult to attribute all achievements to the ADLI, Ethiopia made progress with some of the objectives clearly set in the strategy (Atsbaha and Tessema, 2012). Official reports attribute the sustained annual agricultural growth of 7% over two decades to the ADLI strategy (ATI, 2016). For example, the country's poverty rate declined from around 45% in 1995 to about 23% in 2016 (World Bank, 2015).

Nevertheless, the degree to which the ADLI accomplished its objectives is mixed. While there was a sustained period of agricultural growth, ADLI's success in catalysing industrialization and overall economic development was limited. The ADLI strategy failed to implement several of its key components, which included the introduction of labour-intensive technologies, proper utilization of agricultural land, incorporating indigenous farming technologies, considering agro-climatic zones, integrating agricultural and rural development programmes, building agricultural marketing systems, improving rural finance, encouraging private investments, expanding rural infrastructure, and promoting off-farm activities for people operating in agriculture (Atsbaha and Tessema, 2012).

2.2 *Agricultural Growth Programs (AGP I and II)*

Drawing from the lessons learned through the ADLI, Ethiopia initiated the implementation of two consecutive five-year plans known as the AGP I and AGP II (MOFED, 2010). AGP I was executed from 2011 to 2015, followed by AGP II, which ran from 2016 to 2020 (National Planning Commission, 2016). Unlike its predecessors, these plans had a broader scope and did not concentrate solely on agriculture. The plans outlined a strategic framework that encompassed multiple sectors, including economic expansion, infrastructure development, agricultural advancement, industrialization, social development, human capital, and instilling good governance (ATI, 2016; Atsbaha and Tessema, 2012; Mellor and Dorosh, 2010; MOFED, 2010; National Planning Commission, 2016). These policies aimed to accelerate economic growth, reduce poverty, and improve living standards (World Bank, 2015).

Both AGPs have made a fair contribution towards the transformation of the country's economy (Osh et al., 2018). According to reports from the Ethiopian government, the remarkable growth in average annual GDP of 10.1% in the first half of the 2010s can be credited to AGP I (National Planning Commission, 2016). During the AGP II period, agricultural value added increased by 6.7% and contributed about 2.5 percentage points to the 10.9% growth of the overall economy, which was only slightly short of the 11.1% target for the period (National Planning Commission, 2018).

Despite the substantial economic expansion observed during the AGP eras, which aimed to achieve swift structural transformation and speedy industrialization by prioritizing the growth of the industry sector over agriculture (Mellor and Dorosh, 2010; MOFED, 2010), the actual shift from agriculture to industry fell short of the intended goals in terms of the economic contributions (National Planning Commission, 2018, 2016). By 2015, the manufacturing sector's share of overall GDP remained below 5% (World Bank, 2015). Additionally, agricultural products struggled to meet the quantity, type, and quality requirements of the export market, as well as the demands of growing industries and domestic consumption (National Planning Commission, 2018, 2016; World Bank, 2015). The economic transformation agenda faced limitations such as restricted agricultural input and output markets, unsustainable land and water management practices, inadequate agricultural research and extension services, insufficient infrastructure, limited financial access for smallholders, and a general inability of the economy to undergo structural transformation (Sacko and Mayaki, 2020).

2.3 Recent transformation initiatives

Building upon previous efforts that led to notable improvements in agricultural productivity, the Ethiopian government has demonstrated a strong commitment to agricultural transformation. This dedication is exemplified by recent large-scale government initiatives. The Ethiopian government formulated an Agricultural Sector Plan (MOF, 2020), which was reinforced by the introduction of the "10-in-10" initiative in 2021 (Dessie et al., 2022). This initiative serves as the foundation for the National Agricultural Investment Plan (NAIP), and it focuses on ten commodities over ten years from 2020 to 2030. This plan aims to harmonize the country's priorities with the Comprehensive Africa Agriculture Development Program (CAADP) framework and establishment of Comprehensive National Flagship Programs (Sacko and Mayaki, 2020).

Wheat, which is one of the priority commodities in the "10-in-10" initiative, received considerable attention. The government's objective under the National Wheat Flagship Program was to enhance the productivity of both rainfed and irrigated wheat to attain self-sufficiency (Anteneh and Asrat, 2020). The "10-in-10" initiative encompasses the Ethiopian food system transformation[2] and Agricultural Commercialization Clusters (ACC) programmes. Its objective is to achieve significant increases in cereal production (85%), meat production (496%), fish production (356%), chicken meat production (140%), and milk production (127%) by 2030, compared to the levels of 2020 (African Union, 2021). To accomplish these goals, the agriculture sector is expected to grow at an annual rate of 6.2% between 2020 and 2030 (MOF, 2020; National Planning Commission, 2018). Undoubtedly, such a transformative undertaking necessitates not only unwavering commitment but also exceptional implementation capabilities.

Two questions arise. First, can the required resources (both human and material) be mobilized within the given timeframe? Second, even with such a transformation, will the country produce enough food to adequately feed the growing population? The answer to the first question may potentially be affirmative, considering the historical achievements of previous initiatives. In the subsequent sections, we explore potential answers to the second question.

3 The challenges

3.1 The continuing population explosion

Ethiopia's history is 'punctuated by famine.' This culminated in the 1980s with widespread famine and undernutrition, where over half of the country's population consumed less than 2,100 kcal per day (Webb and von Braun, 1994). In 1985, the number of people requiring food aid reached 7 million. Currently, despite a decrease in hunger rates in Ethiopia, for example, as indicated by a significant reduction in child stunting, the proportion and absolute number of individuals experiencing food insecurity remain disturbingly high. By 2022, the number of Ethiopians experiencing severe food insecurity stood at 22 million (FAO et al., 2022; FAO, 2023; IPC, 2019), nearly one-fifth of the country's population. Approximately 10–15 million required food assistance (FAO et al., 2022).

While the causes of famine and hunger are complex and multifaceted, food availability is a crucial factor. Since 1990, agricultural productivity has doubled (Atsbaha and Tessema, 2012; National Planning Commission, 2018; Osh et al., 2018), while total output grew even faster (Figure 3.1a).

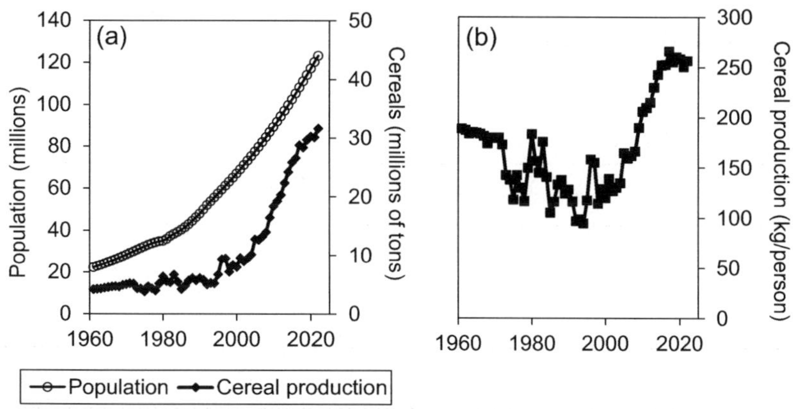

Figure 3.1 Relationship between population growth and change in cereal production between 1961 and 2022 (a) and cereal production per capita (b). Authors' representation (FAO, 2023) for cereal production and (UN-DESA, 2022) for population growth.

Figure 3.1b shows per capita cereal production, which can be categorized into three periods. Prior to the 1970s, the cereal-to-population ratio was relatively stable around 175 kg/person. The ratio declined from the late 1970s to the early 1990s, coinciding with catastrophic famines that hit the country. The third period, from the early 2000s to today, shows a consistent increase in the cereal-to-population ratio, albeit with a slight plateau since 2018. The plateau in recent years could be attributed to ongoing conflicts in the country, the COVID-19 pandemic, and fertilizer shortages following the Ukraine-Russia war. Although cereal production per person has increased since the 1980s, the number and proportion of people experiencing food insecurity have not declined equally fast. This is because food distribution has never been equal among the population, especially as consumption rates among urbanizing communities have been increasing (Mohamed and Weldesilassie, 2021).

3.2 *Persistent food insecurity*

Despite the positive developments, there are concerns that the projected increase in agricultural production may not be sufficient to meet the demand of the rapidly growing population. Figure 3.2 highlights the potential challenges that lie ahead. In 2022, approximately 20% of the country's population was experiencing acute food insecurity (FAO et al., 2022). According to Ethiopian Public Health Institute and Inner City Fund (EPHI and ICF) (2021), 37% of children under the age of five suffer from stunting, 21% are underweight, and 7% are wasted. About 22% of women aged 15–49 have a body mass index (BMI) below 18.5 (the ideal range being 20–24.9), underscoring their vulnerability to malnutrition.

Cereal production plays a critical role in Ethiopia's food security as it accounts for an average of 68% of the total per capita daily caloric intake (Wolle et al., 2020; Worku et al., 2017). Unfortunately, with the current and targeted trends in agriculture and population growth rates, Ethiopia is unlikely to achieve grain self-sufficiency by 2050. Even with efforts to supplement domestic production with imports, the magnitude of the deficit remains concerning (Figure 3.2). It is anticipated that around 2.5 million individuals, constituting 1.7% of the projected population in 2050, will experience extreme food insecurity (IPC, 2019; WFP, 2023).

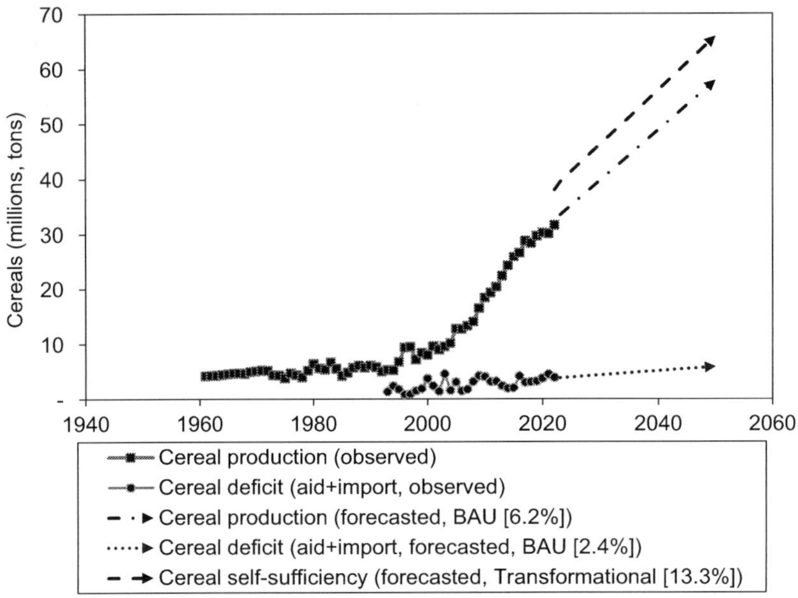

Figure 3.2 Historical and projected cereal production and deficit in Ethiopia (1961–2050). BAU stands for business-as-usual.

Source: Authors' calculation from Central Statistical Agency, Ethiopia (CSA), The World Bank Group (WB), Wold Food Program (WFP), and FAOSTAT and forecast made using the Autoregressive integrated moving average (ARIMA) model.

For Ethiopia to achieve full food self-sufficiency and eliminate the need for imports and food aid, the growth rate in agriculture needs to increase beyond the current target. Figure 3.2 illustrates that Ethiopian agriculture must grow at an annual rate of 13.3% to attain complete independence from food imports and ensure 100% food self-sufficiency through domestic production alone. However, whether domestic production should be considered as the sole path to food self-sufficiency may be debated. Such a significant need for increasing domestic production cannot be accomplished by adhering to business-as-usual practices. Meeting the target growth rate of 13.3% annually calls for a radical shift in the approach to agricultural development. It is a challenging but necessary endeavour to ensure Ethiopia's long-term food security, reduce dependency on imports, and create a sustainable agricultural sector capable of feeding its growing population.

3.3 *Fragmented farms, suboptimal yields, and low labour productivity*

There is a significant literature that reports an inverse relationship (IR) between land size and productivity (Delvaux et al., 2020), although recent findings suggest that the IR is an artefact of measurement error (Abay et al., 2019; Ayalew et al., 2024; Carletto et al., 2013). Contrastingly, agricultural intensification is constrained by the small sizes of farms and the absence of economic incentives (Giller & Andersson, this volume). Figure 3.3 shows that the IR exists only when we consider small farm sizes, less than 5 ha. These farms represent typical smallholder farm sizes in Ethiopia. Fragmentation of land may impede crop intensification due to numerous factors, including lack of economies of scale, limited access to modern agricultural techniques, low labour productivity, and inadequate resource allocation.

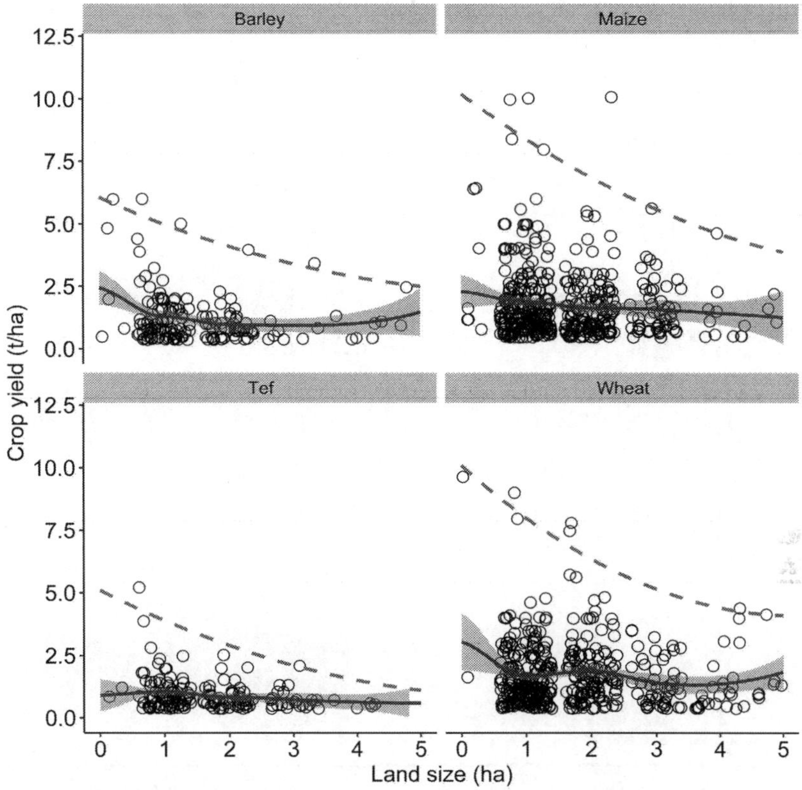

Figure 3.3 Land size per farm and average cereal productivity for non-clustered typical smallholder farms across four crops and from four states in Ethiopia.

Source: Dureti, Tabe-Ojong, and Owusu-Sekyere (2023). Bold lines represent fitted values to the mean, while the broken lines represent values fitted to 99 percentiles.

Upon examining the Agricultural Commercialization Clusters (ACC), an increase in land size for the range of 10–20 ha corresponds to a slight rise in cereal yield (Figure 3.4). The highest cereal productivity appears to be associated with land sizes ranging from 20 to 30 ha. This range potentially allows optimal use of resources, enabling farmers to achieve greater yields per unit of land. Similarly, for land sizes greater than this range, productivity either declines or remains constant, depending on the crop type. Notably, land sizes exceeding 40 ha are associated with a decline in productivity for barley and maize, while productivity remains constant for teff and wheat beyond this range. This suggests that there may be an optimal land holding size for optimizing cereal yield, similar to what is done for other agricultural inputs such as fertilizer. Beyond a certain point, the law of diminishing marginal returns applies, whereby additional land beyond the optimum size may result in reduced efficiency. Data from the ACC initiative indicates that cluster sizes larger than 40 ha do not offer additional productivity benefits. In many instances, clusters larger than this cut-off can actually lead to a decline in productivity. Although the mean land productivity of clustered farms consistently surpasses that of non-clustered farms, it is challenging to solely attribute these gains to land size, given that cluster farms benefit from improved access to inputs and extension advice. In addition, it is important to approach productivity figures resulting from government-led campaigns with caution, as they may require nuanced interpretations.

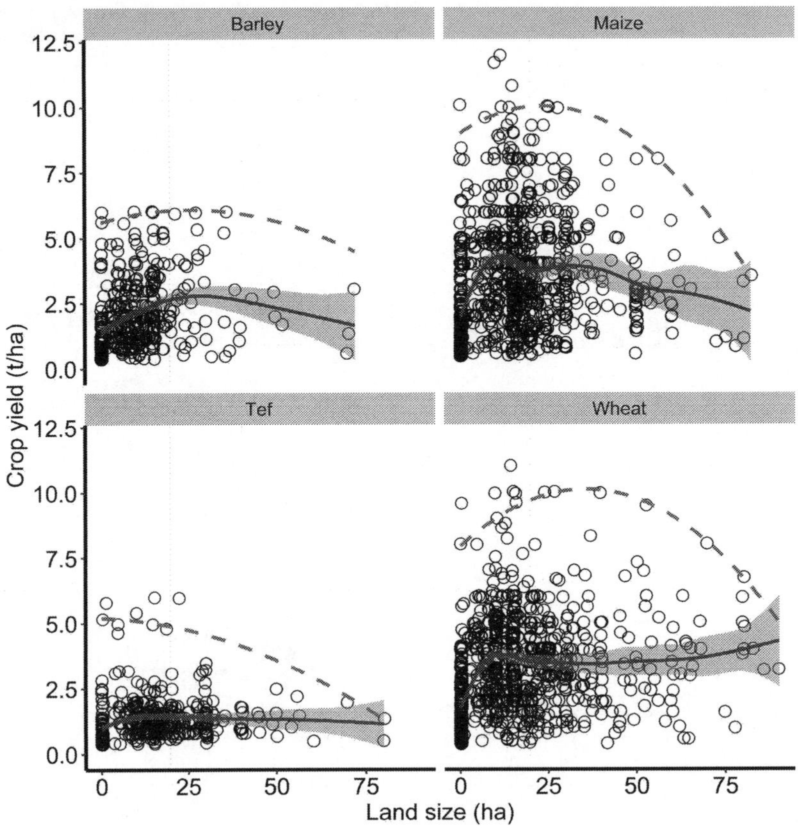

Figure 3.4 Land size per farm and average cereal productivity for clustered farms organized under the Agricultural Commercialization Clusters (ACC) implemented by the Ethiopian government across four crops and in four states in Ethiopia.

Source: Dureti, Tabe-Ojong, and Owusu-Sekyere (2023). Bold lines represent fitted values to the mean, while the broken lines represent values fitted to 99 percentiles.

4 Navigating the future: towards sustained food security for Ethiopia

4.1 Learning and building on past successes

As outlined earlier, Ethiopia has made significant strides in transforming its smallholder agriculture. Nonetheless, two key gaps remain, namely the narrow approach to sectoral change and a lack of implementation capacity (UNCTAD, 2022). In the past, many agricultural transformation efforts focused solely on specific aspects of the sector without considering the broader systemic factors. This often led to disconnected interventions that failed to address the root causes of low agricultural productivity. For example, many initiatives focused solely on boosting productivity by providing farmers with improved seeds or fertilizers without considering other critical factors such as access to credit, markets, infrastructure, or knowledge transfer. By overlooking these interconnected factors, such approaches miss the opportunity to create sustainable agricultural transformation. To overcome this, a more comprehensive approach that involves the entire agricultural value chain, from production to marketing and beyond, is needed. It requires addressing the complex interactions among farmers, input suppliers, traders, processors, and consumers (see Ingenbleek et al., this volume).

Lack of implementation capacity, especially poor human capacity, was another significant barrier to the success of previous approaches. This includes a shortage of staff with the right mindsets, skills, and expertise, which hampers the design, implementation, monitoring, and evaluation of agricultural development programmes. To address this gap, it is crucial to invest in building human capacity at various levels by providing training and professional development opportunities to agricultural extension workers, researchers, and policymakers. Partnerships between academic institutions, research organizations, and governmental agencies can help bridge the knowledge gap and facilitate knowledge transfer. Capacity-building programmes should focus not solely on technical skills but also on fostering the right attitudes and motivations necessary for innovation, collaboration, and adaptive management.

4.2 Towards land consolidation: institutionalizing cluster farming[3]?

The Ethiopian government is taking proactive steps to initiate land consolidation through cluster farming, which involves grouping of smaller landholdings to form larger, more cohesive farming units while maintaining individual household's ownership rights. While cluster farming shows promise (Dureti et al., 2023; Endalew et al., 2024), there is apprehension regarding the associated expense, particularly with prolonged government involvement. The costs related to infrastructure development, resource allocation, and campaign-style support services may impose significant financial and logistical burdens on the government in the long run. To ensure the success and sustainability of this approach, its institutionalization, thereby facilitating permanent land consolidation, becomes paramount. By implementing targeted measures and efficient collaborations, cluster farming can facilitate achieving economies of scale, improved agricultural practices, efficient resource utilization, and expanded market opportunities. It is worth emphasizing that land-related policies and interventions in Ethiopia have been fiercely debated, largely due to the unsuccessful[4] land collectivization effort implemented by the military government between 1974 and 1984 (Omiti et al., 1999). Smallholders are suspicious of government involvement in land consolidation, fearing it could result in land appropriation, a scenario not uncommon during the Dergue era (Ottaway, 1977). There is a high probability that opinions among Ethiopia's present political and academic elites will sharply vary. Additionally, collective memories of the feudal serfdom system may not have entirely faded away among many smallholder communities in Ethiopia. Due to this delicate nature of land-related concerns, it is essential to consider the following points for cluster farming to thrive.

- Granting user rights on the condition of cluster membership: To encourage active participation and responsibility among farmers within cluster farms, user rights should be contingent upon adherence to cluster farming bylaws, conservation practices, adopting environmentally friendly techniques, and participation in capacity-building programmes. Supportive policies should ensure land tenure security without engaging in the controversial subject of land ownership reform in the country.
- Facilitating input delivery: To support the success of cluster farming, it is crucial to ensure timely and reliable access to and delivery of quality seeds, fertilizers, irrigation equipment, machinery, herbicides, and pesticides. A good strategy is to set up dedicated channels for cluster farms to streamline input distribution and lower costs, ideally with significant private sector involvement.

- Establishing storage, aggregation, and grading mechanisms: To maximize cluster farming benefits, efficient post-harvest handling and market integration are essential. Establishing nearby storage and aggregation centres helps preserve produce, reduce losses, and ensure steady supply. Training and infrastructure for grading and packaging can enhance product quality and market competitiveness.
- Facilitating sustained market access: Creating sustainable market linkages for cluster farms is essential for economic viability. By forming partnerships with market actors like wholesalers, retailers, food processors, and exporters, cluster farms can access markets directly, avoid intermediaries, and secure fair prices.
- Integration with agro-processing industries: In the long term, linking cluster farm products with urban agro-processing industries will benefit both parties. Cluster farms will provide raw materials to industries while accessing larger markets and value addition. Encouraging investment in agro-processing, supporting technology transfer, and incentivizing value addition will boost cluster farms' income and create jobs, strengthening rural-urban connections.
- Creating alternative job opportunities for rural labour: Adopting modern farming technologies in cluster farming may decrease the need for manual labour. The surplus rural labour can be absorbed by diversifying the rural economy through the promotion of industries, entrepreneurship, vocational training, and access to microfinance, thereby ensuring sustainable economic development.

4.3 Inclusive structural change and improved labour productivity

Agriculture-led growth has boosted farm productivity in Ethiopia, yet economic growth has lagged behind rapid population increases. This disparity underscores the necessity for a strategy that accelerates labour migration from agriculture to other sectors, emphasizing the need to address food security and create non-agricultural jobs.

It is important to recognize that the viability and sustainability of these approaches may vary across different regions within Ethiopia. Certain areas, such as the Bale plateau, have already experienced successful agriculture-led economic growth, particularly in wheat production. Conversely, in areas like the Central Rift Valley, where such strategies are less effective, exploring and adopting alternative approaches inclusively are essential.

To quickly transition labour from agriculture, it is essential to develop skills that prepare workers for non-agrarian sectors. For instance, Ethiopia's industrial parks had to employ foreign workers due to the local workforce's lack of necessary skills. Short, targeted technical and vocational training programmes, designed for emerging sectors, are crucial for this transition.

4.4 Revisiting cereal production and value chain

Ethiopian cereal production has been increasing but still falls short of its potential due to significant yield gaps (see Schut & van Ittersum, this volume) and very low fertilizer usage, averaging 36.2 kg/ha, although maize in some areas may receive up to 300 kg/ha. A recent study by Falconnier et al. (2023) revealed that the rate of mineral fertilizer required to maximize benefits greatly exceeds the current application rates in East Africa. The soils in Ethiopia have been highly degraded, depleted of nutrients, and mined for centuries, making external nutrient input a necessity rather than an option if the country is to transform its agriculture in a sustainable manner. Supplying appropriate amounts of fertilizers to replenish the depleted nutrients in the soil will support sustainable agricultural practices and increase productivity.

Another challenge is significant post-harvest loss, which ranges from 15.5% to 27.2% for major grain cereals and averages 23% across all crops (Debebe, 2022). This loss not only represents a waste of valuable resources but also contributes to food insecurity and economic inefficiency. Smallholders tend to produce only enough for their own consumption and rely on spontaneous surplus sales to cover small expenses such as basic consumer goods, school, or emergency medical fees (MOFED, 2010). The seasonality of the grain market has been a common disincentive for smallholders to produce surplus for the market (Negede et al., 2024). The current market value chain does not encourage smallholders to produce with commercial intentions, further limiting their economic opportunities. Addressing this issue requires interventions that provide smallholders with access to stable and profitable markets, incentivizing them to increase production and engage in commercial agriculture.

Grain companies are vital in transforming grain production, yet they are either absent or inefficient in the country. Companies involved across the grain value chain—production to processing and marketing—can be facilitated through public-private partnerships and government support. This approach would ensure fair returns for smallholder farmers, enhance production efficiency, and provide reliable market access, thus transforming Ethiopia's agricultural sector.

4.5 *Beyond food self-sufficiency: exploring multiple pathways to food security*

Ethiopia has recognized that achieving food self-sufficiency is not just an economic endeavour but also an important national security agenda. The recent wheat initiative serves as clear evidence of Ethiopia's commitment to achieving self-sufficiency and mitigating potential food security crises that could arise due to disruptions in the global supply chain. However, it is important to consider a broader range of strategies beyond self-sufficiency. Ethiopia possesses differentiated competitive advantages in the production of various commodities. The country, for instance, has a strong competitive advantage in boosting wheat production through the intensification of rainfed wheat. By contrast, large-scale irrigated wheat production does not offer the same competitive advantage due to concerns surrounding the large capital investment required to establish irrigation schemes, extended cost of maintaining the infrastructure, and their long-term environmental sustainability. Lowland irrigated wheat production is likely to face major issues of salinity and sodicity unless well-managed. Precision irrigation schemes may reduce these risks and improve water use efficiency, although establishing and maintaining such irrigation infrastructure are extremely expensive for the government. Smallholders may lack the incentives to pursue irrigated lowland production due to these obstacles. To address these challenges, Ethiopia could allocate its potential for irrigation to the cultivation of high-value commodities. Focusing on the production of crops for cooking oil, fruits, and horticultural crops, as well as other commodities such as cotton, tobacco, and sugarcane, would provide a competitive edge. The country needs to analyse whether achieving food self-sufficiency by replacing imports entirely is better than import of commodities through strategic regional and international trade using resources that can be generated from exporting high-value crops.

5 Conclusion: the way forwards

To feed the burgeoning population, Ethiopia needs to transform its agriculture. This endeavour presents a formidable challenge due to the intricate and multifaceted factors involved. If such a transformation is achievable at all, Ethiopia stands as one of the few countries with a genuine potential to accomplish it, mainly because of the country's present commitment and impressive track record. A comprehensive approach to agricultural transformation is realistic, although challenging. Efforts so far have concentrated narrowly on specific commodities, leading to fragmented interventions and a restricted potential for impact. To overcome this, a holistic perspective that addresses interlinked factors is needed. This involves understanding complex interactions among actors and sectors. Building implementation capacity is vital, including partnerships to facilitate knowledge transfer to bridge existing gaps.

Consolidating fragmented landholdings into cohesive farming units via the institutionalization of cluster farming could significantly accelerate agricultural transformation. This approach promises economies of scale, which can enhance the adoption of productivity-boosting technologies and practices while expanding market opportunities. Optimizing cereal production and value chain management requires careful attention to fertilizer use and minimizing post-harvest losses. Additionally, establishing strong grain producing, marketing and processing companies could improve efficiency and offer smallholders stable market access.

Inclusive structural change is needed to transition the labour force from agriculture and create diverse employment opportunities. This requires investment in education and skills development, promoting technology adoption, improving finance access, developing market linkages, and implementing supportive policies and reforms. Beyond food self-sufficiency, Ethiopia should explore multiple pathways to food security. This includes capitalizing on competitive advantages, intensifying production of commodities where it excels, and considering strategic regional and international trade, the other times.

Notes

1 These GHI calculation scores are made on a 100-point scale, using combination of indicators that include standardized undernourishment, child stunting, child wasting, and child mortality rates, where 0 GHI value is the best score (no hunger) and 100 is the worst. In practice, neither of these extremes is reached. A value of 100 would signify that a country's undernourishment, child wasting, child stunting, and child mortality levels each exactly meets the thresholds set slightly above the highest levels observed worldwide in recent decades. A value of 0 would mean that a country had no undernourished people in the population, no children younger than five who were wasted or stunted, and no children who died before their fifth birthday (GHI, 2022). Ethiopia ranks 101 in the world for the GHI in 2023.

2 The Ethiopia's Food Systems transformation initiative focuses on five action areas that include ensuring access to safe and nutritious food for all, shifting to sustainable consumption patterns, boosting nature-positive production, promoting advanced, equitable livelihoods, and building resilience to vulnerabilities and shocks (Federal Democratic Republic of Ethiopia, 2021). These action areas are designed to align with United Nations Forum on Sustainability Standards Action Tracks.

3 Cluster farming is an agricultural practice that aims to optimize resources and enhance productivity by strategically grouping smallholder farmers within a defined geographical area. The method leverages the proximity of farm plots to foster collaboration, share resources, and achieve economies of scale.

4 With the exceptional case of the Arsi-Bale plateau, where collective farming from the Dergue period succeeded and is still viable.

References

Abay, K.A., Abate, G.T., Barrett, C.B., & Bernard, T. (2019). Correlated non-classical measurement errors, 'Second best' policy inference, and the inverse size-productivity relationship in agriculture. *Journal of Development Economics*, *139*, 171–184.

African Union. (2021). *Ethiopia country food and agriculture delivery compact*. African Union, Addis Ababa, Ethiopia.

Anteneh, A., & Asrat, D. (2020). Wheat production and marketing in Ethiopia: Review study. *Cogent Food & Agriculture*, *6*. https://doi.org/10.1080/23311932.2020.1778893

Assefa, B.T., Chamberlin, J., Reidsma, P., Silva, J.V., & van Ittersum, M.K. (2020). Unravelling the variability and causes of smallholder maize yield gaps in Ethiopia. *Food Security*, *12*, 83–103.

ATI. (2016). *Agricultural transformation agenda progress report covering 2011–15 in the GTP I period.* Addis Ababa.

Atsbaha, G., & Tessema, B. (2012). A review of Ethiopian agriculture: Roles, policy and small-scale farming systems. *Glob Grow Caseb Insights into African Agric, 2010*, 36–65.

Ayalew, H., Chamberlin, J., Newman, C., Abay, K.A., Kosmowski, F., & Sida, T. (2024). Revisiting the size–productivity relationship with imperfect measures of production and plot size. *American Journal of Agricultural Economics, 106*, 595–619. https://doi.org/10.1111/ajae.12417

Belachew, K.Y., Maina, N.H., Dersseh, W.M., Zeleke, B., & Stoddard, F.L. (2022). Yield gaps of major cereal and grain legume crops in Ethiopia: A review. *Agronomy, 12*, 2528.

Carletto, C., Savastano, S., & Zezza, A. (2013). Fact or artifact: The impact of measurement errors on the farm size-productivity relationship. *Journal of Development Economics, 103*, 254–261. https://doi.org/10.1016/j.jdeveco.2013.03.004

CSA. (2020). Population projections for Ethiopia 2007–2037. Central Statistical Agency Population of Ethiopia, 188.

Debebe, S. (2022). Post-harvest losses of crops and its determinants in Ethiopia: Tobit model analysis. *Agriculture & Food Security, 11*, 1–8. https://doi.org/10.1186/s40066-022-00357-6

Delvaux, P.A., Riesgo, L., & Gomez y Paloma, S. (2020). Are small farms more performant than larger ones in developing countries? *Science Advances, 6*, 1–11. https://doi.org/10.1126/sciadv.abb8235

Dessie, T., Zewdie, Y., Yilma, Z., Ayalew, W., & Haile, A. (2022). *Yelemat Tirufat: An overview of the initiative and lessons of experience from selected livestock development interventions in Ethiopia*. International Livestock Research Institute.

Diriba, G. (2020). Agricultural and rural transformation in Ethiopia obstacles. Triggers and Reform Considerations Policy Working Paper.

Dureti, G.G., Tabe-Ojong, M.P., & Owusu-Sekyere, E. (2023). The new normal? Cluster farming and smallholder commercialization in Ethiopia. *Agricultural Economics (United Kingdom), 54*, 900–920. https://doi.org/10.1111/agec.12790

Endalew, B., Elias, A., & Yasunobu, K. (2024). Impact of cluster farming on smallholder farmers teff commercialization in Ethiopia. *CABI Agriculture and Bioscience, 5*, 1–12. https://doi.org/10.1186/s43170-024-00220-7

EPHI & ICF. (2021). Ethiopia mini demographic and health survey 2019: Final report.

Falconnier, G.N., Leroux, L., Beillouin, D., Corbeels, M., Hijmans, R.J., Bonilla-Cedrez, C., ... & Affholder, F., 2023. Increased mineral fertilizer use on maize can improve both household food security and regional food production in East Africa. *Agricultural Systems, 205*, 103588.

FAO. (2018). The state of food security and nutrition in the world 2018: Building climate resilience for food security and nutrition. Food & Agriculture Organization.

FAO, IFAD, UNICEF, WFP, & WHO. (2022). The state of food security and nutrition in the world 2022. https://doi.org/10.4060/cc0639en

FAO. (2021). Global report on food crises: Joint analysis for better decision. Food Security Information Network, FAO, 1–202.

FAOSTAT. (2023). *World food and agriculture – statistical yearbook 2023*. FAO. https://doi.org/10.4060/cc8166en

Federal Democratic Republic of Ethiopia. (2021). Vision 2030: Transforming Ethiopian food systems.

GHI. (2022). Global hunger index 2022: One decade to zero hunger: Linking health and sustainable food systems.

IPC. (2019). IPC acute food insecurity analysis July 2019-June 2020, integrated food security phase classification.

Mellor, J.W., & Dorosh, P. (2010). Agriculture and the economic transformation of Ethiopia. ESSP II Working Paper.

MOF. (2020). A homegrown economic reform agenda: A pathway to prosperity, public version: Edited - March, 2020, 42.

MOFED. (2010). Growth and transformation plan II (GTP II) (2010/11–2014/15). *Ministry of Finance, Economic Development, 1*, 14–120.

Mohamed, J., & Weldesilassie, A.B. (2021). *Food consumption patterns in Ethiopia*. Addis Ababa.

Moreland, S., & Smith, E. (2012). *Modeling climate change, food security, and population: Pilot-testing the model in Ethiopia, MEASURE evaluation*. Addis Ababa.

National Planning Commission. (2016). Growth and transformation plan II (GTP II) 2015/16–2019/20). *World Bank Gr. I*, 236.

National Planning Commission. (2018). The second growth and transformation plan review report. National Planning Commission, 110.

Negede, B.M., De Groote, H., Minten, B., & Voors, M. (2024). Does access to improved grain storage technology increase farmers' welfare? Experimental evidence from maize farming in Ethiopia. *Journal of Agricultural Economics*, *75*, 137–152. https://doi.org/10.1111/1477-9552.12546

Omiti, J.M., Parton, K.A., Sinden, J.A., & Ehui, S.K. (1999). Monitoring changes in land-use practices following agrarian de-collectivisation in Ethiopia. *Agriculture, Ecosystems & Environment*, *72*, 111–118. https://doi.org/10.1016/S0167-8809(98)00162-5

Onoja, A.O., Abraha, A.Z., Gebrekidan, A., & Achike, I.A. (2019). Climate-smart agricultural practices (CSA) adoption by crop farmers in semi-arid regions of West and East Africa: Evidence from Nigeria and Ethiopia. In P. Castro, A. Azul, W. Leal Filho, & U. Azeiteiro (Eds.), *Climate change-resilient agriculture and agroforestry* (pp. 89–113). Springer.

Osh, P., Thurlow, J., KebebeDor, F.W., Ferede, T., & Taffesse, A. (2018). Structural change and poverty reduction in Ethiopia: Economy-wide analysis of the evolving role of agriculture (No. 123), Strategy Support Program, Working Paper. Addis Ababa.

Ottaway, M. (1977). Land reform in Ethiopia 1974–1977. *African Studies Review*, *20*, 79–90. https://doi.org/10.2307/523755

Sacko, J., & Mayaki, I. (2020). Comprehensive Africa agriculture development programme (CAADP). *Biennial Review Report 2015–2018, 2*, 1–87.

Senbeta, A.F., & Worku, W. (2023). Ethiopia's wheat production pathways to self-sufficiency through land area expansion, irrigation advance, and yield gap closure. *Heliyon*, *9*, 1–13.

UN-DESA. (2022). World population prospects 2022: The sustainable development goals report 2022, July 2022. UN-DESA, New York.

UNCTAD. (2022). Workshop on fostering productive capacities in Ethiopia for industrialization, export diversification, and inclusive growth [WWW Document]. https://unctad.org/meeting/workshop-fostering-productive-capacities-ethiopia-industrialization-export-diversification

Van Dijk, M., Morley, T., van Loon, M., Reidsma, P., Tesfaye, K., & van Ittersum, M.K. (2020). Reducing the maize yield gap in Ethiopia: Decomposition and policy simulation. *Agricultural Systems*, *183*, 102828.

Webb, P., & von Braun, J. (1994). *Famine and food security in Ethiopia: Lessons for Africa*. John Wiley & Sons Ltd.

WFP. (2023). Food security, vulnerability and livelihoods across Eastern Africa: Trends and drivers 2015–2022. Research Assessment & Monitoring (RAM).

Wolle, A., Hirvonen, K., de Brauw, A., Baye, K., & Abate, G.T. (2020). *Household food consumption patterns in Addis Ababa, Ethiopia*. International Food Policy Research Institute.

Worku, I.H., Dereje, M., Minten, B., & Hirvonen, K., 2017. Diet transformation in Africa: The case of Ethiopia. *Agricultural Economics*, *48*, 73–86.

World Bank. (2015). Poverty global practice Africa region Ethiopia poverty assessment 2014. AUS6744 v1.

Yigezu Wendimu, G. (2021). The challenges and prospects of Ethiopian agriculture. *Cogent Food & Agriculture*, *7*, 1923619.

4 Past, present and future of food security in Egypt

Racha Ramadan and Antoine Castet

1 Introduction

Egypt is a net food importing country with a population of more than 100 million in 2023, a GDP per capita of USD 4,089 (in constant 2015 prices) and an unemployment rate of 7.1% (according to the data from World Development Indicators). Poverty and food security are major challenges to the development of Egypt. Food demand, mainly for cereals, is expected to increase with population growth, and changes in dietary habits. As local agricultural production is insufficient to meet demand, dependence on imports increases. Within this context, food security in Egypt is challenged by several factors at both food supply (national production and imports) and food demand (access and utilization). These factors are expected to be exacerbated in future with urbanization and climate constraints, in addition to crises such as the COVID-19 pandemic and the Russian invasion of Ukraine.

On the demand side, food security in Egypt provides an economic challenge. Poverty and food price inflation limit access to food. Recently, with currency devaluation and economic slowdown resulting from the pandemic, inflation reached 25.8%, eroding the purchasing power of households. The Egyptian food subsidy system, implemented after the Second World War, plays a significant role in ensuring the availability of and access to basic food items for the Egyptian population. The importance of this system, which covers around 88% of the Egyptian population, increases during periods of crisis.

On the supply side, local agricultural production is constrained by limited natural resources such as water and land. Climate change, reflected in higher temperature, rising sea levels and water scarcity, will adversely impact agricultural production potential in the country. The growing population and increasing food demand resulted in an increasing gap between food demand and food supply. This gap is overcome by increasing dependence on imports. Worldwide, Egypt is the seventh importer of maize (with a self-sufficiency rate of 46%) and the leading wheat importer (with a self-sufficiency rate of 43%). For other products such as rice, potatoes, fresh vegetables, eggs and milk, the self-sufficiency rates exceed 90% (FAO's Food outlook, June 2023; CAPMAS, 2020). The high dependence on imports leaves the country vulnerable to the fluctuation of international food prices.

This chapter aims to discuss different dimensions of food security in Egypt. Section 2 discusses food availability and reviews food access, food utilization and the role played by the Egyptian food subsidy system. Section 3 tackles the main challenges to food security in Egypt. Finally, Section 4 concludes and provides policy recommendations.

DOI: 10.4324/9781032649696-6

2 Egyptian food security: food supply and food demand

Egypt is a net food importer with an average cereal dependency ratio around 48% in 2016–2018. In 2020, 27.3% of the population was considered as moderate or severe food insecure and 6% were considered undernourished (World Development Indicators, 2023). To place Egypt's food security status in a global and African perspective, we can use the Global Food Security Index (GFSI). This index, designed and constructed by Economist Impact and supported by Corteva Agriscience, considers food affordability, availability, quality and safety, and sustainability and adaptation across 113 countries.[1] Egypt ranks 77th on this index, 5th in Africa,[2] with a score of 56. It is a clear score improvement but a downgrading relative to 2012, the first year of the index. The score was 51.6 while the ranking was 52nd on this index, and 2nd in Africa. Improvements were observable mainly in affordability, and in sustainability and adaptation, and less so in availability, and quality and safety.

2.1 Food availability

Food availability is ensured through local agricultural production and imports. For thousands of years, the water of the Nile has been essential for agricultural sector, which employs a third of the population, and contributes 12% of the GDP. Around 60% of total food produced in Egypt originates from the Nile Delta (Fishar, 2016).

Since the 1950s, successive Egyptian governments have tried to expand arable land, most importantly into the dry areas west of the Nile Delta (Acloque-Desmulier, 2014). Those efforts led to a substantial increase of the total cultivated land in Egypt, from 2.6% of total land area in 1961 to 5% in 2020, corresponding to an absolute increase of 1.4 million hectares. The change in covered area since 1961 is shown in Figure 4.1 and is characterized by two phases. The first phase corresponds to the "Tahrir Province" land extension project, launched in 1958, which aimed to experiment a new production model in the desert. The results of this ambitious plan proved

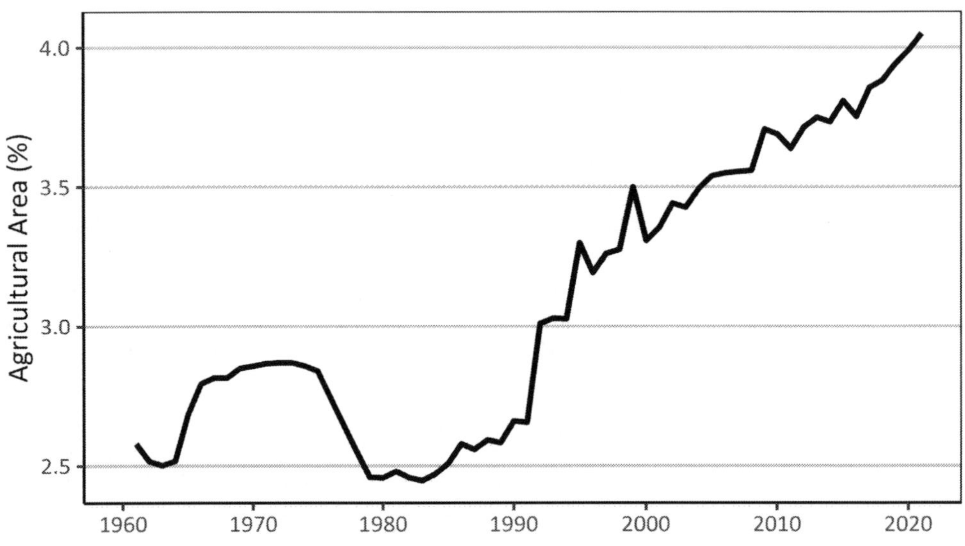

Figure 4.1 Agriculture area in Egypt (% of total land area), 1961–2022.

Source: FAOSTAT (2023).

disappointing, as the cost of land reclamation turned out to be higher than expected. Therefore, following the Six Day War of 1967, the government chose to slow down investment in the "Tahrir Province" in favour of consolidating agricultural production in the already cultivated area. In 1981, Law 143 initiated a new large-scale attempt to extend the land frontier, defining the status of desert land and the terms of its acquisition. This law stipulates that the desert land is owned by the government which decides on its allocation. This has initiated the second phase where the government encouraged the creation of large commercial farms utilizing capital-intensive agricultural technologies at the margins of the Nile Delta, known as the "New Lands" (Acloque-Desmulier, 2019). These farms focused on the production of high-value cash crops, as horticultural products, for export. The use of central-pivot irrigation in the New Lands also enabled the production of cereals such as wheat, and tubers such as barley, potato and sugar beet. New Lands now represent around 15% of the cropped area. The New Lands increasingly also hosted small family farms. Those small producers, in most cases, work with their families on the land and call upon a daily workforce during periods of high activity (Acloque-Desmulier, 2019).

The coexistence of those two models, highly mechanized and small-scale family farming, is quite specific to the New Lands. Indeed, in the "Old Lands", Egyptian agriculture is characterized by the predominance of smallholdings (Tull, 2020; Almas and Usman, 2021). Large farms, above 50 feddans (1 feddan = 0.42 hectares), represent only 0.2% of the landholders, but around 17% of the land area. In 2010, 91.75% of farmers hold 5 feddans or less, but around 47% of the land area (Aboul-Naga et al., 2017). This high land fragmentation inhibits farmers from benefiting from economies of scale and to practise capital-intensive and market-oriented farming. Instead, old lands farmers mainly practise subsistence farming and produce small and irregular amounts for the market (Daburon et al., 2014).

To ensure food availability, increasing agricultural productivity is a key factor. This productivity is historically high in Egypt which relies on irrigation, not rainfall. In 1976, the completion of the Aswan high dam further increased agricultural productivity by putting an end to annual flooding. Until its establishment, crops located at the edge of the Nile benefited, during floods, from its irrigation and fertilizing silt. However, a flood of too high intensity could destroy the crops, while a flood of too low intensity could dry them out. The construction of this dam as well as the creation of numerous canals across the Nile Delta allows farmers to irrigate their land all year round, following a two seasonal pattern, doubling the previous period suitable to agriculture. This allows the cultivation of wheat and clover in winter, and maize, sorghum (subsistence crops), rice and cotton (cash crops) in summer. The increase in productivity, made possible by the dam, may partly explain why the arable land reduction did not affect the production. In addition, the consolidation of old lands agriculture following the disappointment of Tahrir province project around 1967 helped to improve the production, notably through investment in agricultural technology and water management improvement (Fuglie et al., 2021).

Figure 4.2 indicates the production (in tons) of the main agricultural items. It confirms the importance of not only of wheat, vegetables and fruits, but also of milk and meat. Poultry especially is one of the main agricultural industries in Egypt. There are approximately 60,000 intensive poultry farms, which contribute 10% of agricultural value added (ASL2050, 2018).

Despite the increasing agricultural production, there is a gap between local food production and food demand. Import dependency for several key staples has increased steadily to cover this gap. For cereals, the import dependency ratio was 30% in the 1960s. Since the 1970s, dependency on imports, mainly wheat, increased. The import dependency ratio has reached 48% for cereals and 43% for wheat (FAOSTAT, 2023; Tsakok, 2023). The increasing demand for imports and the rising international prices increased the value of imports, mainly during crises period. Figure 4.3 shows fluctuations in the quantity of wheat imported since 1960.

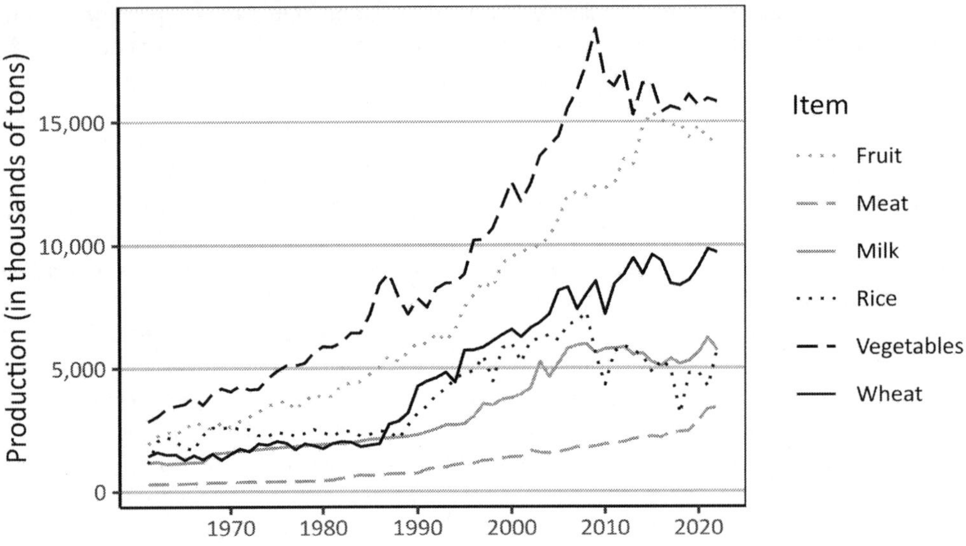

Figure 4.2 Agricultural production of the main crops products (tons).

Source: FAOSTAT (2023).

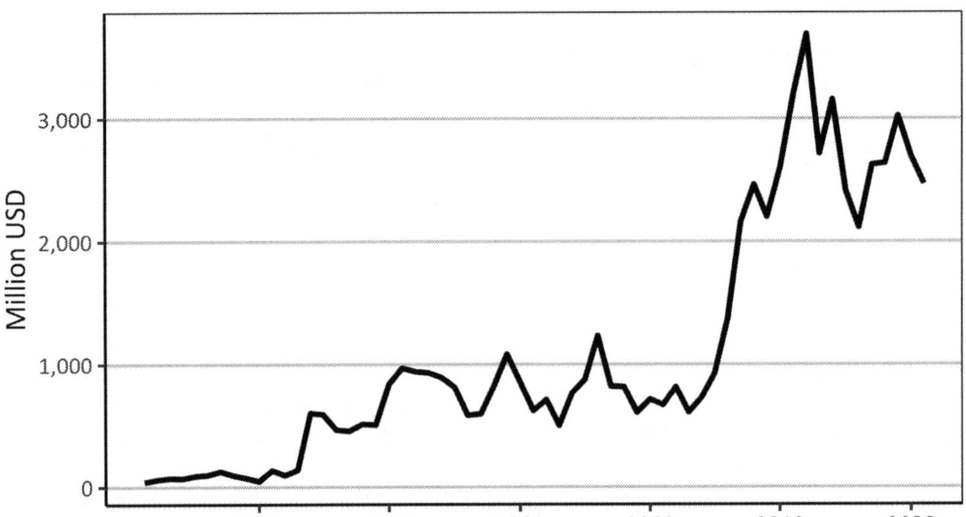

Figure 4.3 The quantity of wheat imports since 1961.

Source: FAOSTAT (2023).

The quantity imported has risen sharply since 1960. As Egyptian wheat production is not sufficient to meet the growing demand due to population growth, Egypt has become the world's largest importer of wheat. The COVID-19 crisis has severely disrupted the international market, leading to a drop in imports in 2020–2021, which seems to be recovering in subsequent years. Recently, the Russo-Ukrainian War of 2022 led to a 100% increase in the world price of wheat; 85% of the wheat imported by Egypt comes from these two countries. As a result, the value of imported wheat has risen, which is worrying given the large quantities imported (Abay et al., 2022).

The climate and soil of the New Lands are particularly favourable to fruit and vegetable production. The government incentivizes its production of fruits and legumes. The government's strategy, since the 1990s, has been to produce high-value agricultural products, sell them on the international market and use the income to import wheat.

Seeking to reduce import dependency, the Egyptian government declares the wheat procurement price in the beginning of each season to encourage its production. Additionally, Egypt doubled the capacity of grain silos to improve grain storage and reduce waste and losses. During the pandemic and the Russo-Ukrainian War, Egypt applied several measures to ensure food availability. These measures include expanding the areas planted with wheat and diversifying the sources of imported wheat by importing from other countries such as India, France, Germany and Romania.

2.2 Food access and food utilization

Having enough food available at the country level, through local production and imports, does not necessarily translate to food security at the household level. The latter is determined by different socio-economic and demographic characteristics of individuals and households, such as individuals' sex, age, education level, employment status, economic activity, access to social security, size of the household, income level and geographical location (Rose and Charlton, 2002; Sultana and Kiani, 2011; Aidoo et al., 2013). In Egypt women, people with low education, households with a high share of children under six years old and the poor are more likely to be food insecure.

Egypt suffers from the triple burden of malnutrition: obesity, stunting and micronutrient deficiencies (leading to health problems such as anaemia). With urbanization and changes in dietary habits, the prevalence of overweight and obesity has increased. The prevalence of overweight among children under five years old increased to 19% in 2022 compared with 12% in 2000. Among the adult population, 32% were obese in 2016 compared with 22% in 2000. Although the prevalence of stunting among children under five years old has declined since 2000, it remained high with 20% in 2022. Similarly, the prevalence of anaemia among women of reproductive age has declined but it remained high with 28% in 2019 (FAOSTAT, 2023).

This specific triple burden may result from the rapid "nutrition transition" that Egypt, like many low- and middle-income countries, has experienced since the mid-1970s. Economic growth leads to an increase in food availability and a shift from a high-fibre plant-based diet towards an energy-dense diet. In Egypt, this phenomenon has been amplified by the food subsidy system. The rationed products, while ensuring sufficient caloric availability for Egyptians, also encourages the consumption of calorie-rich foods at very low prices. Traditionally, subsidizing products such as sugar, cooking oil and rice encourages their overconsumption at the expense of a more diversified diet of micronutrient-rich foods (such as meat, fish, dairy products, pulses, vegetables and fruits).

The Egyptian food subsidy system[3] is one of the social protection programmes that plays a key role in keeping levels of poverty and food insecurity down. The system is financed by the government and represented around 27% of the financial year 2019/2020 budget (Ministry of Finance, 2019/2020). The system is composed of two subprogrammes. The first subprogramme consists of dark country-style bread, known as *baladi* bread. The price of subsidized bread is less than 1 cent a loaf. Bread subsidies used to be a universal subsidy since bread was available to all consumers without restrictions. The second sub-system is a ration card system that provides the cardholders with a specific quota of several items at a subsidized price. Items include rice, sugar and other products that varied over the years.

Instead of being important for food security, especially during crises, the system has been characterized as inefficient because of waste and losses in the supply chain and lack of precise targeting of those who are eligible. Additionally, the locally produced wheat is mainly consumed by rural consumers and is insufficient to sell to the government for the subsidized bread (Ramadan and Thomas, 2011; Almas and Usman, 2021). This means that imported wheat is mainly used to produce subsidized bread, putting more pressure on the government's budget.

Several crises have affected the system over the years. In the 1970s, the recommendations of the International Monetary Fund to remove food subsidies resulted in what are known in Egyptian history as "food riots". As a result, the government did not remove the food subsidies. In 2008, the food, fuel and financial crisis was buffered in Egypt due to the food subsidy system. Since 2015, several reforms have been implemented to reduce the system's budget burden and improve targeting and effectiveness (Hosni and Ramadan, 2018). With the new system, the bread subsidy is no longer universal. This new system provides 150 subsidized loaves of bread per month to each family member using a bread ration card. The new ration card system is considered as a quasi-cash transfer providing a more balanced diet by offering 33 different products. This system allows consumers to choose products that match their preferences and needs. At the same time, it ensures that this quasi-cash amount is spent by households on food products to ensure minimum basic goods. The beneficiaries' database was updated and revised to reduce inclusion and exclusion errors and to ensure that only eligible households have access to the subsidies. This new system allows consumers to convert savings in their bread consumption into points for consuming other subsidized products (Abdallah and El-Shawarby, 2018). Such policy reduces waste of subsidized bread, giving the consumers incentives to buy only the amount required for their own consumption. Subsidized bread used to be sold in the black market or used as feed for animals (Ramadan, 2015).

Other social protection programmes include cash transfer programmes, "Takaful" and "Karama", which means Solidarity and Dignity in Arabic. These programmes were launched in 2015 and cover approximately 10% of Egypt's population. Takaful programme supports poor families with children under 18 years of age, while Karama supports the poor elderly disabled and orphans (World Bank, 2015, 2018). According to an International Food Policy Research Institute (IFPRI) impact evaluation study (2018), beneficiaries of Takaful increased their food consumption and improved the quality of their diets. They consume more fruits, meat and poultry. During the pandemic, the government increased the number of Takaful and Karama beneficiaries.

3 Food security challenges in Egypt

Structural challenges of food security include limited resources such as water and arable land, rising poverty, growing population, and waste in the food supply chain (Mohamed, 2018). Local agricultural production is insufficient to meet the growing demand of the rising population. Climate change puts more pressure on local agricultural production with its impact on water resources and arable land, in addition to the rising temperature. Increasing food import dependency increases vulnerability to external shocks, particularly given the slowing economic growth and devaluation.

3.1 Natural resource constraints and climate change

Limited natural resources, such as land and water, are constraints for local agricultural production in Egypt. Available land for food production is becoming scarce, especially when considering competition with urbanization. Agriculture is mainly based on irrigation water from the Nile and is the major user of clean water in Egypt, accounting for 79% of total annual freshwater withdrawals. With population and urban growth, there is an increasing competition on water between agriculture, residential, energy and industrial uses (Khedr, 2019; Almas and Usman, 2021). Additionally, ground water is slow to recharge given that rainfall in Egypt is low and the flow from the Nile River is expected to vary due to climate change and the Ethiopian Dam.

Climate change is expected to lead to less suitable conditions for crop cultivation. Estimates show that Egypt will become hotter and drier with long-term reductions in agricultural production (WFP and ODI, 2015; Kandeel, 2019; Tull, 2020). Water scarcity will persist through 2050 and climate change is expected to deteriorate water quality and quantity, decrease per capita water availability and jeopardize the food system. By 2030, the delta region is expected to lose a minimum of 30% of its food production. By 2050, crop yields will decline causing an estimated decrease by 6% of total food production, by 15% in wheat production, by 22% in maize production and by 24% in pulses production (Perez et al., 2021; UNFCC, 2022; Abu Hatab, 2023). Kandagatla and Almas (2019) estimate that wheat consumption will reach 40 million tons, while local wheat production will only be 14.5 million tons by 2050. Therefore, the gap between wheat supply and demand will grow, and dependence on imports is projected to be around 53% in 2050 (Le Mouël and Schmitt, 2018).

3.2 *Population growth*

The growing population had been a key challenge for food security and all development goals in Egypt. Since the 1970s, the average population growth rate is 2.2%, which is higher than the global average. Egypt is the most populous country in the Arab World with 111 million in 2022. However, its population growth is lower compared to other African countries and is expected to decrease to less than 2% by 2050 (UN DESA, 2022; also Van Wissen, this volume).

High fertility rate explains population growth. In the 1960s, fertility rate was 6.8 births per women. The rate declined over the years due to the government family planning programmes to reach 3.1 in 2009, then increased again in 2011 to reach 3.4. Recently, the government implemented new policy to raise awareness and control population growth known as *Two are enough*. Fertility rate declined to reach 2.9 in 2021 (World Development Indicators, 2023).

By 2050, Egyptian population is expected to increase to reach around 160.3 million and urban population is expected to reach 75% of the total population (Assaad, 2020; UNFCC, 2022). The population density inside the cities is already high, especially in the capital Cairo. The government's strategy is to promote the establishment of cities in desert land to avoid a too high urban density and at the same time reduce urban sprawl at the expense of agricultural land.

More food supply is required to match increasing future food demand. Using UN (United Nations) and FAO (Food and Agriculture Organization) recent data and assuming the kilocalorie amount remains similar to the 2021 ones (3,127 kcal/day/capita), with an estimated population (160 million), the kilocalorie demand will be 501 billion per day by 2050. All things being equal, the Egyptian government must increase the food quantity available by 75% through local production or importation.

3.3 *Economic access to food*

Food security in Egypt at the household level can be considered as an economic challenge with 29.8% of the population living in poverty. Low incomes and poverty, in addition to inflation, limit the purchasing power of households, threatening economic access to food. The COVID-19 pandemic resulted in loss of income and higher food prices, deepening the existing challenges. Food security of those working in hard-hit sectors, poor households and those who lost their income was jeopardized. Households resorted to coping strategies such as a reduction in the number of meals consumed or consumption of less nutritious cheap food items (World Food Programme, 2020; UNDP, 2021). According to the Economic Research Forum's Middle East and North Africa Household Survey data on COVID-19 (2021),[4] 68% of households in the lowest quartile reported suffering from limited access to food during the pandemic, while 53% had to reduce their usual meals/portions. These percentages were smaller in the highest income quartile group (46% and 26%, respectively) (Ramadan, 2022).

The Russian invasion of Ukraine created a supply-driven crisis resulting in soaring international food prices. The war aggravated the existing challenges resulting from COVID-19, in a period where governments already had limited fiscal space because of the economic problems caused by the pandemic (Ben Hassan and El Bilali, 2022).

4 Concluding remarks

Security is constrained by several factors at both the food supply and food demand levels. These constraints are expected to be exacerbated with the growing population and climate change. Within this context this chapter reviews food security dimensions in Egypt and their challenges.

Based on the four pathways discussed in the introduction of this book, we can conclude that Egypt is in the agriculture stagnation path. But agriculture sector and food security are a priority on the Egyptian government development agenda. The Sustainable Agricultural Development Strategy 2030 of Egypt is based on three main pillars: improving agricultural productivity, raising food security and ensuring the sustainable use of natural resources. And the national structural reform programme is based on the agriculture sector with the manufacturing industry, communications and information technology (Ministry of Planning and Economic Development, 2023).

On the demand side, there is a necessity to slow population growth to ensure the balance between food production and consumption, as discussed above. Social safety programmes such as food subsidy providing diversified products and cash transfer programmes conditional on investment in human capital are required to face malnutrition, ensure food security and reduce poverty and population growth. But, because of the narrow fiscal space, programmes should target the poorest and most affected groups, such as workers in the informal sector, children and women.

In the long term, government's strategies to increase agricultural production and productivity must consider the adaptation to and mitigation of climate change to ensure sustainable agriculture and environment. Developing the agri-food industry to generate higher value-added products and create more non-farm economic opportunities is a necessary condition to provide more economic opportunities to the young, growing population. Egypt benefits of comparative advantages in many products such as fruits and vegetables. Increasing the competitiveness of these products in the international markets and exporting agri-food products with higher value added might be a significant source of foreign currency for the country. Developing regional cooperation and trade between the countries of the African continent would reduce poverty and ensure food security in Egypt and the whole continent.

Notes

1 The index is a dynamic quantitative and qualitative benchmarking model constructed from 68 unique indicators that measure the drivers of food security across both developing and developed countries. The score takes value from 0 to 100, the higher the value the better the food security status of the country. For more details about the index see: https://impact.economist.com/sustainability/project/food-security-index/about.

2 Egypt's rank is below South Africa, Morocco, Algeria and Tunisia, but above all other African countries covered.
3 There is a significant literature tackling the Egyptian food subsidy system and its reform (Ahmed and Bouis, 2002; Kandil, 2010; Ramadan and Thomas, 2011; Hosni and Ramadan, 2018).
4 The Economic Research Forum COVID-19 Monitor Data for Egypt is a phone survey that contains information on basic socio-demographic characteristics of respondents, self-reports on change in income, food expenditure, food bought, care work for women, employment status, education methods and living conditions in June 2021 – compared to February 2020, before the spread of COVID-19. The survey population consisted of mobile phone users aged 18–64 in Egypt. The samples were stratified by mobile operator and weights are used to be nationally representative.

References

Ahmed, A.U., & Bouis, H.E. (2002). Weighing what's practical: proxy means tests for targeting food subsidies in Egypt. Food Policy, 27, 519–540.

Abay, A.K., Abdelradi, F., Breisinger, C., Diao, X., Dorosh, P.A., Pauw, K., … & Thurlow, J. (2022). Egypt: Impacts of the Ukraine and global crises on poverty and food security. International Food Policy Research Institute (IFPRI). Global Crisis Country Brief 20.

Abdallah, M., & Al-Shawarby, S. (2018). The Tamween food subsidy system in Egypt: Evolution and recent implementation reforms. In H. Alderman, U. Gentilini, & R. Yemtsov (Eds.), *The 1.5 billion people question. Food, vouchers, or cash transfers?* World Bank.

Aboul-Naga, A., Siddik, I., Megahed, W., Salah, E., Ahmed, S., Nageeb, R., … & Impiglia, A. (2017). Study on small-scale family farming in the Near East and North Africa region. *Focus Country*. Egypt. FAO. https://www.fao.org/family-farming/detail/fr/c/471489/

Abu Hatab, A. (2023). Egypt's food system under a perfect storm. https://www.siani.se/news-story/egypts-food-system/

Acloque-Desmulier, D. (2014). Entre Delta du Nil and désert- Front pionnier agricole et recomposition territoriale en Egypte. *Carnets de géographes*, *7*. https://cdg.revues.org/499. https://doi.org/10.4000/cdg.499

Acloque-Desmulier, D. (2019). Conquérir le désert: Recomposition des acteurs et des territoires agricoles en Egypte. These, Université Paris-Nanterre.

Aidoo, R., Mensah, J.M., & Tuffour, T. (2013). Determinants of household food security in the Sekyere-Afram plains district of Ghana. 1st Annual International Interdisciplinary Conference, AIIC 2013, 24–26 April, Azores, Portugal – Proceedings.

Almas, L.K., & Usman, M. (2021). Determinants of wheat consumption, irrigated agriculture, and food security challenges in Egypt. *WSEAS Transactions on Environment and Development*. https://doi.org/10.37394/232015.2021.17.67

ASL2050. (2018). *ASL2050 Livestock production systems spotlight* – Egypt. FAO.

Assaad, R. (2020). Prospects for Egypt's Population and Labor Force: 2000 to 2050. Economic Research Forum Working papers series. No.1398.

Ben Hassen, T., & El Bilali, H. (2022). Impacts of the Russia-Ukraine war on global food security: Towards more sustainable and resilient food systems? *Foods*, *11*, 2301. https://doi.org/10.3390/foods11152301

CAPMAS. (2020). Impact of COVID-19 on Egyptian households till May 2020. In Arabic.

Daburon, A., Alary, V., Ali, A., El-Srogi, M., & Tourrand, J. 2014. Agriculture urbaine et périurbaine, les exploitations laitières du Caire, Égypte. In P. Bosc, J. Sourisseau, P. Bonnal, P. Gasselin, E. Valette, & J. Bélières (Eds.), *Diversité des agricultures familiales: Exister, se transformer, devenir.* Éditions Quæ. Retrieved from https://books.openedition.org/quae/29510

FAOSTAT. (2023). https://www.fao.org/faostat/en/#data/FS

Fishar, M.R. (2016). Nile Delta (Egypt). In C. Finlayson, G. Milton, R. Prentice, & N. Davidson (Eds.), *The wetland book*. Springer. https://doi.org/10.1007/978-94-007-6173-5_216-1

Fuglie, K., Dhehibi, B., El-Shahat, A.A.I., & Aw-Hassan, A. (2021). Water, policy, and productivity in Egyptian agriculture. *American Journal of Agricultural Economics*. https://doi.org/10.1111/ajae.12148

Hosni, R., & Ramadan, R. (2018). Food subsidy or cash transfer: Impact of the food subsidy reform on Egyptian households. *New Medit- A Mediterranean Journal of Economics, Agriculture and Environment*, 3. https://doi.org/10.30682/nm1803b

IFPRI. (2018). Impact evaluation study for Egypt's Takaful and Karama cash transfer program. Synthesis Report: Summary of Key Findings from the Quantitative and Qualitative Impact Evaluation Studies.

Kandagatla, R., & Almas, L.K. (2019). *Egypt's reliance on imported wheat: Concerns, challenges and opportunities*. West Texas A&M University.

Kandeel, A.A. (2019). In the face of climate change: Challenges of water scarcity and security in MENA. 11 June 2019. Atlantic Council. https://www.atlanticcouncil.org/blogs/menasource/in-the-face-of-climate-change-challenges-of-water-scarcity-and-security-in-mena/

Khedr, M. (2019). Challenges and issues in water, climate change, and food security in Egypt. In A.M. Negm (Ed.), *The handbook of environmental chemistry. Conventional water resources and agriculture in Egypt*. Springer.

Le Mouël, C., & Schmitt, B. (2018). Food dependency in the Middle East and North Africa region. Retrospective Analysis and Projections to 2050.

Ministry of Planning and Economic Development. (2023). https://mped.gov.eg/singlenews?id=1336&type=previous&lang=en. Accessed on 30th of November 2023.

Mohamed, N.N. (2018). Egyptian food insecurity under water shortage and its socioeconomic impacts. In Negm, A.M. (ed) *Conventional water resources and agriculture in Egypt. The handbook of environmental chemistry*, vol. 74. Springer. https://doi.org/10.1007/698_2018_239

Kandil, M. (2010). *The subsidy system* in Egypt: *Alternatives for reform*. The Egyptian Center for Economic Studies. Policy Viewpoint. Number 25.

Perez, N.D., Kassim, Y., Ringler, C., Thomas, T.S., Eldidi, H., & Breisinger, C. (2021). *Climate-resilience policies and investments for Egypt's agriculture sector: Sustaining productivity and food security*. International Food Policy Research Institute.

Ramadan, R., & Thomas, A. (2011). Evaluating the impact of reforming the food subsidy program in Egypt: A mixed demand approach. *Food Policy*, *36*(5), 638–646.

Ramadan, R. (2015). Where does food subsidy go? In Terre et Mer: Ressources Vitales Pour la Méditerranée. Cosimo Lacirignola. Bibliothèque de l'iReMMO. Ed. L'Harmattan. No.19. ISBN: 978-2-343-05909-9.

Ramadan, R. (2022). Would the food insecure raise their hands? Applying the case of Egypt in the era of COVID-19. EMANES Working Paper No 59.

Rose, D., & Charlton, K.E. (2002). Quantitative indicators from a food expenditure survey can be used to target the food insecure in South Africa. *The Journal of Nutrition*, *132*(11). https://doi.org/10.1093/jn/132.11.3235

Sultana, A., & Kiani, A. (2011). Determinants of food security at household level in Pakistan. *African Journal of Business Management*, *5*(34). https://doi.org/10.5897/AJBM11.1441

Tsakok, I. (2023). Short of water and under increasing pressure to deliver food security: Key policy considerations the case of the Arab republic of Egypt. Policy Center for the new South-Policy paper 01/23.

Tull, K. (2020). The projected impacts of climate change on food security in the Middle East and North Africa (MENA). K4D Knowledge, evidence and learning for development. Helpdesk Report.

UN DESA. (2022). *World population prospects 2022*. United Nations.

UNDP. (2021). Potential impact of COVID-19 on poverty and food insecurity in the Arab region. RBAS Working papers series. https://www.undp.org/arab-states/publications/potential-impact-covid-19-poverty-and-food-security-arab-region

UNFCC (United Nations Framework Convention on Climate Change). (2022). Nationally determined contribution. Egypt.

WFP and ODI (2015). *Food in an uncertain future. The impacts of climate change on food security and nutrition in the Middle East and North Africa.*

World Bank. (2015). Building resilience and opportunity: Social protection reform in Egypt. Retrieved from https://www.worldbank.org/en/news/feature/2015/06/09/building-resilience-and-opportunity-social-protection-reform-in-egypt. Visited 15 October 2015.

World Bank. (2018). The story of Takaful and Karama cash transfer program. Retrieved from https://www.worldbank.org/en/news/feature/2018/11/15/the-story-of-takaful-and-karama-cash-transfer-program

World Food Programme. (2020). WFP global response to COVID-19. https://www.wfp.org/publications/wfp-global-response-covid-19-september-2020

5 Zimbabwe's agriculture and food security: past, present and future (1960–2050)

Mark Nyandoro and Jens A. Andersson

1 Introduction

Once the breadbasket of southern Africa, exporting food to neighbouring countries, Zimbabwe's recent history is one of food imports, recurrent food shortages and hunger. This chapter analyses the country's pattern of agricultural change and its fluctuating food security situation since the 1960s, and will project the identified trends into the future, towards 2050. To understand food system change in Zimbabwe requires a focus on agriculture as the sector was and continues to be the mainstay of the economy (Nyandoro, 2007). We start our analysis with the 1960s, as this decade marks the emergence of institutionalized food relief as a response to droughts. Until the 1960s, recurrent droughts had sometimes caused food shortages – and severe shortages too (like in 1949) – but until then government had not systematically provided food aid to the poor (Iliffe, 1990).

Two processes are key in our understanding of food system change during Zimbabwe's colonial era: rapid population growth and the racial division of land. When settlers of European descent started to appropriate land in the late 19th century, Zimbabwe was a sparsely populated country with an estimated population of about 1 million people (Zinyama & Whitlow, 1986). A racial division of land was institutionalized in the 1930 Land Apportionment Act, which severely restricted land ownership for the indigenous African population. They were increasingly evicted from the large, privately owned farms of the settlers, and relegated into 'reserves'[1] (Nyandoro, 2019; Phimister, 1993).

Thus, by the 1960s, a rapidly growing African population farmed predominantly on sandy soils that are inherently infertile and prone to rapid degradation under continuous cultivation, and where rainfall is often erratic (Andersson, 2007). A much smaller proportion of the African population was granted privately owned, small-scale commercial farms, in the so-called Purchase Areas, a half-hearted attempt of the colonial government to create a yeoman class of African farmers (Green & Nyandoro, 2024). By contrast, (the descendants of the) European settlers – a minority comprising 2% of the population – occupied 40% of all farmland, the most fertile red loam and dark loam soils found on the highveld, where rainfall is generally more stable and higher (Lima and Lessard, 2023).

Agriculture became the settlers' main occupation and the core of colonial Zimbabwe's economy. The white settler farmers initially struggled, but with unrelenting government support – which also fended off competition from African farmers (Keyter, 1978) – a strong commercially oriented large-scale farming sector was formed. The white commercial farmers could draw on government credit, subsidized fertilizer, seed distribution and marketing systems (Tawonezvi & Hikwa, 2006). The sector was further supported by infrastructure development such as road networks, fertilizer industry support (Minde et al., 2010) and the construction of

DOI: 10.4324/9781032649696-7

dams, enabling water and irrigation development (Nyandoro, 2007). Government-funded agricultural research further contributed to the sector's success. For example, already in the 1930s, soon after the development of the first hybrid maize varieties, Southern Rhodesia (as Zimbabwe was then called) initiated its own research into hybrid maize. The research resulted in the release of the high-yielding, SR 52 hybrid maize variety (Musimwa & Derera, 2017; Eicher & Kupfuma, 1998). Similarly, government investment stimulated the expansion of Southern Rhodesia's major export crops: tobacco and cotton. Support was mainly given to white commercial farmers as they dominated politics; it was a white commercial farmer, Ian Smith, who became prime minister and unilaterally declared independence (UDI) from Britain in 1965.

By the early 1960s, our point of departure in this chapter, colonial Zimbabwe had become an early-industrializing economy (Andersson, 2002), dominated by the commercial farming sector that employed thousands of (foreign) workers. Immediately after the Second World War, an urban industrial sector had developed, attracting African labourers from rural areas and abroad. In the 1960s, tobacco production was the country's largest foreign currency generator, accounting for 10%–43% of the country's GDP (Nyambara & Nyandoro, 2019). Cotton was exported but was also an important raw material for the expanding domestic manufacturing industry, for example, textiles, cooking oil and livestock feeds (Nyandoro, 2007). In Zimbabwe's dualistic agricultural sector, it was the highly productive commercial farming sector that dominated the production of tobacco, cotton and food crops such as maize and wheat, although new areas were opened up for the production of cotton by African smallholders in the Zambezi valley in the 1950s and in the Sanyati-Gokwe frontier regions since the 1960s (Nyambara, 1999; Nyandoro, 2022; Baudron et al., 2011). Meanwhile the concentration of Africans on the degradation-prone lands of the 'reserves', and the imposition of destocking and land management policies there, put African farmers' food production in these areas increasingly under pressure. More and more rural families had to rely on wage labour (remittances) to make ends meet and to sustain their farming activities. But as long as the wage labour employment was secure, so was food security for urban and rural families – either directly through food purchases, or indirectly through investments in smallholder farming in the African reserves.

This chapter argues that since the early 1960s, food security in Zimbabwe is dependent on a thriving wage labour sector that supports both (urban) livelihoods and agricultural investments in the Communal Areas (the former reserves). It shows how the locus of food production has shifted between the commercial and smallholder farming sectors, and how this has made food security more dependent on government intervention and vulnerable to climate variability. The remaining discussion is divided into five periods. First, from 1960 to 1980 the country progressively became an inward-looking economy following international sanctions imposed on its white minority regime. Although increasing numbers of people could no longer provide for their own food and required government food relief, the country managed to keep net food imports to a minimum (Figure 5.1). Second, the decade following independence in 1980, the period of Zimbabwe's smallholder production revolution, was characterized by widespread food security. During this decade the locus of food production shifted to the smallholder sector, making food production increasingly dependent on the vagaries of the weather and government intervention – yet the latter's resources were rapidly dwindling. Third, in the 1990s, an emergent crisis in the wage labour sector manifested itself. Increasing numbers of rural and urban households started to face food insecurity, necessitating food imports. Fourth, the deepening economic crisis culminated in a politically motivated fast-track land reform and economic collapse in the 2000s. Zimbabwe's agrarian structure radically changed, but without investment from either government or rural households' wage labour incomes, smallholder production declined. Food imports and donor-funded emergency food relief became the new normal.

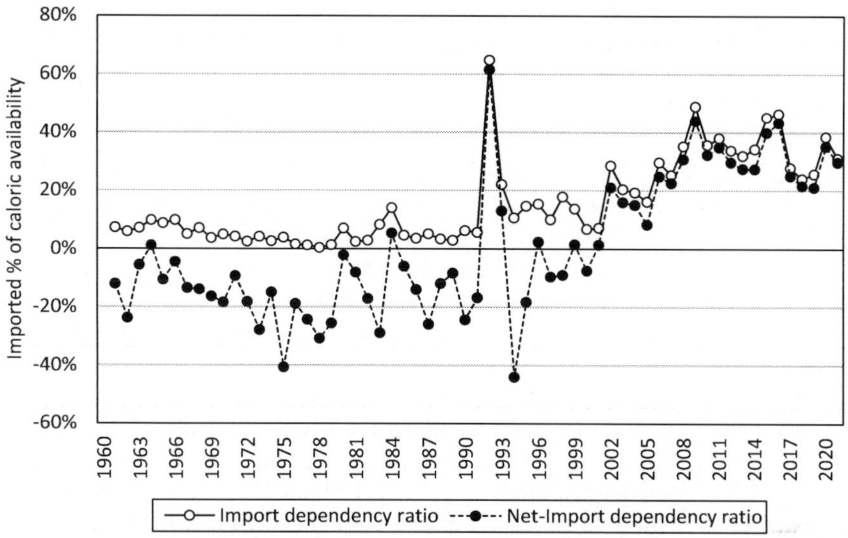

Figure 5.1 Import dependency ratio = imports/(production + imports - exports) and net import dependency ratio = (imports - exports)/(production + imports - exports) for (colonial) Zimbabwe, 1961–2020. Taken from Joshipura (forthcoming).

Lastly, we discuss the past decade and the period towards 2050. Although food insecurity and import dependency remain high, domestic food production is picking up again. Yet, without jobs, urban consumers and smallholder producers will be unable to buy the food and invest in smallholder food production in the Communal Areas. Returning to a path of economic growth is Zimbabwe's major challenge; whether it can or should fuel another smallholder food production revolution is a moot point.

2 Food production in a dualistic agricultural sector: 1960–1980

Although the country recurrently suffered from droughts and famines, Zimbabwe's early colonial period was not marked by massive starvation and hunger-induced deaths (Iliffe, 1990). The drought season of 1959/1960 did, however, mark a turning point. While in previous droughts grain trade had often mitigated food scarcity, in 1960 malnutrition was more widespread and more people needed government food aid, especially the poor.

The 1950s had witnessed agricultural growth in both the African rural areas – where grain production per capita increased – and the settler farmer sector, which more than doubled maize deliveries to the maize control board, while tobacco had become its engine of growth (Iliffe, 1990). Also, the urban sector had grown rapidly after the Second World War. But by 1960 this post-war economic boom was over. Following several seasons of bad rainfall, a looming crisis of smallholder farming in the African 'reserves' became apparent. A growing population concentration in these areas and, in some areas, the imposition of the Land Husbandry Act and its destocking measures (Machingaidze, 1991; Phimister, 1993) had undercut agricultural productivity in these areas.[2] Also, the wage labour economy was doing badly. It could no longer absorb the ever-growing number of job seekers, resulting from rapid population growth and the progressive eviction of Africans from land appropriated from them. The settler-dominated capitalist economy was in crisis, and there was growing urban opposition against the white minority government (van Velsen, 1964). Over the course of several decades, it had created a category of marginalized people who could no longer sustain themselves through farming in the absence of wage labour income (Iliffe, 1990). These people could no longer resolve food scarcity by bartering assets for food. Thus, food shortages, and the need for government intervention to mitigate them, became endemic.

The economic stagnation of the early 1960s was accompanied by growing anti-colonial protest. The Federation (1953–1963) of Northern Rhodesia (now Zambia), Southern Rhodesia (Zimbabwe) and Nyasaland (Malawi) was dismantled (Mlambo, 2014). But in Zimbabwe, this did not result in independence. Instead, the white settler government led by Ian Smith unilaterally declared independence (UDI). Faced with economic sanctions, the Smith government embarked on regulating the financial sector and a policy of import substitution. Aided by instituted low wages and the suppression of labour protests, it was the manufacturing sector that drove economic growth in the period 1965–1975. Meanwhile, government stimulated domestic agricultural production to increase 'imperial self-sufficiency' in raw materials, particularly cotton (Nyandoro, 2007, 2022; Munro, 1976). National food security gained significance in policy making of the white-minority government. In 1975, the Grain Marketing Board (GMB), which had hitherto served the white settler farming sector, extended its buying depots to the African 'reserves'. At the same time, food production was increasingly hampered by guerrilla insurgence from Mozambique and Zambia. White settler farms were sometimes attacked, causing an interruption of operations, while African farmers – in an attempt by the colonial government to stop their support of the guerrillas – were forced into 'protected villages' where farming became impossible. Hence, the liberation war caused widespread food shortages (Nyandoro, 2007).

3 Zimbabwe's smallholder agricultural boom and bust, 1980–2000

At independence in 1980, Zimbabwe inherited an inward-looking economy, with an outdated manufacturing industry that had been starved of productive investment for more than two decades. The government of Robert Mugabe sought to redress the institutionalized racism and socio-economic imbalances of the colonial period, by more fully integrating African farmers and rural areas into the mainstream economy. Massive government investments, supported by the international donor community, were made in schooling, health facilities and infrastructure (electricity, roads, water) in the former reserves – now referred to as Communal Areas. In 1982, Mugabe announced a ten-point policy plan (Rukuni, 1984; Nyandoro, 2007) which meant massive government support for smallholder farming, including irrigation development, agricultural credit, marketing facilities, a re-orientation of agricultural research towards the smallholder sector, an expansion of agricultural extension for smallholder farmers and price support. As a result, Zimbabwe experienced a boom in smallholder maize and cotton production by the mid-1980s (Rukuni and Eicher, 1994). Touted as a 'miracle' that should be replicated elsewhere, the country succeeded in producing food surpluses in most years (Cliffe, 1988) and became known as the 'food basket' of southern Africa. Rural accumulation and affluence were exhibited by smallholder farmers who were able to build brick houses under asbestos roofing, buy cars and accumulate livestock (Nyandoro, 2007).

The large-scale commercial farming sector – as settler agriculture became known after independence – also recovered fast from the disruptions of the liberation war. Although the number of such farms decreased from some 6,000 in 1980 to about 4,000 in 1990 (Muir and Blackie, 1994), the sector flourished through increased tobacco production and diversification into emerging export markets for horticultural products like flowers, fruit and vegetables (Muir and Blackie, 1994). The commercial farmers could remain on the land, as land redistribution was based on a policy of 'willing buyer, willing seller' (Moyo, 2006; Nyandoro, 2012). As part of a 'growth with equity' policy, some 11% of the total farmland was used to resettle over 70,000 families during the 1980s. While well short of the planned 160,000 households, this poverty-focused *minda mirefu* (long fields) resettlement programme was still one of the biggest land redistributions in Africa. A decade later, the resettled farming families often did relatively well as compared to their Communal Area counterparts: they cultivated more land, had higher incomes and accumulated more assets (Hoogeveen & Kinsey, 2001).

Ironically, the smallholder agricultural production revolution of the 1980s laid the foundation for Zimbabwe's increasingly vulnerable food security situation. Massive government investment in smallholder farming had shifted the onus of maize production towards the degradation-prone soils of the Communal Areas, where rainfall is often erratic (Figure 5.2). Maize production thus became more vulnerable to climate variability and to wider macro-economic developments, as crop production on the sandy soils of the Communal Areas requires continued investment in soil fertility – e.g. fertilizers, manure (Andersson, 2007). Meanwhile, the production of food crops like wheat, a winter crop grown by large-scale commercial farmers under irrigation, declined in the 1980s, due to a lack of profitability and increased competition from global markets. These farmers increasingly went for lucrative export crops rather than food crops (Andersson, 2002).

The government's expenditures rose sharply during the 1980s, compromising public investment in rural areas and smallholder farming and government's capacity to buy land for redistribution decreased as land prices rose. By 1990, the land redistribution programme had virtually come to a standstill (Kinsey, 1999) and Zimbabwe's macro-economic ills became increasingly apparent. Unemployment was rising at an alarming rate: on average about 3,000 new jobs were

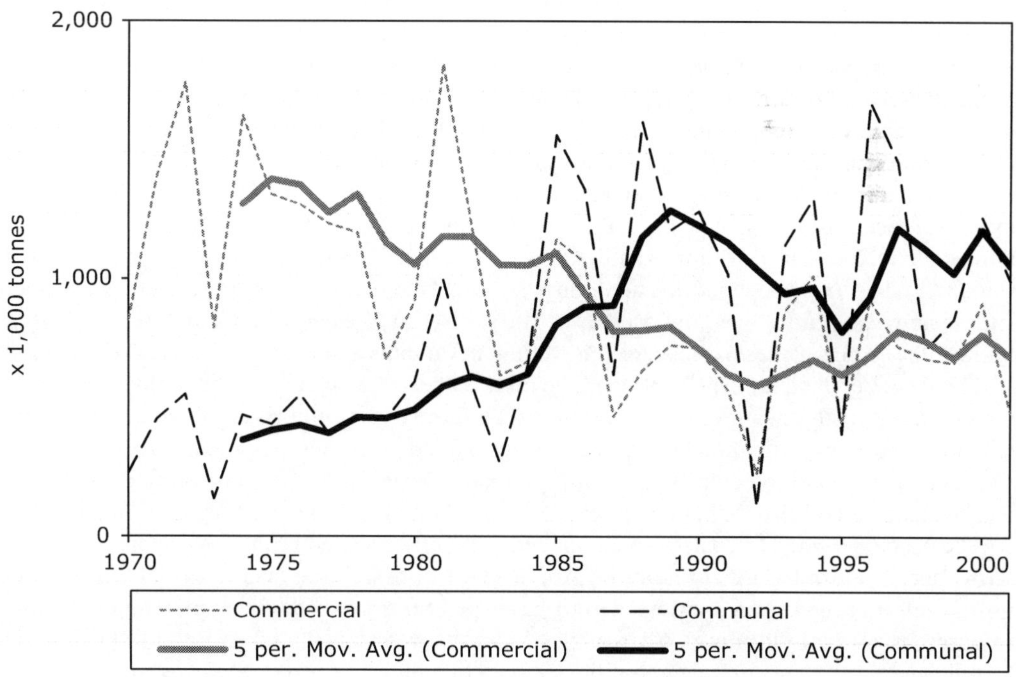

Figure 5.2 Maize production (× 1,000 tonnes) of commercial (grey-dotted line) and smallholder (black-dotted line) farming sectors and five-year moving averages, 1970–2000. Taken from Andersson (2007).

created per annum, while some 300,000 new jobseekers were entering the labour market (Durevall et al., 1999). Under pressure from the International Monetary Fund (IMF), the government launched an economic structural adjustment programme (ESAP) with employment creation as one of its principal aims. However, an outdated technology base (due to isolation during UDI) had undermined the international competitiveness of the manufacturing sector. Industries that had developed based on domestic production of raw materials, such as the textile industry, could not compete with cheap imports, and closed down. Furthermore, high interest rates, exacerbated by large-scale government borrowing on the money market, hampered productive investment. Rather than growing, formal employment declined in the 1990s. With high inflation rates, massive unemployment, a strongly devaluated currency and acute foreign exchange and fuel shortages, Zimbabwe's economy faced a deep crisis.

As ESAP also meant the removal of government subsidies that had bolstered agricultural production in the smallholder farming sector (Muir-Leresche, 2013), a gradual reduction in productivity of this farming sector set in after 1996, when maize production peaked at 2.6 million tons (Maiyaki, 2010). This trend was aggravated by the decline of the (urban) wage labour sector; smallholder farming in the Communal Areas thus lost two major sources of investment. Average maize yields declined, from an estimated 1.3 t/ha in 1986 to approximately 0.8 t/ha in 2004 (FAO, 2007). By the end of the 1990s, Zimbabwe's Communal Area farmers were less food secure than they had been in the 1940s (Nyandoro, 2007).

4 Fast-track land reform, economic meltdown and food insecurity: 2000–2009

The economic crisis of the late 1990s, manifested by fast-growing unemployment, rising food prices and food riots in Zimbabwe's urban centres, demanded a policy response. In early 2000, government supported the invasion of large-scale commercial farms operated by white settlers or their descendants. This was farmland that had been appropriated during the colonial era. This move was politically motivated, as the government of President Mugabe had largely lost its popular support and faced an emergent opposition. The chaotic invasions were hastily formalized into the controversial 'fast-track land reform' policy. Regarded as a 'violation of property rights' by some (Richardson, 2007), and a 'taking back of its land' (Hanlon et al., 2013) by others, the immediate effect of land reform was highly negative (Sachikonye, 2003). Foreign investment stalled, food security declined further, donor support was reduced to merely humanitarian aid and economic sanctions were imposed in 2001 by the USA, the European Union (EU) and Australia on listed companies, some banks, government officials and members of the ruling party.

The fast-track land reforms continued the historical division between smallholders and commercial farms, using two models: A1 and A2. The former was comprised of smallholder farms akin to farms in the Communal Areas. Farm sizes, including communal grazing areas, varied in size, depending on agro-ecological conditions and land quality. The A1 farms often formalized the spontaneous and chaotic land invasions. Most A1 farmers had been Communal Area farmers.

The A2 model sought to establish larger, commercially oriented farms, cutting three to seven farms out of a former large-scale farm. Applicants for these farms had to go through a more formal selection procedure, and needed to have qualifications, a business plan and their own resources to invest in farming. Whereas the A1 farms were quickly taken into production, the lengthy and more competitive and politicized selection process for the A2 farms was marred by irregularities, political patronage and cronyism (Marongwe, 2011) and caused delays (Utete, 2003). A land audit report in 2006 showed that 'nearly half of the A2 farms were underused or not used at all' (Hanlon et al., 2013). It resulted in a drop in total farm output in areas that used to be the commercially oriented and forex earning. As the productivity of the smallholder sector declined further due to lack of investment from outside the sector and sharply rising input prices, Zimbabwe's food security situation rapidly deteriorated (Rukuni et al., 2006).

The controversial fast-track land reform remained at the centre of both popular and policy debates, as if Zimbabwe's economic problems had started with it. Meanwhile, studies of its impacts suggested that those who had acquired land – notably the 146,000 small-scale A1 farmers (Scoones et al., 2010) – were doing relatively well, especially when considering the lack of support and the adverse macro-economic environment. The 23,000 large-scale A2 farmers (Hanlon et al., 2013) were struggling initially, but after the dollarization of the economy in 2009, this sector also became increasingly productive. However, comparisons of the A1 and A2 farmers with existing or historical farming sectors in Zimbabwe are inherently problematic. First, the quality of the land and (irrigation) infrastructure on these farms could differ substantially. For instance, although farm sizes of A1 farmers were sometimes comparable to those of the existing smallholder farmers, the latter were often located on much sandier and more intensively used soils with lower fertility. Second, the performance of the new farms was heavily affected by the rapidly deteriorating economic environment of the mid-2000s. For instance, high-input, export-oriented agricultural production became unprofitable due to the rapidly declining exchange rate of the Zimbabwean dollar and government rules that controlled access to forex. Third, highly diverse situations in different parts of the country could only provide a partial picture of the impacts of the land reform. For instance, early studies of Scoones et al. (2010) focused on Masvingo province, one of the driest areas of the country where crop productivity has always been highly variable and lower than in the high-potential zones of Mashonaland (northern Zimbabwe) that are more suitable for maize production. Whereas political patronage played a limited role in land redistribution in Masvingo, reports on the formation of A2 farms in areas closer to the capital provided a different picture (Marongwe, 2011). Fourth, which benchmark does one use for performance comparisons and what contextual factors does one take into account? For instance, Hanlon et al. (2013), who compared A1 farmers with white settler farmers that had farmed in the same area in the 1970s, point out that the latter had large tracts of un(der)utilized land – it was estimated that in the early 1980s the white settler farmers in Mashonaland only cropped 23% of their arable land. By contrast, the succeeding A1 farmers cultivated on average 34% more land. Yet, yields obtained by these A1 farmers were often low. For instance, Zikhali (2008) reported average maize yields of 2.4 t/ha in the Mazoe district (north of Harare) where farmers growing long duration maize varieties can also participate in the so-called '10+ tonnes clubs' (SeedCo, 2024), akin to similar clubs among white commercial farmers decades earlier. While 2.4 t/ha compares well to the average maize yields obtained on the sandy soils of the Communal Areas (0.8 t/ha) – even when compared with yields obtained during the smallholder agricultural revolution of the 1980s – they are rather low for Zimbabwe's red and black clay soils in high rainfall areas. In the white settler farming sector, which had dominated production on these soils, average maize yields were 3.9 t/ha in the period 1970–2000 (Andersson, 2007) and 4.4 t/ha just before the land reform (Zikhali, 2008).

Crop yields are not merely functions of farmer ability and the quality of the available natural resources. They are also highly dependent on the wider socio-economic environment. As Hanlon et al. (2013) point out, white settler farming was heavily subsidized during the colonial era, yet in the 1970s, about a third of these farms were insolvent. Evaluations of the fast-track land reform thus seem marred by problems of comparison, and often disregard the historical realities of underutilization of land and the varying economic success of the settler farms, as well as the low profitability of commercially oriented farming in general (Giller et al., 2021).

While popular debate has often focused on the land and the controversial land reform, Zimbabwe's largest problem was, however, a broken economy. Government was not only unable to provide support to the newly resettled farmers, but its economic policies often made matters worse. Through state-supported violence against a burgeoning informal sector, threats of seizure of private businesses, corruption and large-scale money printing, the government undermined Zimbabwe's ailing economy. Money printing causing hyperinflation and a plummeting exchange rate were the most visible manifestations of this economic meltdown. Whereas the exchange rate of the Zimbabwean dollar was Z$55 to 1 USD in the year 2000, by mid-2008, the parallel (unofficial) rate for the US dollar was Z$10,000,000,000,000 (Hanlon et al., 2013). The situation became untenable. Agricultural (input) trading was completely disrupted and in many areas farmers reverted to bartering, while many left the country driven by poverty and food insecurity, trying to send money to those remaining behind (Nyandoro, 2011). Zimbabwe's economy became remittance-dependent (see Giller & Andersson, this volume). Smallholder farmers, once supported by government investments and remittances obtained through wage labour incomes, could now no longer access inputs and reverted to recycling hybrid maize seeds and the cultivation of sorghum and millets. Fertilizer production dropped from 505,000 mt in 1999 to 166,000 mt in 2007. Without external inputs (manure, fertilizer), many farmers would not bother sowing their most infertile soils, and rather concentrated on their best plots – often near their houses (Zingore et al., 2007). Food insecurity became the new normal. In the early 2000s, 12% of all cereals consumed in Zimbabwe were imported. By 2009, the imported share of consumption stood at 50% (Mudimu, 2020).

5 Post-reform: agricultural policy (2009 to present)

Since 2009, after Zimbabwe had abandoned its national currency in favour of the US dollar, its macro-economic situation has stabilized somewhat. But more than two decades after the start of its controversial land reform, Zimbabwe is still far from regaining its position as 'the grain-basket of southern Africa'. Government tried to regain some control over the economy, amongst others by trying to introduce a new currency, and by different kinds of agricultural interventions. For instance, after the drought-stricken season of 2015/2016, in which maize production had dropped to about a quarter of the country's needs (UNcomtrade, 2017), a Special Maize Programme for Import Substitution (SMPIS) was launched (Odunze and Uwizeyimana, 2019). Popularly known as 'command agriculture', this government scheme stipulated production targets and funded cereal (maize and wheat) production through contract farming arrangements with large-scale farmers (including the A2 model farmers) in Zimbabwe's high-potential areas. The next season indeed saw a much higher maize (and wheat) production in these areas. While criticized for not reaching its ambitious targets, despite benefiting from an exceptionally good rainfall season, the scheme was expanded to include other crops (Mazvi et al., 2019). The SMPIS programme has continued a development that had already started: a more even distribution of Zimbabwe's food crop production over different farming sectors and agro-ecological zones, which makes food production less vulnerable to adverse climate conditions (Figure 5.3).

Next to the 'command agriculture' programme, which benefited mostly A2 farmers, a Presidential Input Scheme was put in place to support the smallholder farming sector (including A1 farms) – still the main producer of Zimbabwe's maize crop (Dube, 2020). Like the SMPIS, this programme aimed to improve Zimbabwe's food self-sufficiency and reduce its growing food import bill.

Other policies, like the *Pfumvudza* programme that started in 2019, promote forms of agriculture that concentrate input use on small areas of land, as a food security and coping strategy.

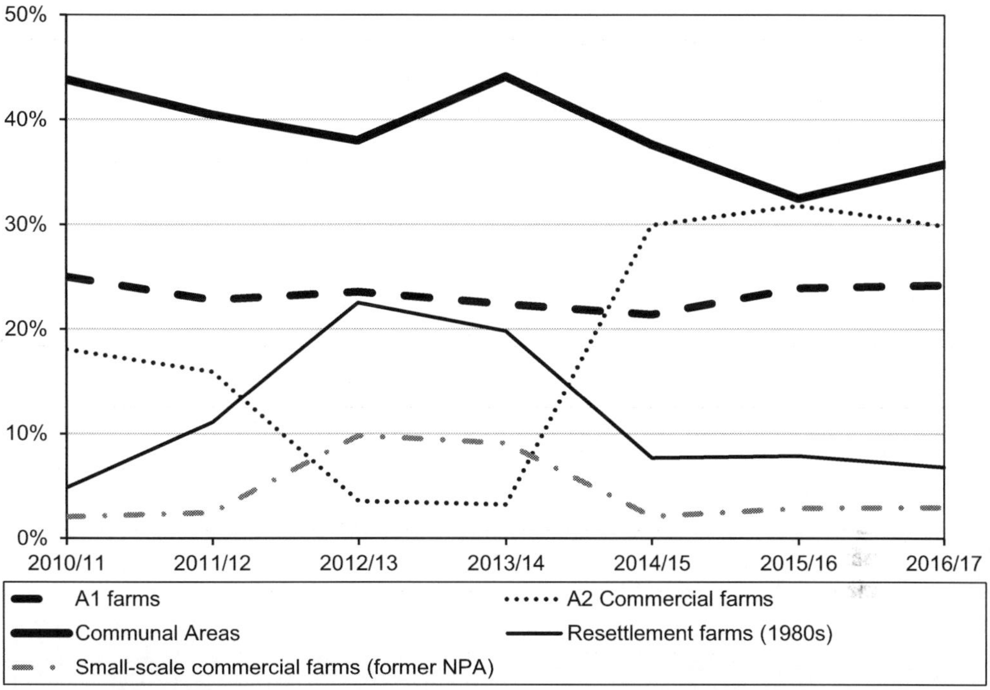

Figure 5.3 Estimated percentage share of different farming sub-sectors in national maize output, 2010–2017.

Data sources: USAID (2016) and Mutenga (2017).

Pfumvudza, which refers to the blooming of new tree leaves that signal the onset of a new farming season, is a form of conservation agriculture (CA) (Mavesere & Dzawanda, 2023). Like earlier promotion of conservation agriculture under the banner of donor-funded humanitarian aid (Andersson & Giller, 2012), this free input support scheme stresses the efficient use of resources on a small area of land (Mavesere & Dzawanda, 2023). Promotion of sorghum and millets cultivation, small-grain crops that are better suited to the semi-arid conditions prevailing in many Communal Areas constitute another attempt to boost food security with very limited resources. However, Zimbabwe's food security situation remains precarious; after the 2022/2023 agricultural season, more than 4.1 million people faced food insecurity (WFP, 2023).

6 From structural transformation in reverse to food security by 2050?

Looking at the sharp economic downfall immediately after the start of the fast-track land reform and the burgeoning literature on the reform and its impacts, it may be tempting to see the land reform as the cause of Zimbabwe's current food (in)security situation. However, as this chapter has shown, the country's food security problems did not start with the land reform programme. Rising food prices, urban food riots and mounting food insecurity had already started in the 1990s (Andersson, 2002). Figure 5.1 shows this; food import growth already started before 2000, when Zimbabwe's employment crisis manifested itself.

During the 1980s, it had been the smallholder farming sector that had become the chief producer of Zimbabwe's main food crop, maize, while the large-scale commercial farming sector increasingly abandoned food production. Government support and the wage labour economy fuelled smallholder farmer productivity, making food production increasingly vulnerable to climate variability as it was re-located to the Communal Areas, which are characterized by low and erratic rainfall (Andersson, 2007). Conversely, when government support came to an end with ESAP and unemployment rapidly increased in the 1990s, investments in smallholder agriculture decreased and productivity of the sector declined. Bad agricultural seasons now had an immediate effect on the country's food security situation, necessitating large-scale food importations and humanitarian aid.

Zimbabwe's economic crisis of the 1990s was thus not rooted in agriculture, but in its (urban) wage labour economy. After independence in 1980, a rapidly growing workforce could not be absorbed into a slowly growing industrial sector. As a result, Zimbabwe's economy witnessed a process of structural transformation in reverse; instead of shifting resources from low to high-productivity sectors of the economy, primary sectors such as agriculture and mining gained in relative importance. The share of manufacturing in the country's GDP fell from 26.9% in 1992 to 11.7% in 2014. Formal employment declined and the share of informal employment in total employment increased from 80.0% in 2004 to 94.5% in 2014 (Kanyenze et al., 2017).

6.1 Towards food security by 2050?

The fast-track land reform and the economic meltdown have had a profound effect on Zimbabwe's agrarian structure. While the racial division of land was ended, the historical division between large-scale and smallholder farming sectors has remained – albeit with fewer very large farms. Yet, the adverse economic circumstances forced farmers, including the beneficiaries of the land reform programme, to concentrate on food crop production – as export-crop production became unprofitable. While dollarization in 2009 made export-oriented agriculture possible again, government's high budget deficit, attempts to re-introduce a national currency, forex problems and inflationary pressures continued to contribute to a volatile macro-economic situation.

The more even distribution of food crop production and command agriculture over the country's different farming sectors and agro-ecological zones is cause for optimism though; if a larger share of Zimbabwe's most productive lands can be used for food production, its production will become less vulnerable to adverse weather and climate change. It would enable Zimbabwe to reduce its import dependency for food. Recent figures show that cereal productivity at national level is increasing (World Bank, 2024). And although Zimbabwe's food import dependency ratio is still high, it appears to be stabilizing (Figure 5.1). However, whether Zimbabwe's food security situation is going to substantially improve in the coming decades is probably more dependent on developments outside the agricultural sector than within it.

Notes

1 Known as 'Native Reserves' until 1962, when they were renamed Tribal Trust Lands (TTLs). At independence in 1980, they became known as Communal Lands or Areas.

2 The Land husbandry Act of 1951 was a comprehensive policy that sought to create individual land and livestock holdings in the 'reserves' and forced African farmers to implement soil conservation measures (like contour bund construction). Lack of implementation capacity and strong opposition (especially against destocking measures) limited the Act's implementation, and led to its abandonment in 1961 (Phimister, 1993; Andersson, 2002).

References

Andersson, J.A. (2002). Going places, staying home: Rural-urban connections and the significance of land in Buhera District, Zimbabwe (PhD thesis). Wageningen University.

Andersson, J.A. (2007). How much did property rights matter? Understanding food insecurity in Zimbabwe: A critique of Richardson. *African Affairs*, *106*(425), 681–690.

Andersson, J.A., & Giller, K.E. (2012). On heretics and God's blanket salesmen: Contested claims for Conservation Agriculture and the politics of its promotion in African smallholder farming. In J. Sumberg & J. Thompson (Eds.), *Contested agronomy: Agricultural research in a changing world* (pp. 22–46). Routledge.

Baudron, F., Corbeels, M., Andersson, J.A., Sibanda, M., & Giller, K.E. (2011). Delineating the drivers of waning wildlife habitat: The predominance of cotton farming on the fringe of protected areas in the mid-Zambezi Valley, Zimbabwe. *Biological Conservation*, *144*(5), 1481–1493.

Cliffe, L. (1988). Zimbabwe's agricultural "success" and food security in Southern Africa'. *Review of African Political Economy*, *43*, 4–25.

Dube, L.M. (2020). Command agriculture and food security: An interrogation of state intervention in the post-fast track land redistribution era in Zimbabwe. *Journal of Asian and African Studies*, *56*, 1626–1643.

Durevall, D., Ncube, M., & Bjurek, H. (1999). *Zimbabwe: Employment, labour market reform and trade.* SIDA.

Eicher, C.K., & Kupfuma, B. (1998). Zimbabwe's maize revolution: Insights for closing Africa's food gap. In C.K. Eicher & J. Staatz (Eds.), *International agricultural development* (3rd ed.; pp. 550–570). Johns Hopkins University Press.

FAO. (2007). FAO/WFP crop and food supply assessment mission to Zimbabwe. Special report. Food and Agriculture Organization of the United Nations / World Food Programme, Rome.

Giller, K.E., Delaune, T., Silva, J. V., Descheemaeker, K., van de Ven, G., Schut, A. G., … & van Ittersum, M.K. (2021). The future of farming: Who will produce our food? Food Security, 13(5), 1073–1099.

Green, E., & Nyandoro, M. (2024). Property rights and labour relations: Explaining the relative success of native purchase area farmers in Southern Rhodesia, 1930–1965. *Journal of Southern African Studies*, *49*(5), 1–18.

Hanlon, J., Manjengwa, J.M., & Smart, T. (2013). *Zimbabwe takes back its land*. Kumarian Press.

Hoogeveen, J.G.M., & Kinsey, B.H. (2001). 'Land reform, growth and equity: Emerging evidence from Zimbabwe's resettlement programme - A sequel', *Journal of Southern African Studies*, *27*(1), 127–136.

Iliffe, J. (1990). *Famine in Zimbabwe 1890–1960*. Mambo Press.

Joshipura, T. (forthcoming). How Sub-Saharan Africa became food import dependent – A long-run perspective (Unpublished doctoral dissertation). Wageningen University.

Kanyenze, G., Chitambara, P., & Tyson, J. (2017). *The outlook for the Zimbabwean economy*. Supporting Economic Transformation.

Keyter, C.F. (1978). *Maize control in Southern Rhodesia 1932-1941: The African contribution to White survival.* Central African Historical Association.

Kinsey, B.H. (1999). Land reform, growth and equity: Emerging evidence from Zimbabwe's resettlement programme. *Journal of Southern African Studies*, *25*(2), 173–196.

Lima, F., & Lessard, T. (2023). *Unlocking Zimbabwe's agricultural potential: Food security*. International Monetary Fund.

Machingaidze, V.E.M. (1991). Agrarian change from above: The Southern Rhodesian native land husbandry act and African response. *International Journal of African Historical Studies*, *24*(3), 557–589.

Maiyaki, A.A. (2010). Zimbabwe's agricultural industry. *African Journal of Business Management*, *4*(19), 4159–4166.

Marongwe, N. (2011). Who was allocated fast track land, and what did they do with it? Selection of A2 farmers in Goromonzi District, Zimbabwe and its impacts on agricultural production. *Journal of Peasant Studies*, *38*, 1069–1092.

Mavesere, F., & Dzawanda, B. (2023). Effectiveness of *Pfumvudza* as a resilient strategy against drought impacts in rural communities of Zimbabwe. *GeoJournal*, *88*, 3455–3470.

Mazvi, F., Chemura, A., Mudimu, G.T., & Chambati, W. (2019). Political economy of command agriculture in Zimbabwe: A state-led contract farming model agrarian south. *Journal of Political Economy*, *8*(1–2), 232–257.

Minde, I.J., Mazvimavi, K., Murendo, C., & Ndlovu, P.V. (2010). Supply and demand trends for fertilizer in Zimbabwe: 1930 to date: Key drivers and lessons learnt. Paper presented at the Joint 3rd African Association of Agricultural Economists (AAAE) and 48th Agricultural Economists Association of South Africa (AEASA) Conference, Cape Town, South Africa, 19–23 September.

Mlambo, A.S. (2014). *A history of Zimbabwe*. Cambridge University Press.

Moyo, S. (2006). The evolution of Zimbabwe's land acquisition. In M. Rukuni, et al. (Eds.), *Zimbabwe's agricultural revolution revisited*. University of Zimbabwe Publications, 143–164.

Mudimu, T. (2020). Food imports, hunger and state making in Zimbabwe, 2000–2009. *Journal of Asian and African Studies*, *55*(1), 128–144.

Muir-Leresche, K. (2013). Agricultural policies for growth and the reduction of poverty and food insecurity. The EC-FAO Food Security Information for Decision Making Programme. https://www.foodsec.org

Muir, K., & Blackie, M. (1994). The commercialization of agriculture. In M. Rukuni & C.K. Eicher (Eds.), *Zimbabwe's agricultural revolution* (pp. 195–207). University of Zimbabwe Publications.

Munro, J.F. (1976). *Africa and the international economy 1800–1960: An introduction to the modern economic history of Africa South of the Sahara*. J.M. Dent & Sons Ltd.

Musimwa, T.R., & Derera, J. (2017). Why SR52 is such a great maize hybrid? II. Heritability, correlation and path-coefficient analyses. Euphytica*,* 213(3), 62.

Mutenga, T. (2017). Command agriculture misses targets. *The Zimbabwe Mail*, 3 August 2017.

Nyambara, P.S. (1999). A history of land acquisition in Gokwe, Northwestern Zimbabwe, 1945–1997 (PhD Dissertation). Northwestern University.

Nyambara, P.S., & Nyandoro, M. (2019). "Tobacco thrives, but the environment cries": The sustainability of livelihoods from small-scale tobacco growing in Zimbabwe, 2000–2017. Special Issue on Sustainable Rural Livelihoods, *Global Environment: A Journal of Transdisciplinary History*, *12*(2), 304–320.

Nyandoro, M. (2007). Development and differentiation: The case of TILCOR/ARDA irrigation activities in Sanyati (Zimbabwe), 1939 to 2000 (PhD thesis). University of Pretoria.

Nyandoro, M. (2011). Implications for policy discourse: The influx of Zimbabwean migrants into South Africa. In E. Guild & S. Mantu (Eds.), *Constructing and imagining labour migration: Perspectives of control from five continents* (pp. 109–134). Ashgate Publishing Company.

Nyandoro, M. (2012). Zimbabwe's land struggles and land rights in historical perspective - The case of Gowe-Sanyati irrigation (1950–2000). *Historia*, *57*(2), 298–349.

Nyandoro, M. (2019). Land and Agrarian policy in colonial Zimbabwe: Reordering of African society and development in Sanyati, 1950–1966. *Historia*, *64*(1), 111–139.

Nyandoro, M. (2022). Emerging smallholder cotton irrigation agriculture and tensions with estate labour requirements in Sanyati, Zimbabwe, 1967–1990. *Journal of Southern African Studies*, *48*(3), 453–472.

Odunze, D.I., & Uwizeyimana, D.E. (2019). Zimbabwe's special maize programme for the import substitution (command agriculture) scheme: A hit-and-miss affair. *Journal of Reviews on Global Economics*, *8*, 1329–1338.

Phimister, I. (1993). Rethinking the reserves: Southern Rhodesia's land husbandry act reviewed. *Journal of Southern African Studies*, *19*(2), 225–239.

Richardson, C. (2007). How much did droughts matter? Linking rainfall and GDP growth in Zimbabwe. African Affairs, *106*(424), 463–478.

Rukuni, M. (1984). An analysis of economic and institutional factors affecting irrigation development in communal lands of Zimbabwe (PhD thesis). University of Zimbabwe.

Rukuni, M., & Eicher, C.K. (Eds.) (1994). *Zimbabwe's agricultural revolution*. University of Zimbabwe Publications.

Rukuni, M., Tawonezvi, P., Eicher, C.K., Munyuki-Hungwe, M., & Matondi, P. (2006). *Zimbabwe's agricultural revolution revisited*. University of Zimbabwe Publications.Sachikonye, L.M. (2003). From "growth with equity" to "fast-track" reform: Zimbabwe's land question. *Review of African Political Economy, War and the Forgotten Continent*, *30*(96), 227–240.

Scoones, I. Marongwe, N., Mavedzenge, B., Mahenehene, J., Murimbarimba, F. & Sukume, C. (2010). *Zimbabwe's land reform: Myths & realities*. James Currey/Boydell & Brewer.

SeedCo (2024). *It is a bumper harvest for Mazowe commercial farmer Pfachi*. https://www.seedcogroup.com/zw/fieldcrops/it-is-a-bumper-harvest-for-a-mazowe-commercial-farmer-pfachi/ (accessed 18 November 2024).

Tawonezvi, P.H., & Hikwa, D. (2006). Agricultural research policy. In M. Rukuni, P. Tawonezvi, C.K. Eicher, M. Munyuki-Hungwe, & P. Matondi (Eds.), *Zimbabwe's agricultural revolution revisited* (pp. 197–216), University of Zimbabwe Publications.

UNcomtrade (2017). *International Trade Statistics Database: Yearbook volume 1: Trade by Country*. United Nations International Trade Statistics Database. https://comtrade.un.org/pb/downloads/ 2017/VolI2017.pdf.

USAID. (2016). Maize production and marketing in Zimbabwe: Policies for a high growth strategy.

Utete, C. (2003). A report of the presidential land review committee on the implementation of the fast track land reform programme 2000–2002. Government of Zimbabwe, Harare.

van Velsen, J. (1964). Trends in African nationalism in Southern Rhodesia. *Kroniek van Afrika* (2).

WFP (2023) Zimbabwe Food Security and Markets Monitoring Report, September 2023. World Food Programme, at: https://docs.wfp.org/api/documents/WFP-0000153348/download/ (accessed 20 December 2023).

World Bank. (2024). Cereal yield (kg per hectare) – Zimbabwe, 1961–2022. FAO data. https://data.worldbank.org/indicator/AG.YLD.CREL.KG?locations=ZW&type=shaded&year=2022

Zikhali, P. (2008). Fast track land reform and agricultural productivity in Zimbabwe, School of business, economics and law. University of Gothenburg.

Zingore, S., Murwira, H.K., Delve, R.J., & Giller, K.E. (2007). Soil type, historical management and current resource allocation: Three dimensions regulating variability of maize yields and nutrient use efficiencies on African smallholder farms. *Field Crops Research*, *101*, 296–305.

Zinyama, L. & Whitlow, R. (1986). Changing patterns of population distribution in Zimbabwe. *GeoJournal*, *13*, 365–384.

Part 2

Challenges

6 Africa's demographic challenge

Leo van Wissen

1 Introduction

Africa's population growth is unprecedented. In 1950 the continent counted 225 million people, and with a population density of only 7.7 per square kilometre, it was among the least populated world regions, and only about a sixth that of Asia (44) (UN DESA, 2022). Since then, more than 1 billion people have been added to the continent's population of 1.41 billion (UN DESA 2022). With an annual rate of 2.5–3% in most years since 1960, population growth has far outpaced all other world regions in relative terms, and since 2020 it is also the largest in absolute terms with more than 30 million annually. Today, almost half of the global annual population growth is taking place in Africa. By 2050 another 1 billion persons will have been added to the African continent, according to the UN (United Nations) medium variant of the World Population Prospects, to reach a total of 2.47 billion. By that time almost all global population growth will be in this continent.

Feeding so many additional mouths in the coming decades is certainly a challenge. This chapter aims to shed light on the demographic dimensions of this issue, with a special focus on the role of the agricultural sector, not only as a provider of food, but also in economic development (which in turn is closely linked to food security, see de Haas et al., this volume). For a good understanding of the population and development discussion two societal trends are important to consider. First, educational expansion is a crucial factor. Will Africa be able to sufficiently increase its human capital to outpace population growth and increase incomes? Second, urbanization is key, as it offers the prospect of a more diversified economy with a diminishing size of the agricultural sector, and a growing importance of the secondary and tertiary sector, to offer a viable perspective to the young population outside agriculture (see also Pieters, this volume).

In this chapter the role of education and urbanization in the population and development nexus in Africa will be discussed. Africa follows the path of the demographic transition: declining mortality, followed by declining fertility. According to this theory, mortality decline, especially infant mortality, precedes fertility decline. Children that survive until adulthood will support their parents in old age. It is therefore rational to have many children if child mortality is high, and the need for many children diminishes if child mortality decreases. This mortality decrease is fuelled by better sanitation, health facilities and medicine, better diets, higher education, urbanization, and related indicators of modernization (see Caldwell, 2006).

It is important that, although in this chapter we take a broad look at the demography of Africa, "Africa is not a country" (Faloyin, 2022). The demographic diversity within Africa is large, which includes the timing of the onset and the pace of the demographic transition among African countries. Nevertheless, the similarities are large as well: 25 of the 30 poorest countries in the world are in Africa (World Bank, 2021a), 22 of the 30 countries with fastest population

DOI: 10.4324/9781032649696-9

growth are African countries (UN DESA, 2022), and 24 out of the 30 countries with the highest share of agriculture in employment are African countries as well (ILO, 2024). While acknowledging the diversity within the continent, we largely analyse Africa at a continental scale. Section 2 focuses on Africa's demographic transition. Section 3 discusses future population growth paths for the continent. In Section 4 some demographic dimensions of agricultural change, introduced in Chapter 1 of this volume, are elaborated. Section 5 concludes.

2 The demographic transition in Africa

Although today Africa's population growth is unprecedentedly high, historically its growth was very low. Precise numbers diverge but estimates show only very slow population growth between 1500 and 1800. Whereas by 1800 the population of Europe had more than doubled in size compared with 1500, Africa had only increased from 87 to 102 million, or by 17% (Livi-Bacci, 2017; Braudel, 1979). Compared to other world regions, the natural environment was not very conducive for population growth, but in addition to this the slave trade between 1500 and 1870 drained the continent, especially West Africa, of a significant share of its young population. Estimates put the number of enslaved people transferred to the Americas between 1500 and 1900 at 12.5 million (Trans-Atlantic Slave Trade Database, 2021), and the flow of enslaved eastwards to the Middle East to at least 16.5 million (Lovejoy, 2011). The indirect effects of enslavement, in terms of warfare, dislocation, forced migration, and insecurity, resulted in further depopulation of the affected areas (Lovejoy, 2011). The West African population, most affected by the slave trade, is estimated to have declined from 50 to 40 million between 1700 and 1900 (Manning, 2014).

After 1900 the situation started to change and Africa's population reached a growth rate roughly the same as Asia's (from 138 to 224 million in 1950, an increase of 62%, or almost 1% annually). Nevertheless, Africa remained among the least populated of the continents, comparable in total population size to Latin America (Figure 6.1). The population balance among the continents has changed dramatically since then. Asia experienced a population explosion,

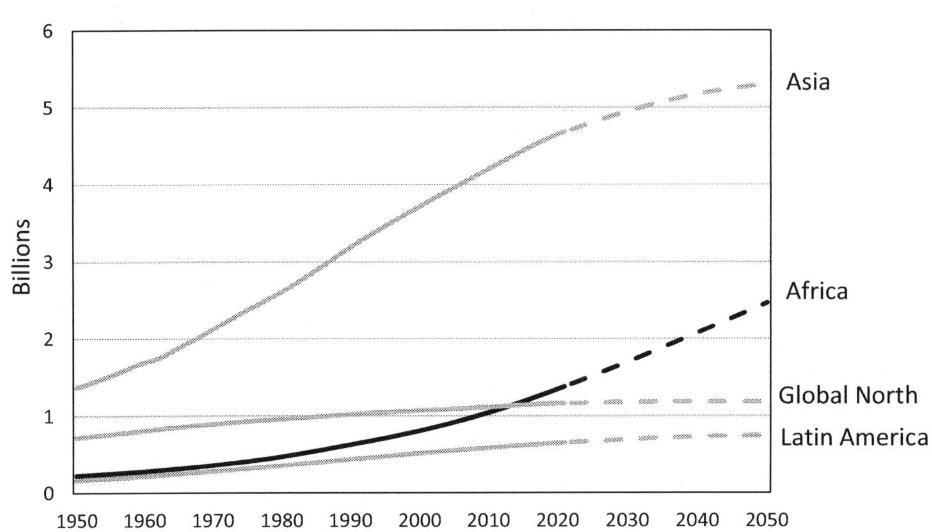

Figure 6.1 Total population estimates and medium variant projections by World Regions 1950–2050.
Source: UN DESA (2022).

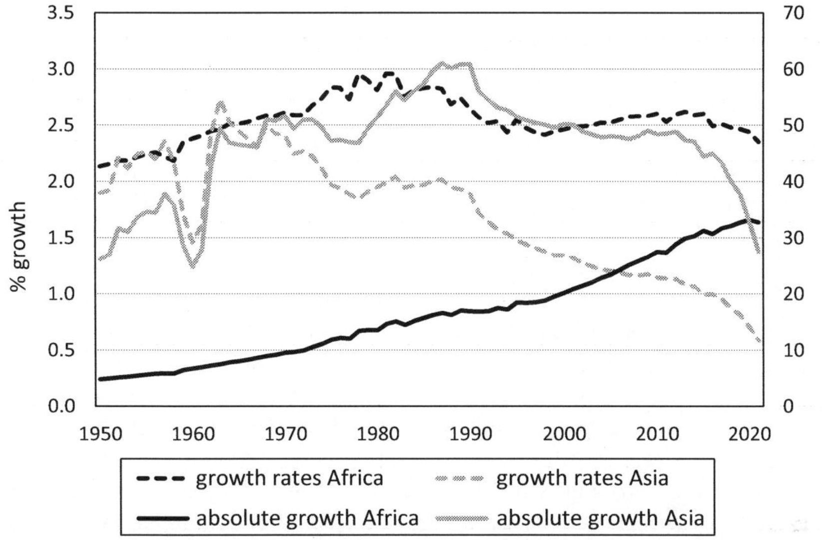

Figure 6.2 Relative (left axis) and absolute (right axis) population growth of Asia and Africa compared.
Source: UN DESA (2022).

especially after 1960. Growth rates in the Global North (Europe, North America, and Oceania) have levelled off and have almost reached zero. By contrast, African growth took off, with growth rates comparable to those of Asia. Whereas Asia's growth rate reached a peak of 2.7% in 1963 and has since steadily levelled off to 0.6% in 2021, the African annual growth rate reached 3.0% in the early 1980s and has stabilized around 2.5% since the mid-1990s (Figure 6.2). In persons added to the population per year, Africa has surpassed Asia since 2019, with a growth of more than 30 million people annually. Of the 134 million live births worldwide in 2021, one third, or 45 million, is in Africa. As a result, Africa is currently the second continent in terms of population size, with about 17% (1.38 billion) of the world population, which is more than the number of people in the Global North (North America, Europe, Japan, Australia, and New Zealand) combined (1.16 billion).

Africa's population development follows the predictions from the demographic transition framework: declining mortality, followed by declining fertility. Africa's child mortality has steadily declined since the 1950s, when one in three children died before reaching the age of five, and only 60% reached maturity. To illustrate what this 60% means: if parents want to be sure to have at least two children to reach maturity, as their old age life insurance, they need at least seven live births, which incidentally was the average number of children a woman had in Africa in 1950. Today only one in 15 children die before the age of five, and 90% reaches mature age (UN DESA, 2022). This means that with four live births there is a 95% chance of at least two mature children surviving. Nevertheless, while the African and Asian life expectancies were almost equal in 1960, at 41.5 and 41.7 years respectively, the increase in longevity in Africa has severely lagged behind that in Asia. Currently an African person can expect to live 61.7 years, while the life expectancy of an Asian person is almost nine years higher, at 70.1 years (UN DESA, 2022). Especially in the 1990s the increase stalled in Africa, mainly due to the impact of the HIV-AIDS virus, although since 2000 the increase in life expectancy of 8.9 years has been faster than in any other world region, due to the catching up after the HIV-AIDS epidemic.

In line with the demographic transition framework, African fertility started to decrease much later than mortality. The number of live births a woman is expected to have during her lifetime (the total fertility rate, TFR) remained at about 6.6 births per woman until 1981, after which it started to decline to 4.3 at present, an average annual decline of 1.1%. There is large variation though, with the lowest TFR values to be found in Southern (2.5) and Northern Africa (3.1), and the highest values in Central (5.6) and West Africa (5.0). The Asian fertility transition started almost 20 years earlier, in 1963, when it peaked at 6.3 births per woman. Since then, it has decreased 2.0% on average annually, to arrive at a value of 1.9 today. Incidentally, this is almost identical to the path of the fertility transition in Latin America. The one-child policy of China was a major factor behind this rapid change, but in Southern and Southeast Asia the fertility decline was still 1.8% per year. If African fertility would have started at the same time and experienced the same rate of reduction as in South or Southeast Asia, its present level would have been 2.3 births per woman, not the present value of 4.3, which would have resulted in a population size today of approximately 900 million, instead of the actual number of almost 1.4 million.

The question why Africa's fertility transition started later and progressed at a much slower rate compared to other developing world has been the focus of a long debate. The dominant explanation is the lagging economic development of the continent, compared with other Global South regions. This was not always the case. In 1960 average income (GDP) per capita in sub-Saharan Africa was about the same as that of Southeast Asia, but in the early 1990s GDP per capita in Southeast Asia was double the average of sub-Saharan countries (Vlasblom, 2013; Maddison Project Database, 2020).

Bongaarts (2017) has provided a more elaborate explanation, taking into account the most important drivers of the demographic transition: GDP per capita, education, and urbanization. These drivers explain about 84% of the variation in fertility in the countries of the Global South in 2010, but for African countries the number of children per woman, given the state of the transition, is still 1.1 child higher after taking these factors into account, which calls for additional explanations of African fertility levels. Bongaarts found the same 'Africa factor' of about one child per woman when analysing desired family size. This would imply that even if all obstacles to family planning practices were removed, African families would still be larger than families in other countries of the Global South. Various explanations have been brought forwards for this difference. According to Boserup (1985) high fertility is encouraged by the structure of the agricultural sector, where traditionally the subsistence farming system necessitated many hands to work the land, especially women and child labour. Under such circumstances it paid off to have a large family. Due to Africa's historical underpopulation, labour scarcity for agricultural work on the vast areas of land was the rule. Resultant pro-natalist values have persisted, even when having a large family is no longer needed for agricultural work and the attainment of wealth and income. Traditionally, for men larger families mean social esteem, and for women, having children enhances social status (Ojakaa, 2022). This culture may also be an explanation for the less-than-optimal implementation of family planning policies in various countries (Bongaarts, 2017). Another factor is African religion (Caldwell & Caldwell, 1987). However, more recent research has supplemented these cultural explanations of the 'African fertility exceptionalism' in favour of an economic one, by pointing at the impact of introducing a social pension system on reducing fertility (Rossi & Godard, 2022; also see Pieters, this volume), which could be in the order of one child per woman. Whatever the causes, contraceptive use in Africa among women in the fertile ages remains low (34%) compared, for instance, with Asia (66%) or Latin America (77%) (UN DESA, 2019; see also Madsen, 2015; Banougnin & Bolarinwa, 2022), although the variation across African counties is large, with values as low as 6% in Niger and Chad, and values close to 60% in Morocco or Kenya (World Bank, 2021b). Contraceptive use is a strong predictor of fertility: it explains 42% of the TFRs of sub-Saharan countries in 2019 (UN DESA, 2020).

3 Population projections for Africa

According to the latest version of the medium variant of the UN Department of Economic and Social Affairs (DESA), the African population will increase from 1.41 in 2021 to 2.47 billion by 2050 (UN DESA, 2022) (Figure 6.1). According to the same projections, another 1.45 billion persons will be added until the end of the century, to reach a total of 3.92 billion (Table 6.1). This projection is made under the assumption that fertility will drop to a value of 2.9 in 2050 and 2.0 in 2100, down from 4.3 today, and that life expectancy will increase to 74.9 in 2100, up from 61.7 today (both sexes combined). How likely is this to occur? What is the uncertainty around these outcomes? And what size will the population be if other fertility and mortality trajectories will be followed?

It is worth to address the assumptions and uncertainties around future population growth, by comparing various projections from the UN with other global population projections.[1] The key assumptions that drive the projections are those that concern the future number of births per woman. Small differences in these assumptions have large consequences. For instance, in the low variant of the UN projections the number of children per woman will be 0.5 child less in 2100 (1.53) than in the medium variant, which results in 1.1 billion fewer persons (2.7 billion) in 2100 (see Table 6.1). The high variant assumes 0.5 child per woman more than in the medium variant in 2100, which will lead to 1.4 billion more persons by the end of this century. In other words, a difference of one child in the fertility assumptions between the low and high variant will result in a difference of 2.5 billion persons in the outcome, which is almost twice the current population size of Africa. These are not 'extreme' projections in the sense that they do not reflect unexpected events such as war, famine, or pandemics. Instead, they should be interpreted as the plausible bandwidth of the projection, given the large variation in the trend in the past. Assumptions about migration and mortality have far fewer consequences for population growth (UN DESA, 2022). For instance, the zero net migration scenario will result in 17 million people more for Africa in 2050, a negligible change in total number compared with the consequences of fertility change.

Other organizations have made global population projections which are cited frequently. The projection outcomes are summarized in Table 6.2. They use a different methodology than the UN projections. Most importantly, they explicitly include one or more key drivers of demographic change, notably education, which in general results in smaller projected populations. The projection made by the Wittgenstein Centre for Demography and Global Human Capital (WCD) (Lutz et al., 2018) is based on the projection of human capital in all countries of the world, i.e. the projection of all populations by level of education. The higher the level of education, the fewer children per woman. For instance, highly educated women in sub-Saharan Africa have on average two children less (3.5 versus 5.5) than those with little education (DHS, most

Table 6.1 Fertility and life expectancy assumptions and resulting projected population size for Africa of UN DESA 2022 population projections

Variant	*2021*			*2050*			*2100*		
	TFR[1]	*Life Exp.*[2]	*Pop. Size*[3]	*TFR*[1]	*Life Exp.*[2]	*Pop. Size*[3]	*TFR*[1]	*Life Exp.*[2]	*Pop. Size*[3]
Medium	4.31	61.7	1.41	2.87	68.3	2.47	1.99	74.9	3.92
Low	4.31	61.7	1.41	2.38	68.2	2.28	1.50	74.8	2.77
High	4.31	61.7	1.41	3.36	68.3	2.66	2.49	75.0	5.40

Source: UN DESA (2022).

Notes: [1] Total fertility rate; [2] Life expectancy; [3] Billions.

Table 6.2 Population projections for sub-Saharan Africa by UN DESA, WCD, and IHME for 2050 and 2100

Source	*Variant/scenario*	*2021*	*2050*	*2100*
UN DESA	Low	1.12	2.09	2.44
	Medium	1.12	1.93	3.44
	High	1.12	2.26	4.72
WCD	SSP1 Sustainability	1.12	1.51	1.44
	SSP2 Middle of the Road	1.12	1.93	2.59
	SSP3 Regional Rivalry	1.12	2.29	4.29
IHME	Faster uptake and education expansion	1.12	2.02	2.45
	Reference	1.12	2.15	3.07
	Slow uptake and education expansion	1.12	2.40	6.27

Source: UN DESA (2022); WCD (2018); IHME (2019).

Notes: Numbers in billions. UN DESA projections are variants. WCD and IHME projections are scenarios.

recent years, own calculations). There is strong evidence that education is a crucial factor in the realization of the so-called 'demographic dividend' (the economic beneficial effects of having a young population in the working age category) (Lutz et al., 2019). The Shared Socio-Economic Paths (SSPs) scenarios produced by WCD (which are used in the IPCC (Intergovernmental Panel on Climate Change) climate scenarios as well, Lee et al., 2023) lead to different distributions of the population by educational attainment in the future. The most important, from a demographic perspective, are the Sustainability scenario (SSP1), with large educational expansion, the Middle of the Road scenario (SSP2), often viewed as the trend- or baseline-scenario, and the Regional Rivalry scenario (SSP3), with the smallest, yet plausible educational expansion (Lutz et al., 2018). The Middle of the Road Scenario results in 1.92 billion people in sub-Saharan Africa in 2050 (Table 6.2), with a bandwidth, defined by the SSP1 and SSP3 scenarios, of 1.51 and 2.29 billion. The SSP1 Sustainability scenario would require an extremely fast educational expansion, whereas current educational investment indicators in African countries show that this goal is very far from achieved at present (UNESCO, 2019; Reiter et al., 2022).

A third set of projections were carried out by the Institute of Health Metrics and Evaluation (IHME, 2020; Vollset et al., 2020). These scenarios also take the role of education explicitly into account, but also the effect of contraceptives on fertility, which may be interpreted as an indicator of the effectiveness of reproductive health and family planning policies. Based on different assumptions regarding these fertility determinants, their reference scenario results in a population for SSA in 2050 of 2.15 billion, with a bandwidth defined by faster and slower uptake of contraceptives and educational expansion scenarios of 2.02 and 2.40 billion. Interestingly, the most plausible variants of the UN, WCD, and IHME are not very different from each other, between 1.93 and 2.15 billion. In summary, we can be reasonably certain about the population size of Africa in 2050. For Africa as a whole, the range is between 2.3 and 2.7 billion around the medium scenario of 2.5 billion, or ±8%. For SSA the range is between 1.9 and 2.3 billion. The uncertainty in the outcomes after 2050 grows exponentially, with a range in 2100 for sub-Saharan Africa from 1.4 billion (SSP 1, assuming a very high expansion of education and leading to population decline by the end of the century) to 6.3 billion in the slow uptake of contraceptives and education expansion scenario of IHME.

A fast-growing population is a young population. Currently, 15% of Africa's population is under the age of five, 40% is under the age of 15, and half is below 20. Such a young age structure will guarantee a large population growth in the future. Even assuming an instant replacement level of 2.1 births per woman would lead to population growth of half a billion people until 2050 (the so-called 'population momentum'; UN DESA, 2022). It is an enormous challenge for African governments to find adequate schooling and housing and for the economy to create jobs for this vast generation (see also Pieters, this volume). To give an idea: between 2017 and 2050 the number of schoolchildren in primary and secondary education in sub-Saharan Africa is predicted to increase from 400 to 650 million (UNESCO, 2019 cited in Paice, 2021), with the largest growth rates in the least developed countries. For many countries it will be very hard to keep real per capita expenditures even at the same level as today, and current levels of education spending in Africa are very low: only a tenth of the average spent per pupil in East Asia, or a quarter of that in Latin America (Paice, 2021).

4 The demographic context of agricultural systems change in Africa

The Introduction to this volume defines four paths that shape the context in which food systems can change: inclusive structural change (Path 1), agri-led growth (Path 2), exclusive structural change or 'enclave growth' (Path 3), or stagnation (Path 4) (de Haas et al., this volume). Each of these paths faces its own challenges when it comes to achieving food security for a growing population. Do these four paths follow distinct demographic trajectories? It is important to highlight that the classification of the countries in the four paths is based on the trajectories of these countries in the period 1991–2019 (see Figure 1.6), does not take into account initial levels in 1991, and is not necessarily predictive for the future. This also applies to their demographic trajectories. The population size of the countries in each group is broadly similar, with about 300 million for Paths 1, 2, and 4 and 400 million for Path 3 in 2020. Structural change and agricultural intensification, the two underlying dimensions of the four paths, are not totally independent of demographic change, although the relations are indirect. Higher outflow from agriculture to other economic sectors is related to urbanization, whereas agricultural productivity growth in principle is an important factor for income growth. Both urbanization and income growth in turn are important drivers of the demographic transition. But in reality, the picture is not as straightforward. The clearest relationship is between the outflow of agriculture and urbanization, as can be seen from Table 6.3, with the strongest urbanization, as expected, taking place in Paths 1 and 4, characterized by large outflows out of agriculture.

In addition, there is some evidence that the low-productivity growth Paths 3 and 4 are associated with a less advanced stage in the demographic transition, as would be expected. They have the highest fertility, and Path 4 has the lowest life expectancy, but the differences are not dramatic.

Table 6.3 Demographic characteristics of the four paths: total, urban, and rural population growth (1990–2020) and total fertility rate (TFR) and life expectancy (2020)

Pathway	*% Population growth (1990–2020)*			*% Urban population*	*TFR*	*Life expectancy*
	Total	*Urban*	*Rural*	*2020*	*2020*	*2020*
Agri-inclusive change	111	212	69	49.9	4.37	64.5
Agri-led growth	116	203	134	36.0	4.35	63.7
Agri-stagnation	113	196	89	39.7	4.80	64.1
Agri-exclusive change	101	229	26	50.9	4.68	57.7
Africa total	111	194	76	43.7	4.68	62.2

Source: UN DESA (2022) (own calculations).

Notes: Statistics for population growth and TFR for category 3 are without outlier country Egypt. Despite being an agri-stagnation country, Egypt has one of the lowest growth and fertility rates of Africa in 1990 (4.5 births per woman in 1990 and 2.96 in 2020).

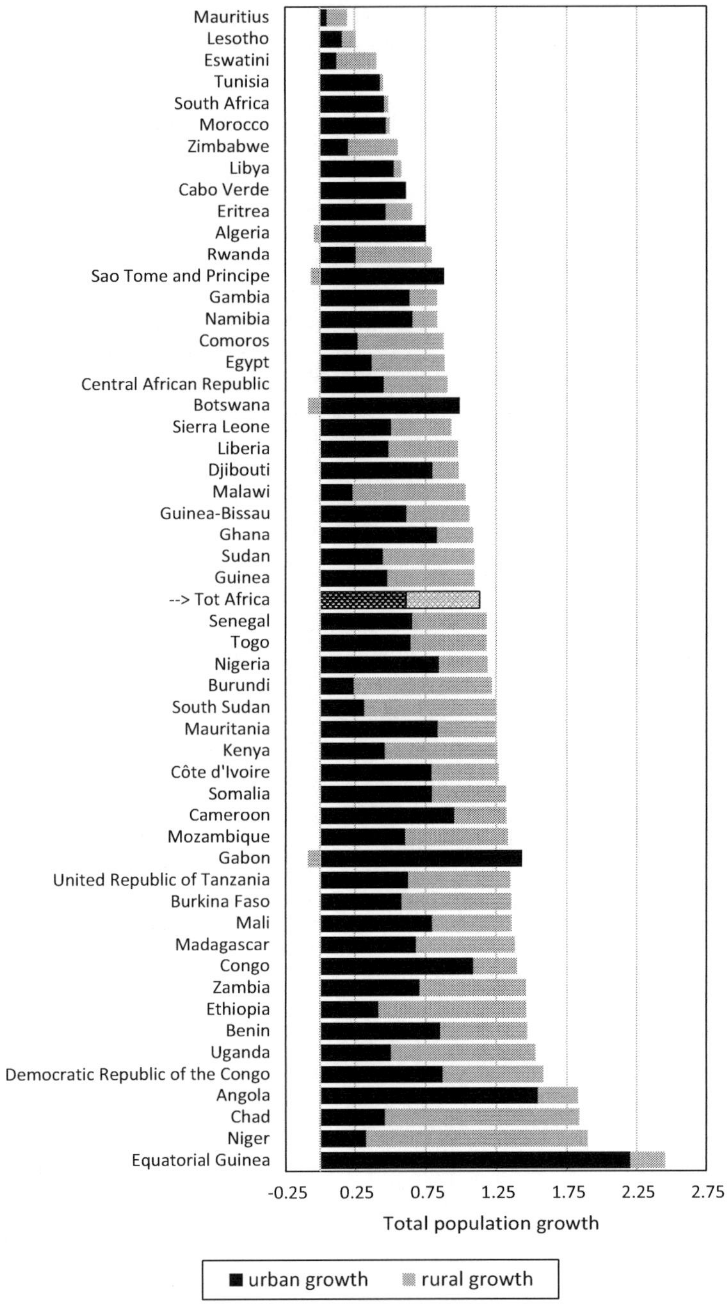

Figure 6.3 Total, urban, and rural relative population growth (1990–2020) of African countries.
Source: UN DESA (2018, 2022).

Figure 6.3 shows that the diversity within Africa in total, urban, and rural growth is large. The countries are sorted according to total population growth in the period 1990–2020. Each bar shows the total relative population growth, and the urban and rural contribution to this growth. Clearly, urban population growth has taken the largest share of population growth in the last 30

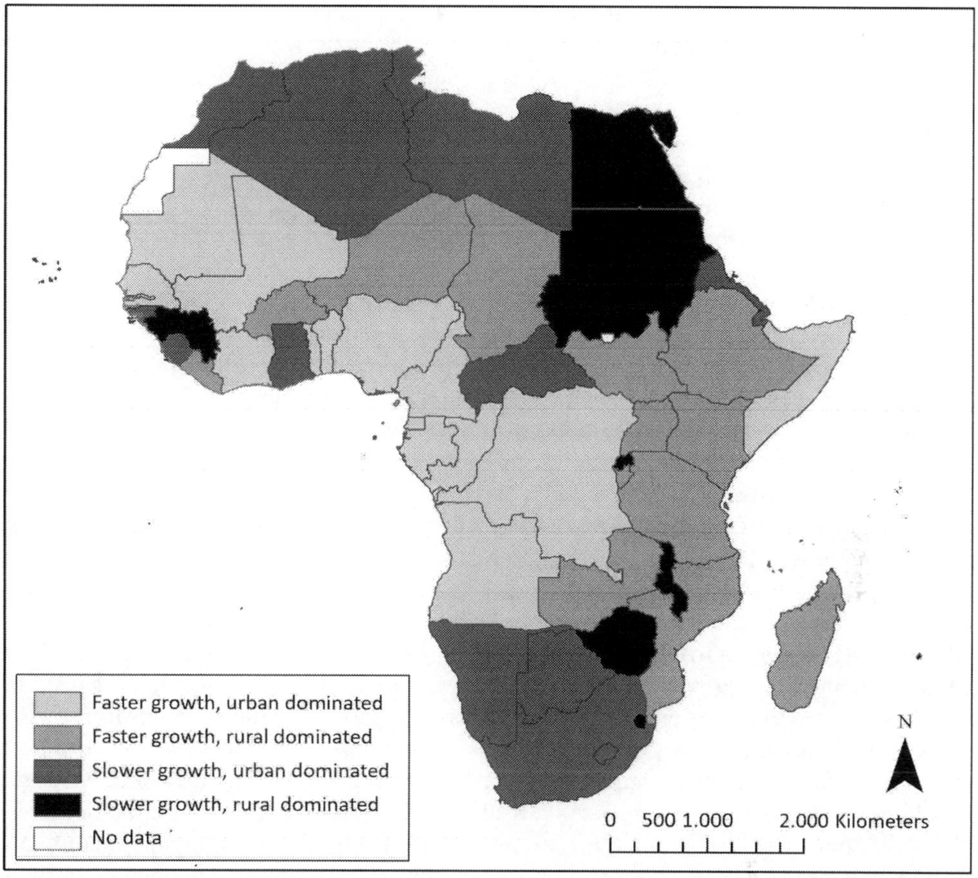

Figure 6.4 Population growth and urban and rural contributions to population growth (1990–2020) of African countries.

Source: UN DESA (2018).

years in most countries. For Africa in total, the total population growth factor was 113% (i.e. the population more than doubled), with 61% of this growth in urban regions, and 51% in rural expansion. Total growth rates vary between 20% (Mauritius) and 245% (Equatorial Guinea). In some countries (for example Equatorial Guinea, Angola, Gambia, Botswana, Libya, Algeria, Tunisia, Morocco, South Africa) total growth is almost completely due to urban growth, whereas in some other countries the growth is predominantly rural (for example Niger, Chad, South Sudan, Malawi).

Figure 6.4 shows that the combination of population growth and the contribution of urban and rural growth has an interesting geographical distribution across the continent. In this figure a distinction is made between higher-than-average African population growth (111%) in the period 1990–2020, and whether urban or rural growth dominated in this total growth. The map shows a clear pattern. West Africa is dominated by fast-growing countries, predominantly driven by urban growth (category 1). Around this group of.countries is a band of countries, stretching from the Sahel to Eastern Africa, which are dominated by fast growth, and predominantly driven by rural growth (category 2). Both North Africa and Southern Africa are characterized by slower growth, mostly due to urban growth, but with notable exceptions (Egypt, Sudan, Zimbabwe, Malawi), where rural growth dominates.

The future of African population growth is predominantly urban. According to the medium variant of the 2018 World Urbanization Prospects of the UN (UN DESA, 2018) of the total population growth of Africa in the period 2020–2050 of 83% (1.1 billion), 63 percentage points of this growth will be realized in urban regions, and 20 percentage points in rural regions. In other words, three out of four additional people in this 30-year period will be living in urban regions, resulting in an urbanization ratio of 59%. Nevertheless, even in 2050 Africa will be less urbanized than Asia (68%). Some countries (e.g. Mauritius, Tunisia, Morocco, South Africa, Libya, Algeria, Botswana, Namibia) will even experience a decline in their rural population (Figure 6.5). There are only four countries where rural population growth exceeds urban population growth: Niger, Burundi, Malawi, and Rwanda. This rapid pace of urbanization will have significant consequences for the demand for food and for the required infrastructure to bring food to the growing urban consumer markets.

Population density is another crucial demographic indicator for understanding the dynamics of the agricultural sector. As remarked above, for a long time Africa was very sparsely populated, which restricted the buildup of adequate infrastructure, or the creation of markets for a favourable development of the agricultural sector. The current population density of Africa, 46 persons per square kilometre, was reached by Asia in 1954 (Table 6.4), where it enabled the take-off of the 'Green Revolution'. The historical low population density has hindered such a development in Africa (Platteau and Hayami, 1998). But at the same time, current exceptional population growth in Africa, which can be seen as catching up with the rest of the world, has also resulted in severe population pressure in many regions, and especially those regions with historically the most advantageous agricultural conditions. This has resulted in ever-declining farm sizes, accompanied with reduced yields per hectare (Giller & Andersson, this volume) and lower productivity. As a result, high population density can be detrimental to the development of the agricultural sector as well. Both outcomes, although conflicting, may be true, depending on the specific regional context. The exceptional population growth of Africa will result in a continent with the second highest population density in 2050, although with 84 persons per square kilometre it is still only half that of Asia (UN DESA, 2022), but higher than the world average of 74. This will create both opportunities and constraints on agricultural development.

Table 6.4 Population density (persons per km^2) by continent, 1950, 2020, and 2050

Continent	*1950*	*2020*	*2050*
World	19.2	60.1	74.1
Africa	7.7	46.0	83.5
Asia	44.0	148.7	168.8
Latin America	8.3	32.2	36.0
Europe	24.9	33.8	31.8
North America	8.7	20.0	22.9
Oceania	1.5	5.2	6.8

Source: UN DESA (2022).

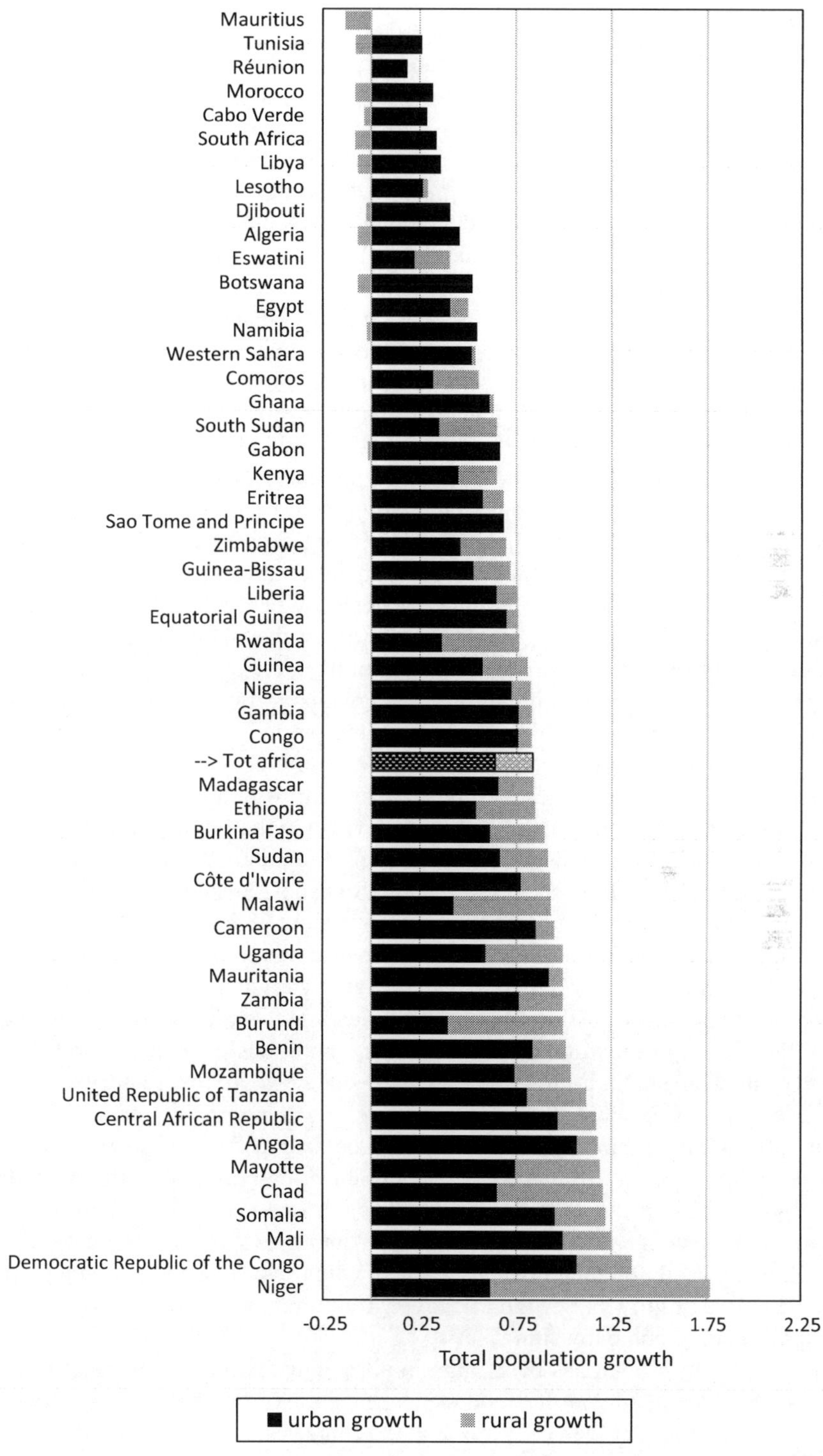

Figure 6.5 Total, urban, and rural relative population growth (2020–2050) of African countries.

Source: UN DESA (2019, 2022).

5 Conclusions

This chapter has given a bird's-eye view of the past and future demography of Africa. From a global demographic perspective, the 21st century will be dominated by Africa. In absolute terms the largest growth in population will take place in this continent, whereas population growth in most other parts of the world has stalled already or will be stalling in the course of this century. The African demographic transition is not a copy of that of other regions in the world. From the most underpopulated continent, it has developed into the most rapidly growing region, with most global population growth concentrated there. We can be reasonably certain that the 2050 population of Africa will be between 2.3 and 2.7 billion, around the medium scenario of 2.5 billion, or ±8%. By then, it will have the second highest population density among the continents. The crucial component is fertility, and relatively small differences in fertility assumptions have large consequences in terms of future population size. These assumptions remain a highly debated issue, both academically and in the policy arena. The move from a discussion of macro-level population numbers to the micro-level of the individual human rights and needs has been beneficial to stimulate the introduction of policies to protect sexual and reproductive rights in many countries. Often these policies are part of a larger package that also includes general health issues, education, and jobs, and taken together women and families have benefited from them. But the issue whether African population growth is sustainable is not solved automatically by the success of these policies. Whether Africa will follow the high or the low future growth path relates to a large extent to the future of the 'Africa factor' discussed before. The African fertility exceptionalism, whether culturally determined, or the result of lacking old-age pensions, may disappear in the future. The large educational gradient in fertility and the rural-urban fertility differential of almost two children per woman (DHS, multiple years, own calculations) may point in this direction. Favourable future economic and societal conditions may be supportive of introducing a social pension system, which will remove the necessity of having many offspring. Creating such conditions may bring the future population growth close to the low-growth scenarios just discussed, and an onset of population decline before the end of this century. If African societies fail to significantly expand education and introduce reproductive health and family planning policies, and if they are not able to generate economic growth for jobs outside agriculture and to support a viable pension system, the African population will most likely follow the high-growth path. Given the large variation across countries, some countries will be able to create the favourable conditions leading to slower growth, while other countries will follow the high-growth pattern. Moreover, conditions may change in the future, and countries may switch from high-growth to low-growth paths over this century.

In line with the broader objective of this volume, the demographic features were sketched of the four paths of agricultural systems change introduced in the Introduction to this volume. Two features are clear from the analysis. First, urbanization correlates with the paths characterized by high outflow of the agricultural workforce. Moreover, there is a clear geographical pattern across Africa, with particularly West-African countries showing high growth, fuelled by urbanization, and a band of Sahel plus East-African countries characterized by high population growth fuelled by rural growth. Second, the low agricultural productivity growth paths are less advanced in the demographic transition.

Overall, population growth has been large in all countries, and will be high in the coming decades, with huge demands on the food system. This may be challenging everywhere, but particularly for the highest growth countries and the countries that show agricultural stagnation. High urban growth countries will face the problem of feeding the ever-growing urban population, with large demands on concentrated infrastructure to the cities. Countries with access to ports have an advantage in terms of the potential of importing food, if local supply is not sufficient. High rural growth countries, especially the landlocked countries, will need to develop a finer infrastructural net, to bring local supply to the local markets. Nevertheless, in assessing the demographic context of agricultural transformation, one should consider the large diversity across countries.

Note

1 One of the projections, by IHME, does not produce separate projections for Africa as one region, but projections for sub-Saharan Africa are available.

References

Banougnin, B.H., & Bolarinwa, O.A. (2022). Contraception. In C.O. Odimegwu & Y. Adewoyin (Eds.), *The Routledge handbook of African demography* (pp. 198–216). Routledge.

Bongaarts, J. (2017). Africa's unique fertility transition. *Population and Development Review*, *43*(1), 39–58. https://www.jstor.org/stable/26622873

Boserup, E. (1985). Economic and demographic interrelationships in sub-Saharan Africa. *Population and Development Review*, *11*(3), 383–397. https://doi.org/10.2307/1973245

Braudel, F. (1979). *Civilisation materiélle, Economie et Capitalisme XVe-XVIIIe Siècle*. Armand Colin.

Caldwell, J.C., & Caldwell, P. (1987). The cultural context of high fertility in sub-Saharan Africa. *Population and Development Review*, *13*(3), 409–437.

Caldwell, J.C. (2006). *Demographic transition theory*. Springer.

Faloyin, D. (2022). *Africa is not a country*. Harvill Secker.

IHME (Institute for Health Metrics and Evaluation). (2020). *Population forecasting*. IHME, University of Washington, 2020. Available from https://vizhub.healthdata.org/population-forecast/ (Accessed Feb 25th 2023).

ILO (International Labour Organization). (2024). "ILO modelled estimates database." ILOSTAT. Accessed February 07, 2024.

Lee, H., Calvin, K., Dasgupta, D., Krinner, G., Mukherji, A., Thorne, P., … & Park, Y. (2023). IPCC, 2023: Climate change 2023: Synthesis report, summary for policymakers. In Core Writing Team, H. Lee, & J. Romero (Eds.), *Contribution of working groups I, II and III to the sixth assessment report of the intergovernmental panel on climate change*. IPCC.

Livi-Bacci, M. (2017). *A concise history of world population*. John Wiley and Sons.

Lovejoy, P. (2011). *Transformations in slavery: A history of slavery in Africa*. Cambridge University Press.

Lutz, W., Crespo Cuaresma, J., Kebede, E., Prskawetz, A., Sanderson, W.C., & Striessnig, E. (2019). Education rather than age structure brings demographic dividend. *Proceedings of the National Academy of Sciences*, *116*(26), 12798–12803.

Lutz, W., Goujon, A., Kc, S., Stonawski, M., & Stilianakis, N. (2018). *Demographic and human capital scenarios for the 21st century: 2018 assessment for 201 countries*. Publications Office of the European Union.

Maddison Project Database. (2020). (Bolt & van Zanden, 2020) – With minor processing by Our World in Data. "GDP per capita – Maddison project database – Historical data" [dataset]. Bolt and van Zanden, "Maddison project database" [original data]. Retrieved March 5, 2024 from https://ourworldindata.org/grapher/gdp-per-capita-maddison.

Madsen, E.L. (2015). What's behind west and central Africa's youthful demographics? High desired family size. *NewSecurityBeat*. Wilson Centre. https://www.newsecuritybeat.org/2015/05/whats-west-central-africas-youthful-demographics-high-desired-family-size/

Manning, P. (2014). African population, 1650–2000: Comparisons and implications of new estimates. In E. Akyeampong, R.H. Bates, N. Nunn, & J. Robinson (Eds.), *Africa's development in historical perspective* (pp. 131–150). Cambridge University Press.

Ojakaa, D.I. (2022). Trends and patterns in fertility and its proximate determinants in sub-Saharan Africa. In C.O. Odimegwu & Y. Adewoyin (Eds.), *The Routledge handbook of African demography* (pp. 198–216). Routledge.

Paice, E. (2021). *Youthquake. Why African demography should matter to the world.* Bloomsbury Publishing.

Platteau, J-P., & Hayami, Y. (1998). Resource endowments and agricultural development: Africa versus Asia. In Y. Hayami & M. Aoki (Eds.), *The institutional foundations of East Asian economic development* (pp. 357–412). Palgrave MacMillan.

Reiter, C., Goujon, A., & Samir, K.C. (2022). The demography of human capital formation in sub-Saharan Africa, 1950–2100. In C.O. Odimegwu & Y. Adewoyin (Eds.), *The Routledge handbook of African demography* (pp. 591–615). Routledge.

Rossi, P., & Godard, M. (2022). The old-age security motive for fertility: Evidence from the extension of social pensions in Namibia. *American Economic Journal: Economic Policy*, *14*(4), 488–518.

Trans-Atlantic Slave Trade Database. (2021). SlaveVoyages. https://www.slavevoyages.org. Accessed December 20, 2023.

UNESCO. (2019). *Meeting commitments: Are countries on track to achieve SDG 4?* UNESCO Institute for Statistics.

UN DESA (United Nations Population Division Department of Economic and Social Affairs). (2018). *World urbanization prospects 2018*. United Nations.

UN DESA (United Nations Population Division Department of Economic and Social Affairs). (2019). *Estimates and projections of family planning indicators 2019*. United Nation, DESA.

UN DESA (United Nations Population Division Department of Economic and Social Affairs). (2020). *World fertility and family planning 2020*. Highlights (ST/ESA/SER.A/440).

UN DESA (United Nations Population Division Department of Economic and Social Affairs). (2022). *World population prospects 2022*. United Nations.

Vlasblom, D. (2013). *The richer harvest: Economic development in Africa and Southeast Asia compared. The "Tracking Development Study" 2006–2011*, African Studies Centre Leiden Occasional Paper.

Vollset, S.E., Goren, E., Yuan, C.W., Cao, J., Smith, A.E., Hsiao, T., … & Murray, C.J. (2020). Fertility, mortality, migration, and population scenarios for 195 countries and territories from 2017 to 2100: A forecasting analysis for the Global Burden of Disease Study. *The Lancet*, *396*(10258), 1285–1306.

World Bank. (2021a). *National accounts data: GDP per capita (current US$)*. Accessed March 5, 2024.

World Bank. (2021b). *Contraceptive Prevalence, any modern method (% of married women ages 15-49)*. Accessed March 5, 2024.

7 From yield and nutrient gaps to assessing future food self-sufficiency in Africa

Antonius G.T. Schut and Martin K. van Ittersum

1 Introduction

Food security in Africa at macro level can be achieved by a balance between food demand, food production and food import (de Haas et al., this volume). International and regional trade will remain important (van Berkum and de Steenhuijsen Piters, this volume), but finance for imports requires export of other products and services and may not provide food for the poor without sources of income. A large dependency on food imports is undesirable not only from a geopolitical and economic perspective, but also from a sustainability perspective as large amounts of nutrients are added to the food system, limiting options for circularity. The green revolution intensified agriculture and increased local food production and labour productivity and is a major catalyst of structural transformation in early stages of development (Jayne, Chamberlin, et al., 2018). Agricultural productivity drives non-agricultural employment (Kamdem & Nembot, 2020) as it reduces real food prices compared to wages (Dorward, 2013). Achieving food self-sufficiency in Africa for the rapidly growing population (van Wissen, this volume) is a major challenge, even more so when considering that this needs to be done within planetary boundaries. Limiting demand growth by preventing food loss and waste (Sheahan & Barrett, 2017; Totobesola et al., 2022) and producing more on existing farm land are key to limit agricultural expansion with loss of biodiversity and increased emissions of greenhouse gases (GHGs) as a consequence. We show that the currently large yield gaps provide opportunities to produce much more on existing farm land with sustainable intensification while some cropland area expansion and regional food trade will also be needed. First, some background is given on the African farming context before discussing the opportunities and limitations of sustainable intensification with a reflection on climate change and opportunities to reduce dependency on fertilizer imports.

2 The African agricultural landscape

The old and strongly weathered landscapes that prevail across much of the continent provide little natural soil fertility to support a large human population. The African Rift valley and the large river deltas of the Niger and Nile are exceptions where fertile soils developed on relatively young volcanic or river deposits. In these areas, the population density is in general far higher than in other parts of the continent, although local population density is influenced by many more factors than soil fertility alone (Iliffe, 2017). Intensive cropping with net nutrient offtakes over a long period of time has strongly reduced soil fertility in these fertile landscapes. Soil fertility constraints for crops are most prevalent in the savannas and the tropical regions. But also in the drier areas towards the Sahelian climate zone, nutrients limit crop yields. In these zones, soil fertility constraints are strongly exacerbated by lack of soil water (Breman et al., 2001) and

DOI: 10.4324/9781032649696-10

temporal mismatches between crop demand and soil nutrient supply as it takes quite some time for the soil to wet at depth and before soil nutrients are plant available. Although there are opportunities to extend irrigation in Africa (Veldwisch et al., this volume), current irrigated areas are small. Yet, seasonal rainfall is abundant in many parts with high rainfed yield potentials (van Ittersum et al., 2016). For example, the tropical rainforest and Guinea savannah climate zones cover a large area with long seasons that often support two crops per year. Overall, it has been known for many decades that rainfall less frequently limits yields than nutrients (Silva et al., 2023; van Keulen & Breman, 1990).

2.1 Nutrient limitations

African soils have long been mined: nutrient offtakes with harvested products are on many fields larger than nutrient inputs resulting in negative soil nutrient balances at the national scale (Lassaletta et al., 2014; Stoorvogel & Smaling, 1998). Farmers relied for long on animal manures for nutrient supply, exhausting the surrounding landscape (de Ridder et al., 2004). Animals recycle nutrients that plants have taken up from the soil. Weathering of soil minerals and biological N fixation in the wider landscape determine the nutrients that are available for plant growth and hence the amount of feed and nutrients that can be concentrated in animal manures. Only limited amounts of feed and thus manure are available and this scarce resource is typically concentrated on the best fields, while poor fields receive little or none. Organic fertilizers are a key component of integrated soil fertility management (Vanlauwe et al., 2010; Vanlauwe et al., 2015). Yet, farmers have limited opportunities to address nutrient deficiencies with organic manures and other organic sources alone (Vanlauwe & Giller, 2006; Vanlauwe et al., 2014). Evidently, there is insufficient quantity and quality of feed available to increase the number of animals and manure in the system. To address nutrient deficiencies and grow more food and feed locally, mineral fertilizers are required.

The legacy of animal manures in soils is reflected in large differences in soil P and K contents and the ensuing crop yields (Njoroge, 2019). Nitrogen is typically the most limiting element when crops are fertilized solely with animal manures, because N contents are low when compared to P and K contents relative to crop demand. The use of animal manures on selected fields created a patchwork of soil fertility in African landscapes. Concentrating manures on the best fields is a better strategy for farmers than evenly distributing manure over all fields (Rowe et al., 2006), yet also mines other fields of macro and micro nutrients. Such depleted fields consequently have a poor yield response to the standard (N only or N with P) fertilizers as other nutrients are limiting, such as the macronutrients K and S that are typically not supplied. For these fields with strongly depleted soil nutrient pools, larger investments are needed with animal manures (Zingore et al., 2008) and/or balanced fertilizer use for two to three seasons (Njoroge, 2019) to ensure a good yield response.

Soil fertility is directly related to the farmers' socio-economic conditions. More endowed farmers usually own more livestock and can manure their better fields while poorly endowed farmers typically cannot. As a consequence, the less fertile fields are more often managed by poor farmers with few resources. On such fields investments in fertilizer carry more risk, and risk-averse smallholders are reluctant or unable to invest, creating poverty traps as soil fertility and crop yields further decline (Tittonell & Giller, 2013).

2.2 Opportunities for intensification

There are ample opportunities for more food production through intensification on existing farm land. Yield gaps are very large in most African countries; generally farmers achieve only

20–40% of the rainfed potential yield (van Ittersum et al., 2016; Yuan et al., 2024). The yield gaps are mainly due to a nutrient gap (ten Berge et al., 2019), although efficiency and technology gaps are also important (Silva et al., 2021). Africa thus has a large potential to increase yields as the use of mineral fertilizers is still very limited, despite strong pledges by African and international politicians. The green revolution did not yet transform African agriculture (Frankema, 2014). Most countries have a negative soil P balance, with the exception of Morocco, Mauritania, Egypt, Kenya and South Africa (Zhang et al., 2020), but even here the surpluses are very small (>3.5 kg P ha^{-1}). As a result, only small yield increases have been observed over the last decades, estimated at about 30 kg ha^{-1} y^{-1} for cereals in Africa (Grassini et al., 2013; Van Ittersum et al., 2016). Large yield increases in Egypt and South Africa are the exceptions (Grassini et al., 2013), while yields in Ethiopia (Abate et al., 2015; Sida et al., this volume) and Zambia strongly increased in more recent years.

While Africa lacks local N fertilizer production capacity, it has substantial P deposits and fertilizer production facilities: Morocco is the world's most important P exporter but also Tunisia, Egypt and South Africa have substantial capacity to mine and process rock phosphate. Smaller P deposits are found in Algeria, Tanzania, Senegal and Togo. There is hardly any K fertilizer production in Africa (FAO, 2022). Only Ethiopia has a large K deposit and smaller deposits are present in Botswana, the Republic of the Congo and Eritrea.[1] Although N and P fertilizer production in Africa increased and potential capacity exceeds demand (FAO, 2022), many countries rely on N and P fertilizer imports (Chianu et al., 2012). Nitrogen production is concentrated near large natural gas fields, present in a few countries (Egypt, Nigeria, Algeria and Libya). Large fertilizer companies are now investing in green ammonia, derived from air and water with wind or solar energy to produce N fertilizer, although many practical issues remain (Ojelade et al., 2023). This new technology offers opportunities for local urea production (Srivastava et al., 2023), important for especially isolated African countries where transport costs of fertilizers are very high but where large areas of unproductive land are available that may be used for wind and/or solar energy production. Investment in these green technologies enables countries to become self-sufficient in N production and less vulnerable to global market disturbances.

3 Challenges for food self-sufficiency

In this section and the next, we address four major and interrelated challenges for food self-sufficiency in Africa: (1) yield increments need to keep pace with increasing food demand; (2) fertilizers need to be produced and distributed; (3) fertilizers need to be used by farmers; and (4) this needs to be done while minimizing GHG emissions and adapting to climate change.

3.1 Increasing yields of cereals and other staple crops

To illustrate the grand challenge of keeping up with the enormous increases in food demand, we focus on cereals. Note that the outcomes of this analysis also apply to other crops needed for a nutritious diet. Cereals account for approximately 50% of caloric intake and total crop area in Africa. Cereal demand in the region has been projected to nearly triple between 2015 and 2050 due to rapid population growth and, to a lesser extent, dietary change (Van Ittersum et al., 2016). As indicated above, increases in cereal yields are very slow in most African countries; at current rates it would take more than 30 years to achieve a 1 t/ha maize yield increase (van Ittersum et al., 2016; FAOSTAT). To assess what is possible, van Ittersum et al. (2016; updated for this chapter) evaluated four different intensification scenarios for ten selected countries which make up more than half of the population and cropland of sub-Saharan Africa (SSA). In 2050:

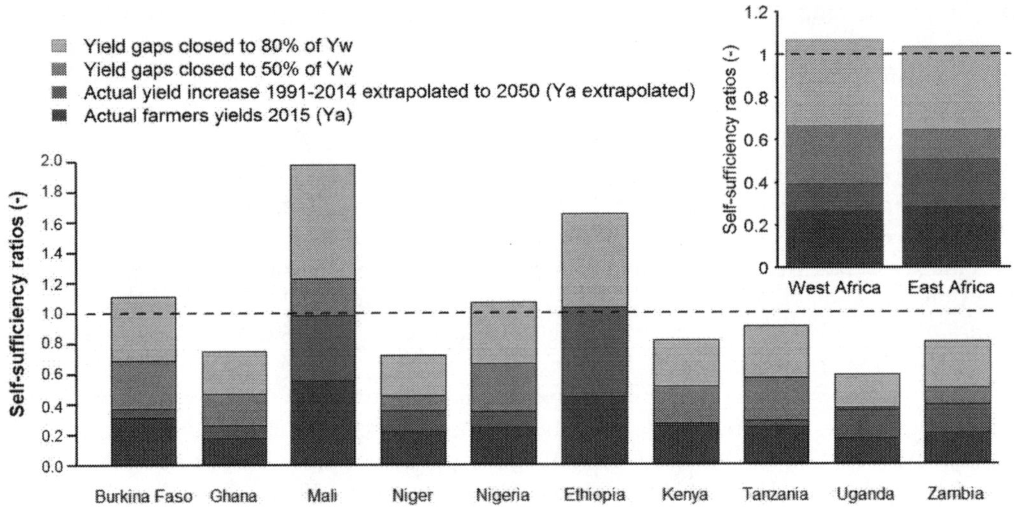

Figure 7.1 Cereal self-sufficiency ratios of ten selected countries in 2050 in relation to current farmers' yields (Ya) for scenarios where current yield trends are extrapolated and where yield gaps are close to 50 or 80% of water-limited yield (Y_W) under rainfed conditions. Figure updated from van Ittersum et al. (2016) (*PNAS*) kindly prepared by Marloes van Loon (2023).

(1) per-ha cereal yields are the same as in the year 2015; (2) current cereal yield trends over the period 1991–2014 continue up to 2050; (3) rainfed cereal yields are 50% of water-limited potential yield (Y_W) or irrigated cereal yields are 50% of the potential yield (Y_P); and (4) cereal yields are 80% of Y_W or Y_P. Estimation of Y_W and Y_P in this study did not include expected impacts of ongoing climate change. While for some countries (Mali and Ethiopia) current trends in cereal yields would suffice to achieve self-sufficiency in 2050, most countries will not be self-sufficient by 2050 even if farmers achieve 80% of the yield potential (Figure 7.1). Under the scenario with the highest intensification, cereal self-sufficiency for the whole SSA region in 2050 is just possible with the current (2010) cereal area. Current yields (on 2010 crop areas) of ca. 20–40% of yield potential will have to increase to ca. 80% of the potential, the latter being even more than that observed in Europe, USA and China, which requires an unprecedented steep and continuous increase in production for rainfed systems. The analysis above is currently being updated with the 2020 data on crop areas and yields and yield trends, potential yield estimations and projected cereal demand changes. Preliminary results indicate slightly more optimistic scenarios, in particular for (parts of) East Africa, assuming that the recent steep increases in yields in several countries (including Ethiopia) will continue.

There are also large opportunities to increase yields of other important staple crops, such as banana, cassava, sweet potato or yams (Schut & Giller, 2020b). Especially cassava has an enormous yield potential at almost ten times greater than current yields (Adiele et al., 2021; Adiele et al., 2020). To realize these required yield increments in a range of crops, the development of extension and support services and targeted fertilizer subsidy programmes combined with credit, insurance and market regulation programmes are needed to support local food production.

3.2 Fertilizer needs

The amounts of fertilizer needed to achieve food self-sufficiency on the African continent are very large. Sustaining 2015 average yields into the future alone will require at least three times more N and P input than the amounts currently used to prevent further soil depletion (Ten Berge et al., 2019). To increase yields to 50% or even 80% of Y_w requires large inputs of all macronutrients. It is estimated that minimum N requirements for maize crops with ca. 80% of rainfed yield potential in the ten countries in SSA (mentioned above) will be ca. 140 kg N/ha, while current average fertilizer inputs are between 10 and 50 kg N/ha (www.yieldgap.org; Ten Berge et al., 2019). This minimum N requirement assumes that other nutrients in the soil are not limiting. However, availability of P and occasionally K are often far below desired levels and require substantial inputs. Current phosphorus and potassium inputs are on average very low to negligible (less than 5 kg/ha). The required amounts to reach optimal soil fertility are so large that the strategy to fertilize the soil as used in high-income countries is unlikely to succeed. The alternative strategy is to fertilize the crop (Njoroge, 2019). Balanced fertilizer applications including N, P and K ensure a good agronomic efficiency (Schut & Giller, 2020a) and will build up soil pools steadily.

Markets for required crop inputs and products function poorly in many African countries, due to a raft of reasons (Holden, 2018). There are large challenges in the supply chain and the extensive distribution network needed to reach the large numbers of smallholder farmers, directly related to the large distances and poor road networks to connect farms to supply and produce markets (Ingenbleek et al., this volume). Fertilizer subsidy programmes attempt to reduce costs of fertilizer, yet often do not serve their purpose of enhancing access to fertilizers by smallholders (Jayne, Mason, et al., 2018) or address underlying market problems. For example, distributors servicing many farmers need simple, one-size-fits-all solutions to keep costs down while the large variability in soil fertility requires a more tailored solution. Overall, earlier input subsidy programmes resulted in modest yield (20%) and income (18%) increases (Balthzer & Hansen, 2011), although econometric evidence is thin (Jayne, Mason, et al., 2018). Many subsidy programmes work with vouchers, reducing costs on available fertilizer with little attention for fertilizer composition tailored to the crop needs, leading to a poor nitrogen use efficiency (NUE). A better NUE is identified as one of the key factors in sustainable intensification (Jayne & Sanchez, 2021; ten Berge et al., 2019) which can be realized by balanced fertilizers with macronutrients in the right proportions (Schut & Giller, 2020a). Subsidy programmes should ensure balanced fertilizer use, for example by promoting the production and use of blended fertilizers adapted to needs in the local area (Jayne & Sanchez, 2021) reflecting typical soil fertility and specific crop demand. For example, increasing the proportion and productivity of legumes in the system must be the first step in intensification, which requires P not N. Subsidy programmes should, also for cereals and root crops, first focus on P and K and maybe S and Mg because this ensures that N is effectively used. These nutrients are also less mobile and remain longer in the system and hence contribute to increasing production for a longer period of time. The stronger focus on other nutrients is also justified because N fertilizers are already known and used by many farmers.

3.3 Overcoming barriers to intensification

The majority of smallholders are risk-averse and reluctant to invest in agriculture because prices of products strongly fluctuate and seasons strongly vary (Jindo et al., 2020). Increasing yields may not be economically rational when gross margins per unit of labour do not increase due to low labour productivity (Silva et al., 2019). Many smallholder farmers are food insecure and are net food consumers (Frelat et al., 2016), and may be farming for lack of other options (Giller et al., 2021). About 40–70% of the food-secure smallholders in Africa are just "hanging in", while only 11–13% are "stepping up" and investing in intensification (Thornton et al., 2018). An important factor enabling farmers to step up is credit availability. For example, the One Acre Fund (1AF, https://oneacrefund.org) is popular as they provide a standard package including hybrid maize seeds, one 50 kg bag of diammonium phosphate (DAP) and calcium ammonium nitrate (CAN) per acre on credit. Credit needs to be repaid only after the season when the crop has been harvested. The 1AF operates in nine countries and has served 1.5 million smallholder farmers and successfully increased yields by 26% and profits by 16% in Kenya (Duru et al., 2021). However, such initiatives strongly depend on donor funds to support costs of staff, limiting their commercial viability.

Urbanization continues and populations in urban areas grow faster than rural areas (van Wissen, this volume). Hence, the required yield increments to provide food for this growing urban demand from existing farm land exceed all realistic scenarios for yield increases on smallholder farms. Hence, some of the required food may need to be produced on well-managed larger farms (Schut & Giller, 2020b) that have access to credit and can use mechanized equipment with direct yield benefits (Aune et al., 2019). Such large farms do not have to be owned by a single farmer, but can also be consolidated farms, cooperative farms or farms on rented land. For example, younger farmers with an entrepreneurial attitude may be able to rent land from other farms and produce for markets (Holden & Tilahun, 2021).

4 Food security and greenhouse gas emissions

Next to the daunting task of producing sufficient and nutritious food, GHG emissions need to be kept as low as possible to mitigate climate change. Emissions from global crop production are mostly caused by the use of peatlands (Leifeld & Menichetti, 2018; Lin et al., 2022) and by the conversion of forest and grassland to cropland (area expansion) which results in a net loss of carbon stocks in soil and vegetation. SSA has already seen a continuous increase in emissions from agriculture-driven deforestation. Yet, intensification, i.e. higher yields per hectare with sufficient and judicious use of inputs, will also lead to larger emissions per hectare. Estimations suggest that intensification of cereal production with sufficient *and* efficient use of fertilizers will lead to lowest GHG emissions overall (Van Loon et al., 2019), as area expansion comes with large GHG emissions. But this will be possible only when combined with strongly improved agronomy, including the use of well-adapted cultivars; proper planting densities; good nutrient management; and crop protection against weeds, pests and diseases (Vanlauwe et al., 2015). Production and use of fertilizer N accounts for 10.6% of agricultural and 2.1% of global GHG emissions (Menegat et al., 2022). The GHG emissions from N fertilizer production can be reduced drastically. The biophysical conditions in Africa are generally favourable for solar energy production, providing new options for green fertilizer N production. Elsewhere, new production facilities for green ammonia production are already approved. In future, local green ammonia production can contribute strongly to reduce GHG emissions, reducing transport costs and infrastructural hurdles that currently complicate intensification efforts.

5 Conclusions and outlook for the future

Given the above analysis, a scenario of agricultural intensification without cropland area expansion seems highly unlikely, especially when considering that intensification is needed across a wide range of crops to supply a nutritious diet. Until now, increase in African food production resulted largely from area expansion (Giller, Delaune, Silva, Descheemaeker, et al., 2021; Yuan et al., 2024). In the short term, agricultural area expansion remains an important means to keep up with the growing food demand, causing losses of forests or grazing lands, thereby reducing carbon stocks and accelerating loss of biodiversity. Agricultural expansion driven by market demand is clearly an option in countries with large areas of suitable but unused land, reliable seasons and access to inputs such as Tanzania (Palmas & Chamberlin, 2020) and Zambia. However, when incentives to produce are strong, farmers may also start using less suited land with low productivity and land erosion and degradation as a consequence, as observed for example in Senegal (Tappan et al., 2004) and Malawi (Li et al., 2021). Agricultural expansion can also be driven by land degradation where farmers are forced to move to new areas for cropping. Soils already low in fertility are most vulnerable to degradation without fertilizer input, for example as a result of the economic collapse of Zimbabwe (Chagumaira et al., 2015). This is a worst-case scenario with undesired outcomes for food self-sufficiency, loss of biodiversity and large GHG emissions from land conversions (Brink et al., 2023).

Several additional factors ought to be considered in the above analysis. First, rainfall variability is a complicating factor as it increases the risks associated with investments in inputs by farmers. Agricultural intensification in dry and drought-prone areas is much more difficult and farmers in such areas should not target high yields, but first focus on investment in water retention technologies or use of agroforestry systems to enhance resilience (Sheil, this volume) before investing in fertilizer use. The assumption of attainable yields at a level of 80% of water-limited yield potential in harsh rainfed regions with substantial year-to-year variation in rainfall is too optimistic. At the same time, it is likely that there will be further genetic progress in yield potentials in the decades to come thanks to breeding, which would lead to more optimistic assessments. Second, the required yield increments are extremely difficult to achieve across the agricultural landscape when only a small proportion of farmers is willing to invest. The highly variable yield responses to standard inputs (that do not match the crop needs) and the larger investments needed for fields with a poor fertility mean that required yield increases will probably only be realized on a small proportion of the land that is currently in use. Third, climate change is likely to affect future yields and their variability. However, it is estimated that such effects are relatively modest until 2050 and adaptation of crops, cultivars and planting windows may mitigate negative yield effects (Alimagham et al., 2024). Yet, present climate variability, aggravated by climate change with changes of average conditions but also more frequent extreme events (Descheemaeker et al., this volume), is a valid reason to aim for more than full food self-sufficiency to be on the safe side.

The different potentials among African countries for intensification and expansion emphasize that food trade among countries within (and perhaps between) West, Central and East Africa will be essential to achieve regional food self-sufficiency and food security. This is even more true considering the fact that relatively land-abundant and less populated countries in especially Central Africa have a larger scope for area expansion (Chamberlin et al., 2014; Giller et al., 2022). Densely populated subregions with a high female fertility and a limited area of suitable land, such as Niger and Nigeria (Van Ittersum et al., 2016), will partly depend on food imports, also in 2050.

The analysis presented in this chapter underpins why it is unlikely that yield increments alone can keep pace with population growth in many African countries. Policies which are good for food security and for the climate must therefore seek to strike an optimum balance between intensification and controlled expansion of the agricultural area. Strong incentives to produce more food are needed, combining price policies and subsidies on inputs, especially P, K and where needed S. A key component is improved extension services to support farmers and promotion of tailored fertilizer blends. This will increase the yield response to N thereby reducing emissions of GHG related to N production and application. Hence, policies to support agriculture need to be combined with environmental policies to protect natural resources, quality of land, air and water. Supporting local fertilizer production is needed to reduce transport costs. SSA can strongly benefit from local green N fertilizer production technologies with lower transport costs. Investment in such green production facilities has large benefits for food self-sufficiency and GHG emissions and reduces its current dependency on international markets.

Note

1 https://www.afdb.org/sites/default/files/2019/10/21/cross-border_study.pdf.

References

Abate, T., Shiferaw, B., Menkir, A., Wegary, D., Kebede, Y., Tesfaye, K., … & Keno, T. (2015). Factors that transformed maize productivity in Ethiopia. *Food Security*, *7*(5), 965–981. https://doi.org/10.1007/s12571-015-0488-z

Adiele, J.G., Schut, A.G.T., van den Beuken, R.P.M., Ezui, K.S., Pypers, P., Ano, A. O., … & Giller, K.E. (2021). A recalibrated and tested LINTUL-Cassava simulation model provides insight into the high yield potential of cassava under rainfed conditions *European Journal of Agronomy*, *124*, paper 126242. https://doi.org/10.1016/j.eja.2021.126242

Adiele, J.G., Schut, A.G.T., van den Beuken, R.P.M., Ezui, K.S., Pypers, P., Ano, A.O., … & Giller, K.E. (2020). Towards closing cassava yield gap in West Africa: Agronomic efficiency and storage root yield responses to NPK fertilizers. *Field Crops Research*, *253*, 107820. https://doi.org/10.1016/j.fcr.2020.107820

Alimagham, S.M., van Loon, M.P., Ramirez-Villegas, J., Adjei-Nsiah, S., Baijukya, F., Bala, A., … & van Ittersum, M.K. (2024). Climate change impact and adaptation of rainfed cereal crops in sub-Saharan Africa. *European Journal of Agronomy*, *Online*.

Aune, J.B., Coulibaly, A., & Woumou, K. (2019). Intensification of dryland farming in Mali through mechanisation of sowing, fertiliser application and weeding. *Archives of Agronomy and Soil Science*, *65*(3), 400–410. https://doi.org/10.1080/03650340.2018.1505042

Balthzer, K., & Hansen, H. (2011). Agricultural input subsidies in Sub-Saharan Africa. (2011/2). https://www.oecd.org/derec/49231998.pdf

Breman, H., Groot, J.J.R., & Van Keulen, H. (2001). Resource limitations in Sahelian agriculture. *Global Environmental Change*, *11*(1), 59–68. https://www.scopus.com/inward/record.url?eid=2-s2.0-0035108098&partnerID=40&md5=059b36254eef62d9b8902550f051afd4

Brink, B.t., Giesen, P., & Knoope, P. (2023). Future responses to environment-related food self-insufficiency, from local to global. *Regional Environmental Change*, *23*(3), 87. https://doi.org/10.1007/s10113-023-02069-4

Chagumaira, C., Rurinda, J., Nezomba, H., Mtambanengwe, F., & Mapfumo, P. (2015). Use patterns of natural resources supporting livelihoods of smallholder communities and implications for climate change adaptation in Zimbabwe. *Environment, Development and Sustainability*, *18*(1), 237–255. https://doi.org/10.1007/s10668-015-9637-y

Chamberlin, J., Jayne, T.S., & Headey, D. (2014). Scarcity amidst abundance? Reassessing the potential for cropland expansion in Africa. *Food Policy*, *48*, 51–65. https://doi.org/10.1016/j.foodpol.2014.05.002

Chianu, J.N., Chianu, J.N., & Mairura, F. (2012). Mineral fertilizers in the farming systems of sub-Saharan Africa. A review. *Agronomy for Sustainable Development: Official Journal of the Institut National de la Recherche Agronomique (INRA)*, *32*(2), 545–566. https://doi.org/10.1007/s13593-011-0050-0

de Ridder, N., Breman, H., van Keulen, H., & Stomph, T.J. (2004). Revisiting a 'cure against land hunger': Soil fertility management and farming systems dynamics in the West African Sahel. *Agricultural Systems*, *80*(2), 109–131. https://doi.org/10.1016/j.agsy.2003.06.004

Dorward, A. (2013). Agricultural labour productivity, food prices and sustainable development impacts and indicators. *Food Policy*, *39*, 40–50. https://doi.org/10.1016/j.foodpol.2012.12.003

Duru, M., Siegal, K., Tjernström, E., & Deutschmann, J.W. (2021). Can smallholder extension transform African agriculture? https://oneacrefund.org/sites/default/files/2022-10/Teso_RCT_Results_2021_0.pdf

FAO. (2022). World fertilizer trends and outlook to 2022. https://www.google.com/url?sa=t&rct=j&q=&esrc=s&source=web&cd=&ved=2ahUKEwiZit2gosH7AhWN2aQKHRW6CjYQFnoECAwQAQ&url=https%3A%2F%2Fwww.fao.org%2F3%2Fca6746en%2Fca6746en.pdf&usg=AOvVaw3P4OtBPIzAou7P6HsbFwDk

Frankema, E. (2014). Africa and the green revolution a global historical perspective. *NJAS - Wageningen Journal of Life Sciences*. https://www.scopus.com/inward/record.url?eid=2-s2.0-84895631455&partnerID=40&md5=60ca97d9a339d7737bfce40ab108d976

Frelat, R., Lopez-Ridaura, S., Giller, K.E., Herrero, M., Douxchamps, S., Djurfeldt, A.A., … & Van Wijk, M.T. (2016). Drivers of household food availability in sub-Saharan Africa based on big data from small farms. *Proceedings of the National Academy of Sciences of the United States of America*, *113*(2), 458–463. https://doi.org/10.1073/pnas.1518384112

Giller, K.E., Delaune, T., Silva, J.V., Descheemaeker, K., van de Ven, G.W.J., Schut, A.G.T., … & van Ittersum, M.K. (2021). The future of farming: Who will produce our food? *Food Security*, *13*(5), 1073–1099. https://doi.org/10.1007/s12571-021-01184-6

Giller, K.E., Delaune, T., Silva, J.V., Descheemaeker, K., van de Ven, G.W.J., Schut, A.G.T., … & Van Ittersum, M.K. (2022). *The future of farming: Who will produce our food?* (IFAD Research Series, Issue. https://www.ifad.org/en/web/knowledge/-/research-series-83-the-future-of-farming-who-will-produce-our-food-?p_l_back_url=%2Fen%2Fweb%2Fknowledge%2Fseries

Giller, K.E., Delaune, T., Silva, J.V., van Wijk, M., Hammond, J., Descheemaeker, K., … & Andersson, J.A. (2021). Small farms and development in sub-Saharan Africa: Farming for food, for income or for lack of better options? *Food Security*, *13*, 1431–1454. https://doi.org/10.1007/s12571-021-01209-0

Grassini, P., Eskridge, K.M., & Cassman, K.G. (2013). Distinguishing between yield advances and yield plateaus in historical crop production trends. *Nature Communications*, *4*, 1–11. https://doi.org/10.1038/ncomms3918

Holden, S.T. (2018). Fertilizer and sustainable intensification in Sub-Saharan Africa. *Global Food Security*, *18*, 20–26. https://doi.org/10.1016/j.gfs.2018.07.001

Holden, S.T., & Tilahun, M. (2021). Are land-poor youth accessing rented land? Evidence from northern Ethiopia [Article]. *Land Use Policy*, *108*, Article 105516. https://doi.org/10.1016/j.landusepol.2021.105516

Iliffe, J. (2017). *Africans: The history of a continent* (3rd ed.). Cambridge University Press.

Jayne, T.S., Chamberlin, J., & Benfica, R. (2018). Africa's unfolding economic transformation. *The Journal of Development Studies*, *54*(5), 777–787. https://doi.org/10.1080/00220388.2018.1430774

Jayne, T.S., Mason, N.M., Burke, W.J., & Ariga, J. (2018). Review: Taking stock of Africa's second-generation agricultural input subsidy programs. *Food Policy*, *75*, 1–14. https://doi.org/10.1016/j.foodpol.2018.01.003

Jayne, T.S., & Sanchez, P.A. (2021). Agricultural productivity must improve in sub-Saharan Africa. *Science*, *372*(6546), 1045–1047. https://doi.org/10.1126/science.abf5413

Jindo, K., Schut, A.G.T., & Langeveld, J.W.A. (2020). Sustainable intensification in Western Kenya: Who will benefit? *Agricultural Systems*, *182*, 102831. https://doi.org/10.1016/j.agsy.2020.102831

Kamdem, C.B., & Nembot, L.N. (2020). Moving off Agrarian societies: Agricultural productivity to facilitate economic transformations and non-agricultural employment growth in sub-Saharan Africa. *Journal of International Development*, *32*(3), 324–341. https://doi.org/10.1002/jid.3455

Lassaletta, L., Billen, G., Grizzetti, B., Anglade, J., & Garnier, J. (2014). 50 year trends in nitrogen use efficiency of world cropping systems: The relationship between yield and nitrogen input to cropland. *Environmental Research Letters*, *9*(10), 105011, Article 105011. https://doi.org/10.1088/1748-9326/9/10/105011

Leifeld, J., & Menichetti, L. (2018). The underappreciated potential of peatlands in global climate change mitigation strategies. *Nature Communications*, *9*(1), 1071. https://doi.org/10.1038/s41467-018-03406-6

Li, C., Kandel, M., Anghileri, D., Oloo, F., Kambombe, O., Chibarabada, T.P., … & Dash, J. (2021). Recent changes in cropland area and productivity indicate unsustainable cropland expansion in Malawi. *Environmental Research Letters*, *16*(8), Article 084052. https://doi.org/10.1088/1748-9326/ac162a

Lin, F., Zuo, H.C., Ma, X.H., & Ma, L. (2022). Comprehensive assessment of nitrous oxide emissions and mitigation potentials across European peatlands. *Environmental Pollution*, *301*, Article 119041. https://doi.org/10.1016/j.envpol.2022.119041

Menegat, S., Ledo, A., & Tirado, R. (2022). Greenhouse gas emissions from global production and use of nitrogen synthetic fertilisers in agriculture. *Scientific Reports*, *12*(1), 14490. https://doi.org/10.1038/s41598-022-18773-w

Njoroge, S.K. (2019). *Feed the crop, not the soil! Explaining variability in maize yield responses to nutrient applications in smallholder farms of western Kenya* [Wageningen University]. Wageningen. https://edepot.wur.nl/503185

Ojelade, O.A., Zaman, S.F., & Ni, B.-J. (2023). Green ammonia production technologies: A review of practical progress. *Journal of Environmental Management*, *342*, 118348. https://doi.org/10.1016/j.jenvman.2023.118348

Palmas, S., & Chamberlin, J. (2020). Fertilizer profitability for smallholder maize farmers in Tanzania: A spatially-explicit ex ante analysis. *PLoS One*, *15*(9), e0239149. https://doi.org/10.1371/journal.pone.0239149

Rowe, E.C., van Wijk, M.T., de Ridder, N., & Giller, K.E. (2006). Nutrient allocation strategies across a simplified heterogeneous African smallholder farm. *Agriculture, Ecosystems and Environment*, *116*(1–2), 60–71. https://www.scopus.com/inward/record.url?eid=2-s2.0-31444455337&partnerID=40&md5=adaabaa13ea36bc3ae1471f9de7226a8

Schut, A.G.T., & Giller, K.E. (2020a). Soil-based, field-specific fertilizer recommendations are a pipe-dream. *Geoderma*, *380*, 114680. https://doi.org/10.1016/j.geoderma.2020.114680

Schut, A.G.T., & Giller, K.E. (2020b). Sustainable intensification of agriculture in Africa. *Frontiers of Agricultural Science and Engineering*, *7*(4), 371–375. https://doi.org/10.15302/j-fase-2020357

Sheahan, M., & Barrett, C.B. (2017). Review: Food loss and waste in sub-Saharan Africa. *Food Policy*, *70*, 1–12. https://doi.org/10.1016/j.foodpol.2017.03.012

Silva, J.V., Baudron, F., Ngoma, H., Nyagumbo, I., Simutowe, E., Kalala, K., … & Thierfelder, C. (2023). Narrowing maize yield gaps across smallholder farming systems in Zambia: What interventions, where, and for whom? *Agronomy for Sustainable Development*, *43*(2), 26. https://doi.org/10.1007/s13593-023-00872-1

Silva, J.V., Baudron, F., Reidsma, P., & Giller, K.E. (2019). Is labour a major determinant of yield gaps in sub-Saharan Africa? A study of cereal-based production systems in Southern Ethiopia. *Agricultural Systems*, *174*, 39–51. https://doi.org/10.1016/j.agsy.2019.04.009

Silva, J.V., Reidsma, P., Baudron, F., Laborte, A.G., Giller, K.E., & van Ittersum, M.K. (2021). How sustainable is sustainable intensification? Assessing yield gaps at field and farm level across the globe. *Global Food Security*, *30*, 100552. https://doi.org/10.1016/j.gfs.2021.100552

Srivastava, N., Saquib, M., Rajput, P., Bhosale, A.C., Singh, R., & Arora, P. (2023). Prospects of solar-powered nitrogenous fertilizers. *Renewable and Sustainable Energy Reviews*, *187*, 113691. https://doi.org/10.1016/j.rser.2023.113691

Stoorvogel, J.J., & Smaling, E.M.A. (1998). Research on soil fertility decline in tropical environments: Integration of spatial scales. In P.A. Finke, J. Bouma, & M.R. Hoosbeek (Eds.), *Soil and water quality at different scales: Proceedings of the workshop "Soil and water quality at different scales"* held 7–9 August 1996, Wageningen, The Netherlands (pp. 151–158). Springer. https://doi.org/10.1007/978-94-017-3021-1_15

Tappan, G.G., Sall, M., Wood, E.C., & Cushing, M. (2004). Ecoregions and land cover trends in Senegal. *Journal of Arid Environments*, *59*(3), 427–462. https://doi.org/10.1016/j.jaridenv.2004.03.018

ten Berge, H.F.M., Hijbeek, R., van Loon, M.P., Rurinda, J., Tesfaye, K., Zingore, S., … & van Ittersum, M.K. (2019). Maize crop nutrient input requirements for food security in sub-Saharan Africa. *Global Food Security*, *23*, 9–21. https://doi.org/10.1016/j.gfs.2019.02.001

Thornton, P.K., Kristjanson, P., Förch, W., Barahona, C., Cramer, L., & Pradhan, S. (2018). Is agricultural adaptation to global change in lower-income countries on track to meet the future food production challenge? *Global Environmental Change*, *52*, 37–48. https://doi.org/10.1016/j.gloenvcha.2018.06.003

Tittonell, P., & Giller, K.E. (2013). When yield gaps are poverty traps: The paradigm of ecological intensification in African smallholder agriculture [Crop Yield Gap Analysis – Rationale, Methods and Applications]. *Field Crops Research*, *143*(0), 76–90. https://doi.org/10.1016/j.fcr.2012.10.007

Totobesola, M., Delve, R., Nkundimana, J.d.A., Cini, L., Gianfelici, F., Mvumi, B., … & Rolle, R.S. (2022). A holistic approach to food loss reduction in Africa: Food loss analysis, integrated capacity development and policy implications. *Food Security*, *14*(6), 1401–1415. https://doi.org/10.1007/s12571-021-01243-y

van Ittersum, M.K., van Bussel, L.G.J., Wolf, J., Grassini, P., van Wart, J., Guilpart, N., … & Cassman, K.G. (2016). Can sub-Saharan Africa feed itself? *Proceedings of the National Academy of Sciences*, *113*(52), 14964–14969. https://doi.org/10.1073/pnas.1610359113

van Keulen, H., & Breman, H. (1990). Agricultural development in the West African Sahelian region: A cure against land hunger? *Agriculture, Ecosystems and Environment*, *32*(3–4), 177–197. https://doi.org/10.1016/0167-8809(90)90159-b

van Loon, M.P., Hijbeek, R., ten Berge, H.F.M., De Sy, V., ten Broeke, G.A., Solomon, D., & van Ittersum, M.K. (2019). Impacts of intensifying or expanding cereal cropping in sub-Saharan Africa on greenhouse gas emissions and food security. *Global Change Biology*, *25*(11), 3720–3730. https://doi.org/10.1111/gcb.14783

Vanlauwe, B., Bationo, A., Chianu, J., Giller, K.E., Merckx, R., Mokwunye, U., … & Sanginga, N. (2010). Integrated soil fertility management: Operational definition and consequences for implementation and dissemination. *Outlook on Agriculture*, *39*(1), 17–24. https://www.scopus.com/inward/record.url?eid=2-s2.0-77953556615&partnerID=40&md5=d4d43603542699ad4a4adfc72b167cbf

Vanlauwe, B., Descheemaeker, K., Giller, K.E., Huising, J., Merckx, R., Nziguheba, G., … & Zingore, S. (2015). Integrated soil fertility management in sub-Saharan Africa: Unravelling local adaptation. *Soil*, *1*(1), 491–508. https://doi.org/10.5194/soil-1-491-2015

Vanlauwe, B., & Giller, K.E. (2006). Popular myths around soil fertility management in sub-Saharan Africa. *Agriculture, Ecosystems and Environment*, *116*(1–2), 34–46. https://www.scopus.com/inward/record.url?eid=2-s2.0-33744931880&partnerID=40&md5=314e419fbe4554f0d120e51be057900d

Vanlauwe, B., Wendt, J., Giller, K.E., Corbeels, M., Gerard, B., & Nolte, C. (2014). A fourth principle is required to define Conservation Agriculture in sub-Saharan Africa: The appropriate use of fertilizer to enhance crop productivity. *Field Crops Research*, *155*, 10–13. https://www.scopus.com/inward/record.url?eid=2-s2.0-84888198563&partnerID=40&md5=23a92e30f81bf4c74f98f135a3681de6

Yuan, S., Saito, K., van Oort, P.A.J., van Ittersum, M.K., Peng, S., & Grassini, P. (2024). Intensifying rice production to reduce imports and land conversion in Africa. *Nature communications*, *15*(1), 835. https://doi.org/10.1038/s41467-024-44950-8

Zhang, X., Davidson, E.A., Zou, T., Lassaletta, L., Quan, Z., Li, T., & Zhang, W. (2020). Quantifying nutrient budgets for sustainable nutrient management. *Global Biogeochemical Cycles*, *34*(3), Article e2018GB006060. https://doi.org/10.1029/2018GB006060

Zingore, S., Delve, R.J., Nyamangara, J., & Giller, K.E. (2008). Multiple benefits of manure: The key to maintenance of soil fertility and restoration of depleted sandy soils on African smallholder farms. *Nutrient Cycling in Agroecosystems*, *80*(3), 267–282. https://www.scopus.com/inward/record.url?eid=2-s2.0-39849100899&partnerID=40&md5=75fe9e297bda239ea74d0a0ea95ccec6

8 Global market and food security in Africa: more trade or more domestic production?

Siemen van Berkum and Bart de Steenhuijsen Piters

1 Introduction

Many African countries depend heavily on imports of wheat, rice and other food staples. Recent price increases in international food markets and price volatility like in 2008/2009, 2011/2012 and in 2021/2022 threaten food security,[1] especially the poorest segment of the population who spend often more than half of their income on food (FAO, IFAD, UNICEF, WFP and WHO, 2022). The Covid pandemic and the war in Ukraine have illustrated that many food systems depending on international markets lack the resilience to protect vulnerable households against price spikes. A growing population, stagnating agricultural productivity and climate change are just some of the factors leading to concerns about Africa's future food security, especially in the longer term (e.g. OECD-FAO, 2023:95–113). What strategy can be deployed by African governments to strengthen their population's food security? There is a large literature and ongoing debate about food self-sufficiency and food and nutrition security (e.g. Clapp, 2017) in times of globalized food markets. Reducing import dependencies by expanding domestic production can ensure market stability, but may not lead to low consumer prices. Specializing in the production of food commodities in which a country has a comparative advantage, combined with import of cheap foods, is a much followed strategy that can lead to greater general welfare, which may, however, not be distributed equally nor lead to resilience of the food system (Van Berkum, 2021). This essay analyses how international market developments affect domestic food prices in Africa (Sections 2–5) and discusses some strategies to help strengthen domestic food security in countries currently dependent on international markets (Section 6). Key to countervailing food import dependency is diversification of markets (in terms of expanding the number of suppliers of imports), commodities (increasing the variety of sales markets for various products) and consumption patterns. Besides diversification, bottom of the pyramid consumers need more and better employment/income-generating activities, resulting in improved purchasing power.

2 International price developments

As a continent, Africa is a net importer of grains (mainly of wheat, maize and rice) sourced in particular from the Americas (the USA, Canada, Argentina), Europe (Russia and Ukraine) and Asia (rice from India and Thailand) (FAO, 2022). Grain imports contribute directly to food security as imports increase the availability of these staple foods for domestic consumers and reduce domestic grain prices. Import costs are highly dependent on international price developments (combined with the country's exchange rate against the US dollar). Since the 1960s, international agricultural prices have declined in real terms, as a result of improvements in productivity worldwide, which in turn resulted from the development, adoption and widespread use of new

DOI: 10.4324/9781032649696-11

Figure 8.1 Wheat price developments 2002–2023 (nominal terms).

Source: www.indexmundi.com; accessed on 29/09/2023.

Notes: The figure shows wheat (U.S.), no. 2 hard red winter, ordinary protein, export price (US$ per ton) delivered at the US Gulf port for prompt or 30 days shipment.

technological innovations such as fertilizers and mechanization (OECD-FAO, 2023:67). Falling marginal costs led to lower prices despite growing demand driven by global population and per capita income growth. Deviations from this general trend have been observed, such as the price spikes in the 1970s and in the period 2007–2014. In 2020–2023, grain prices increased substantially worldwide (see Figure 8.1 for wheat), although in real terms (that is, the price in US$ adjusted for inflation in the USA) prices were higher in early 2008 (see GuruFocus.com). These price increases have been caused by tight global supply (due to smaller than expected harvests in 2021 in Canada, the USA and Russia, all three major wheat-exporting countries) and rising production and transport costs related to the global Covid pandemic. The Ukrainian-Russian war further pushed up international prices and disrupted price formation on global grain markets in 2022, also because a number of relatively large exporters of food commodities announced export restrictions to stabilize their domestic prices in response to the sharp increase in international prices (IFPRI, 2023). These export restrictions have eased and wheat prices have fallen to pre-war level, but as the war continues, markets remain volatile indicating continued uncertainty.

For the coming decade OECD-FAO (2023) projects a return to the long-term downward trend in global food prices after the disruptions of 2022. Such projections must be interpreted with caution as they assume ‘normal’ weather conditions and predictable macroeconomic and policy developments. However, food systems worldwide are under increasing pressure from the increased frequency and severity of weather extremes. Moreover, there are general fears of new plant and animal disease outbreaks, whether or not influenced by climate change, which already affect food production and consequently food security in several regions in the world. Armed conflicts such as the war between Russia and Ukraine demonstrated that unexpected events in major exporters of agricultural commodities and inputs like nitrogen fertilizers can greatly distort international food markets, resulting in rapid price hikes. The duration of the Ukrainian-Russian war is unknown, while risks of regional or international conflict are on the rise (FAO, 2022), suggesting that disruptions of international food markets will be more common in future than model projections such as those of OECD-FAO suggest.

3 A history of Africa's dependency on imported grains

Over the past 50 years, food consumption patterns worldwide have become more similar in composition due to an increased supply of some globally important cereal and oil crops, and a decline of other cereal, oil and starchy root crops. This increase in homogeneity worldwide highlights the establishment of a global standard food supply, which is relatively species-poor (Khoury et al., 2014). Natural advantages and large-scale bulk production of such foods in exporting countries resulted in historically low international food commodity prices, favouring import over domestic production in many recipient countries. This has increased dependence of countries on international food markets for their national food security.

The growing consumption of wheat by African consumers is a reflection of this homogeneity in global food supply, but is also accompanied by an increasing dependence on imported staples. For instance staple grain consumption in sub-Saharan Africa (SSA) is rising at the same time that the region is becoming more dependent on imported staples as this region consistently produces less than half, and often as little as one-third, of the wheat it consumes (FAO, 2022).

Consumption of wheat in much of Africa took off during the second half of the 20th century (Mason et al., 2012). Wheat consumption was stimulated by the abundant supply of wheat from countries such as the USA, Canada and France where technological development has boosted production per hectare since the mid-19th century. Considering that grain production in many African and Asian countries lagged behind population growth, it was clear to policymakers in wheat-exporting countries that the most promising markets were to be found in developing countries. At the same time, price-competitive high-quality wheat imports found a ready market into African countries, where they were favoured by the relatively well-off urban populations and the politically influential urban elites to secure food availability at reasonable and stable prices (De Janvry and Sadoulet, 2022).

A crucial feature of international wheat markets over the second half of the 20th century has been wheat aid shipped to developing countries. The USA began to ship subsidized wheat to a great number of developing countries and soon after World War II wheat aid was institutionalized on a global scale. Although the programme was said to pursue humanitarian goals, it was crucially motivated by a desire to dispose of growing wheat surpluses (see Friedmann, 1992; González-Esteban, 2017, and De Janvry and Sadoulet, 2022 for a discussion on impacts of food aid on domestic food sector in recipient countries).

4 How food import-dependent is Africa?

The costs of food import by African countries have increased dramatically over time. For instance, in 2021 Africa imported agricultural products worth US$111 billion, against US$21 billion in 2000 (FAOSTAT, values in nominal terms). However, to indicate import dependency of the continent, more important than overall import costs is how imports of key food items relate to what is consumed. Figure 8.2 shows for Africa's five most important food items in terms of daily per capita calorie intake the share of imports in consumption of that product. These five staples together make up 50% of the daily per capita calorie intake in the African countries included. Maize – 18.2% of the daily per capita intake – and wheat – 15.4% – are the staple products with largest shares of intake, yet the net import dependency of Africa of these two staples is starkly different.

There is great variation in diet composition and import dependency in Africa. Table 8.1 shows the share of total food consumption (in calories) and the share of consumed amounts imported for the five most important staple foods. Despite the heterogeneity among countries, reliance on domestic production is broadly low for wheat, rice and palm oil, reflected in relatively

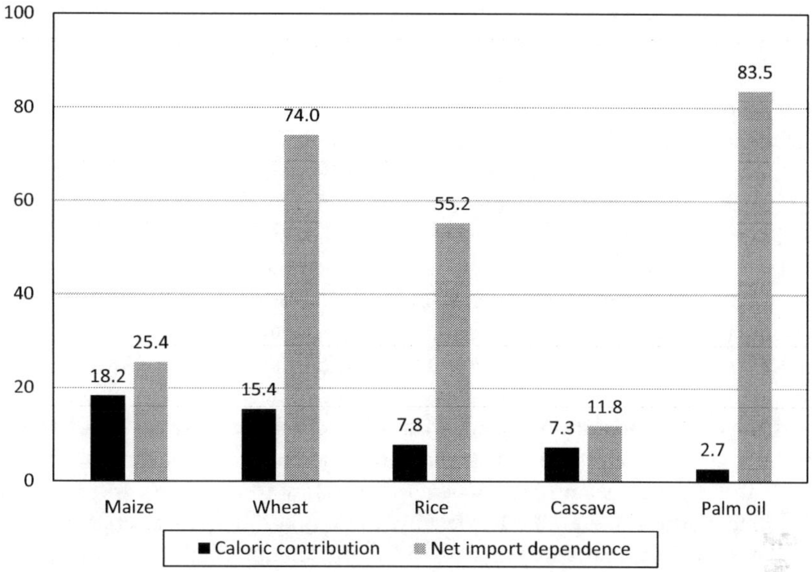

Figure 8.2 Daily caloric contribution and net import dependency (%) of the five most important staple foods in 19 African countries.

Source: FAOSTAT Food Balance Sheets data from 2019 and authors' calculations.

Notes: See Table 8.1 for the countries included; these countries are home to approximately 70% of Africa's population.

Table 8.1 Consumption share (CS) and net import dependence (NID) of the top five staples

Country	*Maize*		*Wheat*		*Rice*		*Cassava*		*Palm oil*	
	CS	*NID*	*CS*	*NID*	*CS*	*NID*	*CS*	*NID*	*CS*	*NID*
Algeria	3.6	**99.9**	40.1	63.6	0.7	**100**	N.a.	**100**	0.3	**100**
Angola	16.9	8.1	10.2	**99.8**	5.5	**98.3**	25.4	0	7.8	**80.1**
Côte d'Ivoire	5.4	2.5	6.3	**100**	26.7	53.8	14	0	6.6	−73.5
Egypt	17.2	51.2	35.3	52.4	11	11.6	N.a.	0	0.1	**100**
Ethiopia	19.3	0.7	13.3	21.1	1.6	**78.9**	N.a.	0	2	**100**
Ghana	7.5	0.5	4.1	**100**	9.5	59.9	26.4	−0.1	1.7	36
Kenya	27.6	6.3	12.2	**84.6**	6.2	**85.5**	2.1	0.1	1.7	**100**
Malawi	44.1	0.1	1.9	**99.2**	1.9	6.3	7.6	0	0.5	**100**
Morocco	10.3	**98.6**	40.7	49	0.4	46.3	N.a.	0	0.1	**100**
Mozambique	28.7	11.4	7.6	**97.3**	12.2	**81**	14.6	0	4.3	**100**
Namibia	16.9	**68.8**	17.3	**95.8**	3.1	**100**	N.a.	0	0	**100**
Nigeria	12.3	2.6	7.9	**98.7**	10.9	1.1	11.4	0	7.2	51.2
Rwanda	8.8	8.1	3.4	**88.5**	3.5	36.2	11.8	14.7	3.2	**100**
Senegal	7.6	41.8	12.9	**100**	29.8	49.7	3.3	1	4.5	**91.1**
South Africa	28.9	−10.5	16.9	51.7	4.8	**99.8**	N.a.	**100**	2.9	**100**
Tanzania	19.7	−1.7	2.8	**95.4**	3.1	31.5	15.6	−0.3	5.5	**100**
Tunisia	N.a.	**100**	45.3	54.1	0.4	**100**	N.a.	0	2	**100**
Uganda	21.4	−3.4	5.6	**93.5**	13.3	7.1	7.8	−3.2	3.7	**99**
Zambia	50.4	−2.1	3	−13.4	1.1	45.5	16.4	0	2.7	**100**

Source: FAOSTAT New Food Balances and authors' calculations.

Notes: Consumption shares are calculated as percentage of average daily per capita caloric contribution data from 2019. Net import dependence = 100*(imports - export)/(production + imports − exports), data from 2019. N.a. = not available. Bold indicates net import dependence greater than 65%.

high import dependency. Regarding import trading parties, Russia is the dominant source of wheat, while palm oil and rice are mainly procured from Malaysia and Indonesia respectively (FAO, 2022). By contrast, maize and cassava are mainly domestically sourced, and have low net import dependency.

We conclude from these data that countries in Africa differ much regarding their dependency on international food markets. Of all countries on the African continent, only South Africa is a net exporter of staple food, though many countries, such as Ghana and Ivory Coast, are exporters of agricultural commodities, such as cocoa and coffee (note that Ivory Coast is a net exporter of palm oil, too – see Table 8.1). Moreover, dependencies can be observed on a single commodity, while other commodities are produced domestically, or on multiple commodities, making such a food system more profoundly reliant on international markets with potential negative effects on its resilience to supply and economic shocks. For instance, North Africa's dependence on the import of wheat is considerable, while wheat also makes a major contribution to the calorie intake of the population. Besides variations among countries, the effects of dependency on food markets are not equally distributed over segments of consumers within countries. Households with little purchasing power benefit from low food prices in times of stability on international food markets, but are hit hardest by price spikes in times of crisis (see for example reports on FEWS.NET that provide early warning information and analysis of food insecurity and its drivers worldwide).

5 What drives food prices in Africa?

Food prices are determined by the processes and powers of trade that ultimately link supply and demand on a specific market, be it domestic or international. Increases in demand for food are largely determined by population and income growth, the latter affecting not only increased consumption but also shifts in preferences for food. In Africa, population growth determines the demand for food to a greater extent than per capita income growth. Population growth is expected to continue at an average 2.5% per year until 2035 and 2% annually in 2035–2050 (UNDESA, 2023). However, food supply via domestic production in Africa is projected to only increase moderately over the coming decade, which also means that Africa will become increasingly reliant on food imports to close the gap between domestic production and consumption (OECD-FAO, 2023). This greater dependency on world markets means that the influence of international developments on domestic markets will also grow, something that, due to the major uncertainties in international prices, entails the risks of fluctuations in foreign supply and thus increases domestic food security risks.

Okou et al. (2022) show for 15 SSA countries that domestic food price variations are strongly related to net import dependency and the share of food staples in overall food consumption. The 15 SSA countries in this study are in fact highly vulnerable to global price shocks as any change in the international price of the imported staples can bring about an almost one-to-one increase in the local price of the imported staples. Moreover, higher domestic production – reflected in lower net import dependency – and lower consumption shares of the imported staple correspond to lower imported food price inflation. This implies that vulnerability to international price fluctuations can be reduced by expanding domestic production as it reduces the share of imported staples in overall food consumption.

Okou et al. (2022) also show that monetary policy framework, fiscal management, income per capita and geographical challenges (e.g. countries that are landlocked) are the main factors explaining differences in staple food prices among SSA countries. Countries with stronger monetary policy frameworks and income growth tend to have lower relative food prices, reflecting better control of inflation and a smaller share of basic food spending, respectively, when household incomes rise. In contrast, food prices are higher in countries with high debt ratios, as weaker fiscal management and transportation challenges tend to drive up the cost of food staples.

6 Building resilience to food market shocks

The above sections indicate the importance for African countries that rely heavily on staple food imports to consider strategies to increase the resilience of their food systems. In addition to macroeconomic policy aimed at sound fiscal policy (keeping government expenditure in order) and monetary policy (aimed at price stability, mainly via the interest rate instrument), sector-specific policy is also required to increase resilience to food market shocks. We have three suggestions: (1) diversify markets and commodities; (2) strengthen competitiveness through inclusive value chain development; and (3) change consumer patterns towards more healthy crops and livestock products. So far, we have focused our analysis on staple foods which provide consumers with daily carbohydrates. Yet, their dominance in African diets has given rise to the increase of many non-communicable diseases on the continent, as the intake of micronutrients is generally far below recommended WHO standards (de Steenhuijsen Piters et al., 2021, Melse-Boonstra et al., this volume).

6.1 Diversification of markets and commodities[2]

Food import dependency becomes a severe problem when countries are less able to finance food imports.

A significant number of the African countries with food deficits depend heavily on export of few primary commodities such as coffee, cocoa or tea, and may face food security risks from a deterioration in those products' terms of trade (FAO, IFAD, UNICEF, WFP and WHO, 2019). Food import dependency is more risky for countries that have small or unstable foreign reserves, which are, in turn, dependent on the diversity and size of the export sector. To reduce these risks, it is vital to promote commodity and market supply diversification (Van Berkum, 2021). Commodity diversification implies a shift towards the production of more and different high-value crops such as fruits, vegetables and livestock products. Market supply diversification means that exports target several markets in different countries, and, in the case of imports, that a country maintains trade contacts with several countries so that it can relatively easily obtain imports from different suppliers. The potential for agricultural diversification differs across regions due to wide heterogeneity in agro-climatic and socio-economic conditions, the latter affected by technological change in crop and livestock production, improved rural infrastructure and diversification in food demand. Based on an extensive analysis of export diversification options in Mali, Chad, Niger and Guinea, López-Cálix (2020) concluded that targeted investments in market infrastructure (for efficient logistics) and in human capital (for building skills that enhance people's productivity and employability) can benefit the diversification of an economy.

In addition to structural adjustments on the supply side, opportunities on the demand side in domestic and foreign markets should be exploited to achieve the transition to a more productive and differentiated agriculture, improve access to food and create jobs outside agriculture. Strengthening regional trade relations with more or less equally developed neighbouring countries is a way to leverage trade to make the transition to a more diversified economy. By focusing on regional trade and improving infrastructure, countries can further exploit and develop their comparative advantages on nearby markets, using the generally existing social connections through which consumer needs and preferences are understood, and that can be served by short supply chains in which small-scale farmers and traders participate. Allen (2017) has shown that food prices are lower in West African countries that participate in the Economic Community of West African States (ECOWAS), suggesting that this free trade zone already has an impact (Frankema, this volume).

Yet regional trade within Africa remains fairly limited and accounts for less than 20% of all exports (UNECA, 2018). One reason is that existing regional trade agreements, such as the Common Market for Eastern and Southern Africa (COMESA), the East African Community (EAC) and the Southern Africa Development Community (SADC), frequently exclude free trade in foods, since their product portfolio is rather similar and countries consider each other as competitors.

The recently established African Continental Free Trade Area (AfCFTA) may stimulate intra-Africa trade, accelerate export diversification and diversify export destinations and types of goods produced on the continent (Echandi et al., 2022). In particular, the AfCFTA promises to increase intra-regional trade in food products that, if accompanied with the right measures, can greatly help boost smallholder farmers' productivity growth and prospects for integration into profitable food value chains. Moreover, more intensified regional trade relations add great potential for buffering against price spikes and against drought – which rarely affects all countries and regions at the same time. To make the most of these opportunities, governments will need to reduce transaction costs by improving trade facilitation – such as import customs clearance procedures and port handling at the border – and invest in physical infrastructure, including roads, railways and harbour facilities (Echandi et al., 2022).

6.2 Strengthening the competitiveness of domestic food production through inclusive value chain development

A second strategy for increasing resilience of food systems to external food market shocks is to strengthen the competitiveness of the domestic agriculture and food sectors by increasing their productivity and efficiency (Van Berkum, 2021). Higher crop and livestock productivity provides farmers with higher incomes and contributes to assuring the supply of food at lower prices to consumers. Increasing the productivity of Africa's agriculture will require the sustainable intensification of production through increased, but rational use of inputs (improved seeds, fertilizer, irrigation, mechanization) as well as better access to (high-value) markets. Achieving this, while not trespassing the planetary boundaries, is a huge challenge in light of Africa's dominant small-scale agricultural structure (Giller et al., 2021).

Value chain development (VCD) is increasingly advanced as an organizational solution to the above challenge (Ingenbleek et al., this volume). In VCD arrangements, public and private sector parties cooperate, assuming each specific roles and responsibilities (Swinnen and Kuijpers, 2020). Public VCD interventions generally aim to increase the efficiency in value chains by improving the business environment. The essence of the more integrated VCD models is to involve all actors in the value chain and "link" smallholder producers with input companies, processing factories or marketing agents. Based on these arrangements, higher volumes of more consistent quality can be procured by the stakeholders in the chain in return for access to credit, agronomic knowledge, price premiums and a reduction of production and market risks for the producers (Swinnen and Kuijpers, 2020). Farmers will become more productive through better access to inputs and more certainty about sales opportunities, while buyers can purchase at lower prices.

Small farmers are not automatically included in value chains, which increasingly have to take into account the demand for high-quality and safe food (Ton et al., 2018; Swinnen and Kuijpers, 2020; Kuijpers et al., 2021). If buyers have a choice between larger and smaller farmers, they will in many cases enter into contracts with the larger farmers because transaction costs are lower.

Governments can enhance the process of involving small farmers in VCD by developing policies that focus on reducing transaction costs for smaller and less resourceful producers for entering more commercial value chains. Such policies include, for example, managing foreign direct investment (FDI) so as to integrate smallholders, and investing in rural infrastructure (roads, storage facilities, energy, information and communication technology [ICT] networks) to connect small-scale farmers in remote areas with markets. Moreover, farmers need to be empowered to obtain a better bargaining position in the supply chain (Kuijpers et al., 2021). Government policies may support the establishment of producer organizations with proper legislation, with information and knowledge transfers enabling them to operate such organizations, sometimes using financial support measures (such as tax exemptions). Also helpful for integrating smallholders into value chains are policies that invest in institutions for independent quality and food safety control, certification, public extension and market information services (Swinnen and Kuijpers, 2020).

6.3 Influencing food consumer patterns and enhancing purchasing power

A third strategy to become less dependent on import of wheat, maize and rice is to influence the demand for these food staples. Two tracks can be distinguished here. One track is to change the relative prices of domestically produced grains to imported grain, and a second track is to promote a healthy diet that is affordable to most consumers. Both tracks are explained below.

6.3.1 Change the relative prices of domestically produced grains to imported grains

Wheat consumption in Africa has been rising faster than in any other region around the world, fuelled by relatively cheap imports. For example, in countries such as Kenya and Nigeria – both large wheat importers – an important driver of increasing wheat demand has been the falling price of imported wheat compared to other staples such as maize, millet and sorghum.

To reduce import dependency on imported wheat and rice, Africa will have to grow more grains and do so more efficiently. In doing so, it could focus on the more than 2,000 native grains and fruits which hold great potential for the food security question to millions of people in Africa (Raheem et al., 2021). The African continent is a major producer of several cereals like sorghum, pearl millet, finger millet, teff and African rice; some may need to be rediscovered and made more attractive to consumers, for example by reduced retail prices or their promotion in menus and recipes. Food security and sovereignty can be improved by revisiting these indigenous food crops and their diversified forms to improve nutrition. Governments and donors should prioritize investments in breeding of local cereals and other efficient farming techniques, to increase productivity per hectare (taking into account climate change effects on the continent's production capacity) and to adapt attributes such as taste, nutritional value and ease of use in products to consumer preferences (low yields and little nutritional value are major constraints to sorghum production as an example from Uganda shows; see Andiku et al., 2021). Costs of production can be further reduced through fiscal subsidies, for example on the use of inputs such as fertilizer and credit (interest discount), and other subsidies aiming at promoting local production to be an attractive alternative to imported grains.

6.3.2 Promote healthy diets and enhance purchasing power

Africa is a region where grains, roots and tubers have a relatively large share in the diet and nutrient-rich foods, such as meat, dairy, fish, fruits, vegetables and pulses, only make up a small part (Willett et al., 2019; Melse-Boonstra et al, this volume). This means that diets are often high in calories and that people consume too little micronutrients (vitamins and minerals). An important reason is that an estimated 80% of the Africans cannot afford a healthy diet. A global study on fruits and vegetable sectors (de Steenhuijsen Piters et al., 2021) confirmed that in Africa fruits and vegetables are more expensive compared to other food groups such as cereals, oils and fats, and drive a large part of the costs of a healthy diet. By promoting more domestic production and efficiency in the value chain (e.g. improved logistics that lowers transport costs) the retail price of these healthy foods should decline, making them more affordable for larger segments of consumers.

Besides promoting the supply of more healthy foods, produced domestically, consumers in SSA will need to achieve more purchasing power to access such foods. Even if the costs of a healthy diet in SSA would decline to average global levels then still the majority of households would not be able to afford these due to low levels of income.

Demographic trends in SSA mean that annual newcomers to the rural workforce will rise from 20 million today to 50 million by 2050 (IFAD, 2021). The rapidly growing numbers of rural youth in Africa present a huge employment challenge. Without employment opportunities, a whole generation will not escape poverty, with significant implications for people's well-being and access to healthy foods. Off-farm opportunities in food value chains and supporting services can provide options for young people and those exiting primary production to engage in entrepreneurial activity (IFAD, 2021). This challenge can only be met by general economic growth, job creation in all economic sectors and improving the working conditions of labourers and other workers at lower socio-economic strata.

7 Conclusions

Recent spikes in world food prices have shown the high vulnerability of African consumers for their supplies of carbohydrates on international food trade.

Strategies to reverse the heavy dependency on international grain markets include diversification of markets and food production, promotion of domestic production in terms of productivity and diversity of supplies, and policies to promote more nutritious food consumption patterns.

Governments have an important role to play in ensuring the most vulnerable groups in society will benefit from these strategies. Governments will have to support diversification of markets and production and strengthening of competitiveness by investing in trade costs reducing market infrastructures (e.g. storage facilities, trading places), information and knowledge exchange, and the building of institutions such as independent quality control organizations. For agricultural VCD to play a meaningful role in poverty alleviation for smallholders, governments must ensure that chain relationships (e.g. in the form of contracts) meet conditions for social equality and inclusiveness. To improve the availability, affordability and accessibility of high-quality nutritious foods for the less well off, governments may need to use subsidies and taxes to entice producers to produce more and consumers to demand more. This means active involvement of African governments in domestic food markets, with policies aimed at strengthening food security being central.

An active food and nutrition security policy in Africa may also require restrictive measures such as import tariffs. This is a sensitive issue because protection of one's own market can go against trade agreements made in the World Trade Organization (WTO). Further, it may only result in more stable, lower domestic food prices in the longer run, as it takes time for domestic production to substitute imports. Achieving more food and nutrition security through increased food system resilience is not a binary choice between domestic production and food import. National policies towards such goals must include all possible measures and optimize domestic production capacities alongside imports, depending on the economic, political and ecological context. As Jennifer Clapp (2017) argues:

> A more nuanced approach based on the real-world application of food self-sufficiency policies does not view the concept as an either/or proposition, but rather sees it in relative terms. Such an approach could potentially create room for a more productive policy dialogue on this issue at the international level.

Notes

1 Food security is used as in the FAO definition which emphasizes accessibility, at all times, to sufficient food in a quantitative and qualitative (that is, nutritional) sense (FAO, 1996).
2 Sections 1 and 2 draw heavily on Van Berkum (2021).

References

Andiku, C., Shimelis, H., Laing, M., Shayanowako, A.I.T., Ugen, M.A., Manyasa, E., & Ojiewo, C. (2021) Assessment of sorghum production constraints and farmer preferences for sorghum variety in Uganda: Implications for nutritional quality breeding. *Acta Agriculturae Scandinavica, Section B — Soil & Plant Science*, *71*(7), 620–632. https://doi.org/10.1080/09064710.2021.1944297

Allen, T. (2017). *The cost of high food prices in West Africa*. OECD Publishing.

De Steenhuijsen Piters, B., Dijkxhoorn, Y., Hengsdijk, H., Brouwer, I., Tichař, T., & Carrico, C. 2021. Synthesis report of the global fruits and vegetables scoping study; Assessing opportunities for philanthropic investment. Wageningen, Wageningen Economic Research, Report 2021-113.

De Janvry, A., & Sadoulet, E. (2022). The puzzle of lagging sub-Saharan Africa agriculture: Toward a theory of connectedness. In H. de Gorter et al. (Eds.), *Modern agricultural and resource economics and policy*, Natural Resource Management and Policy 55. https://doi.org/10.1007/978-3-030-77760-9_12

Clapp, J. (2017). Food self-sufficiency: Making sense of it, and when it makes sense. *Food Policy*, *66*, 88–96.

Echandi, R., Maliszewska, M., & Steenbergen, V. (2022). *Making the most of the African continental free trade area: Leveraging trade and foreign direct investment to boost growth and reduce poverty*. World Bank and AfCFTA Secretariat.

FAO. (1996). Rome declaration on world food security and world food summit plan of action: World food summit 13–17 November 1996. FAO, Rome.

FAO. (2022). The state of agricultural commodity markets 2022. The geography of food and agricultural trade: Policy approaches for sustainable development. Rome, FAO. https://doi.org/10.4060/cc0471en

FAO, IFAD, UNICEF, WFP, & WHO. (2019). The state of food security and nutrition in the world 2019. Safeguarding against economic slowdowns and downturns. Rome, FAO.

FAO, IFAD, UNICEF, WFP, & WHO. (2022). The state of food security and nutrition in the world 2022. Repurposing food and agricultural policies to make healthy diets more affordable. Rome, FAO. https://doi.org/10.4060/cc0639en

FAOSTAT. https://www.fao.org/faostat/en/#data/TCL. Accessed 21/09/2023.

Friedmann, H. (1992). Distance and durability: Shaky foundations of the world food economy. *Third World Quarterly*, *13*, 371–383

Giller, K., Delaune, T., Vasco Silva, J., Van Wijk, M., Hammond, J., Descheemaeker, K., … & Andersson, J.A. (2021). Small farms and development in sub-Saharan Africa: Farming for food, for income or for lack of better options? *Food Security*, *13*, 1431–1454. https://doi.org/10.1007/s12571-021-01209-0.

González-Esteban, A.L. (2017). Why wheat? International patterns of wheat demand, 1939–2010. *Economic History of Europe: New Approaches, New Topics*, *13*(3), 135–150.
IFAD (International Fund for Agricultural Development). (2021). Rural development report 2021. Transforming food systems for rural prosperity. IFAD. Rome.
IFPRI. (2023). Food export restrictions have eased as the Russia-Ukraine war continues, but concerns remain for key commodities. IFPRI (International Food Policy Research Institute), Washington.
Khoury, C.K., Bjorkman, A.D., Dempewolf, H., Ramirez-Villegas, J., Guarinof, L., Jarvis, A., … & Struik, P. (2014). Increasing homogeneity in global food supplies and the implications for food security. *PNAS*, *111*(11), 4001–4006.
Kuijpers, R., Rappoldt, A., Petrutiu, S., Vos, A., Kruijssen, F., Verschoor, R., & Maitland, A. (2021). Value chain approaches for social change. KIT Working paper 2021-01.
López-Cálix, J.R. (2020). *Leveraging export diversification in fragile countries: The emerging value chains of Mali, Chad, Niger, and Guinea*. International Development in Focus. World Bank.
Mason, N.M., Jayne, T.S., & Shiferaw, B. (2012). Wheat consumption in sub-Saharan Africa: Trends, drivers, and policy implications. MSU International Development Working Paper. Michigan State University.
OECD/FAO. (2023). *OECD-FAO agricultural outlook 2023–2032*. OECD Publishing. https://doi.org/10.1787/08801ab7-en
Okou, C., Spray, J., & Unsal, D.F. (2022). Staple food prices in sub-Saharan Africa: An empirical assessment, IMF Working Papers, WP/22/135.
Raheem, D., Dayoub, M., Birech, R., & Nakiyemba, A. (2021). The contribution of cereal grains to food security and sustainability in Africa: Potential application of UAV in Ghana, Nigeria, Uganda, and Namibia. *Urban Science*, *5*, 8. https://doi.org/10.3390/urbansci5010008
Swinnen, J., & Kuijpers, R. (2020). Inclusive value chains to accelerate poverty reduction in Africa. World Bank. Working Paper Issue no.37.
Ton, G., Vellema, W., Desiere, S., Weituschat, S., & D'Haese, M. (2018). Contract farming for improving smallholder incomes: What can we learn from effectiveness studies? *World Development*, *104*, 46–64.
UNDESA (United Nations of Economic and Social Affairs). *World population prospects 2023*. https://population.un.org/wpp/
UNECA (United Nations. Economic Commission for Africa). (2018). *An empirical assessment of the African Continental Free Trade Area modalities on goods*. Addis Ababa. UN.ECA. https://knowledge.uneca.org/ATPC/sites/default/files/PDFpublications/brief_assessment_of_afcfta_modalities_eng_nov18.pdf
Van Berkum, S. (2021). How trade can drive inclusive and sustainable food system outcomes in food deficit low-income countries. *Food Security*, *13*, 1541–1554. https://doi.org/10.1007/s12571-021-01218-z.
Willett, W., Rockstrom, J., Loken, B., M. Springmann, T. Lang, S. Vermeulen, … & C.J.L. Murray (2019). Food in the anthropocene: The EAT–lancet commission on healthy diets from sustainable food systems. *The Lancet*, *393*(10170), 447–492. https://doi.org/10.1016/S0140-6736(18)31788-4

9 Why is food expensive in sub-Saharan African cities?

Ewout Frankema

1 Introduction

African households devote a major part of their income to food. In 2022, the median household in sub-Saharan Africa (SSA henceforth) spent an estimated 55.8% of its budget on food, with national averages ranging from c. 28% in Mauritius to 73% in Burundi, one of the poorest countries in the region. Benin, Mozambique and Uganda are close to the regional median. In the median country, households in the lowest quintile of the income distribution spent c. 60% of their budget on food, which compares to 49% in the upper quintile (World Bank, 2022, pp. 30–31). In contrast, in the USA, households in the lowest income quintile spend c. 31% on food (USDA, 2023a). These figures indicate that food prices and (unexpected) changes therein have a large bearing on the living standards enjoyed by the great majority of Africans, and can make the difference between hunger and satiety.

Several studies have shown that retail prices of food products in SSA are higher than expected given the region's low average per capita income levels (Gelb & Diofasi, 2015; Allen, 2017; Okou et al., 2022). These high prices concern major staples as well as other fresh and processed products that are consumed in smaller quantities, but are important ingredients in a healthy and varied diet. This chapter sets out to disentangle this food price puzzle. Why is food expensive, and why does this apply specifically to African cities south of the Sahara? What can be done about it towards 2050, when the total number of SSA's urban consumers who depend on markets to procure their daily meals will have more than doubled from an estimated 555 million in 2025 to 1.26 billion? (UN DESA, 2023).

The statement that "urban" food prices are "high" requires some further qualification. The "high" means relative to per capita income levels. According to the Balassa-Samuelson theorem, consumer prices in poor countries tend to be lower than in rich countries (Asea & Corden, 1994). To understand this, a distinction is warranted between price setting mechanisms in tradable commodities and services (e.g. textiles, electronics, oil, wheat, gold or ocean-shipping), which are primarily dictated by international markets, and non-tradable commodities and services which are for various reasons place-bound or limited in supply through national regulations (e.g. taxis, hairdressing, tailors, heavy bulk products (e.g. cement, beer), health care, banking and insurance). For the category of non-tradables, prices are set in local markets and are partially a derivative of local wage levels. Since wages are lower in poor countries, consumer prices of non-tradables tend to be lower as well.

Food prices in SSA are "higher" than one would expect on the basis of *the general relationship between food prices and income levels observed in the rest of the world*. To put this in numbers, across SSA urban food prices in 2017 (pre-Covid) were on average about 16% points above what one would expect based on average GDP per capita. Conversely, food prices

DOI: 10.4324/9781032649696-12

in North Africa were about 20% points lower (for details see Section 2). Food price gaps also existed between SSA and Southeast Asian economies. The fact that a considerable part of the food products that are consumed in Africa are internationally traded and, indeed, imported is an important factor in the explanation of "high" food prices, but it cannot as such explain the difference between North Africa and SSA. After all, North African countries such as Egypt and various Asian economies are net-importers of food as well.

This chapter focuses on "urban" prices as the data available for international price comparisons are mostly collected in major cities. Moreover, food prices in SSA's vast rural areas vary enormously, which makes it harder to claim that food is "expensive" across the board. In general, food prices of locally cultivated and traded food items tend to be (far) cheaper than in the city, while imported food or food produced in urban areas (e.g. canned meat, beer) is often more expensive in rural areas depending on how isolated the area in question is.

This chapter explores the question why food prices in SSA cities are high. Section 2 presents and discusses some of the data that underpin international food price comparisons. Section 3 discusses some implications of high food prices and presents five categories of explanation for high urban food prices, paying special attention to their interconnections. This chapter concludes with a discussion of policy interventions that can curtail (urban) food prices in order to improve food and nutrition security in SSA.

2 African food prices in a global perspective

What is the evidence for the claim that food is expensive in urban SSA? Figure 9.1 plots GDP per capita levels for 171 countries scattered across the world (x-axis) against the relative price of a standardized food basket (y-axis) for the year 2017. Prices are scaled relative to the United

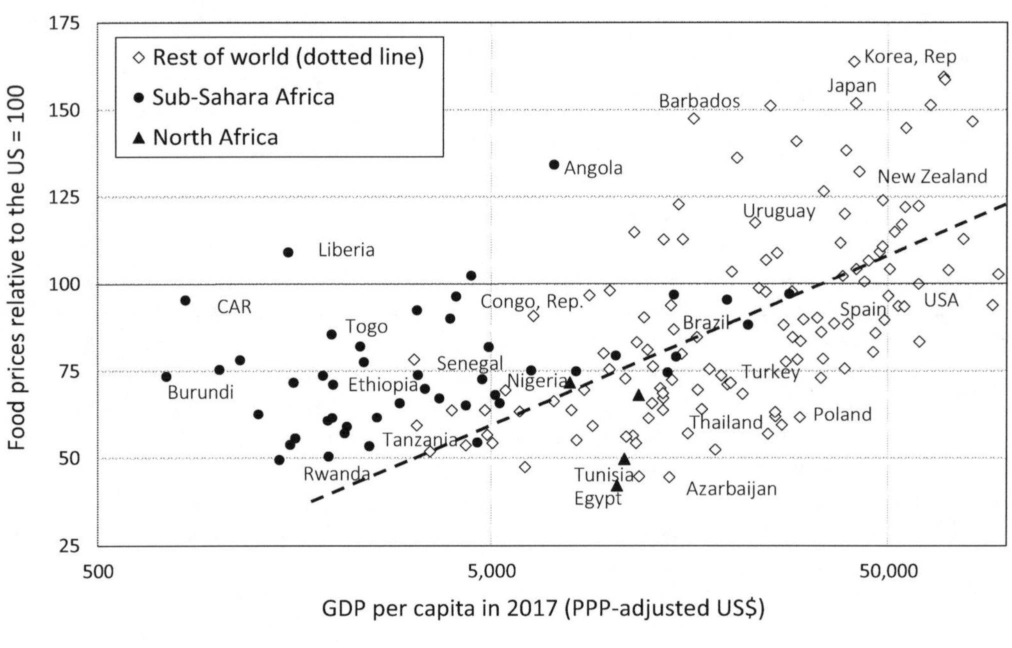

Figure 9.1 Relative food prices versus per capita GDP, 2017.

Source: Food price data from World Bank (2020); GDP per capita from World Bank (2023b).

States (US = 100). The graph distinguishes countries in SSA, North Africa and the rest of the world. Per capita GDP levels (x-axis) are in logarithms, so that the linear regression line, which pertains to the 121 countries in the "rest of the world", is derived from a non-linear function. GDP per capita levels are expressed in purchasing power parity (PPP)-adjusted US$. This means that instead of using official exchange rates, local income levels are converted into US$ using PPP estimates which, like the food price indices, are derived from the World Bank's latest round of the International Comparison Program (ICP) (for details see World Bank, 2020).

Figure 9.1 demonstrates that the great majority of SSA countries are above the regression line. Despite being much poorer countries, food prices in Burundi, Ethiopia and Senegal are broadly comparable with those in Turkey. Food is even much cheaper in Poland or Thailand than in Liberia or Angola, despite higher income levels in the former two countries. The food basket is equally expensive in the Republic of Congo as it is in the USA, which may not be a huge problem for the upper income classes in this oil-exporting country, but severely constrains food access among the urban poor. The graph also shows large variation in food prices across SSA. For instance, the food basket is almost twice as expensive in the Central African Republic, than in Rwanda or Tanzania. In contrast to SSA, three North African countries in our sample are positioned far below the line, while Morocco is positioned on the line (for Libya data is lacking). One of the reasons for this difference is that North African countries have adopted significant food subsidy schemes that SSA countries cannot easily afford (see Section 4 and Ramadan & Castet, this volume).

There are two caveats. First, the food basket is composed of a weighted collection of food items covering nine broad categories: (1) bread and cereals; (2) meat; (3) fish and seafood; (4) milk, cheese and eggs; (5) oils and fats; (6) fruit; (7) vegetables; (8) sugar, jam, honey, chocolate and confectionery; and (9) products not classified elsewhere. Obviously, these items are consumed in varying quantities by people depending on their purchasing power as well as personal and cultural preferences, local availability of particular food items and substitution possibilities. A poor city-dweller in Nairobi will make different choices than a middle-class household in Sao Paulo, Tokyo or Amsterdam. The ICP uses a food basket that better reflects consumption preferences in middle- and higher-income countries than those of the global poor. Prices of yam, millet and cassava, typically African staples, are absent, while luxury items such as cheese, exotic fruits, honey or chocolate are factored in. Second, especially in countries where wet, informal markets compete with supermarkets which operate at the outer end of more extensive value chains, prices for standardized units of food will differ. Supermarket products tend to be more expensive.

These caveats do not render the comparison useless though, if only because the data do bear out that retail prices in a rapidly growing and important segment of SSA food markets are "high" compared to the rest of the world, *controlling for average income levels*. This points to serious inefficiencies (i.e. high costs or exceptional profit margins) in specific segments of the agri-food value chain. Moreover, it is possible to disentangle the food basket per item to obtain more precise comparisons. Doing so, Allen (2017) found for West African countries that dairy products and oils and fats were exceptionally expensive, followed by meat products and fruits and vegetables, while bread and cereals and fish products were more affordable, but still priced high. For securing a healthy and nutritious diet, people need to be able to afford a combination of these items (Melse-Boonstra et al., this volume).

To underscore this point, Table 9.1 compares urban retail prices of a series of staple foods in SSA with those in South and Southeast Asia (SSEA henceforth) for 2011 and 2017. The ICP rounds of 2011 and 2017 offer particularly detailed price information with a good coverage of African and Asian economies. While these regions had more or less comparable income levels up to the early 1970s, the unweighted average income levels in SSEA are now roughly twice as

Table 9.1 Staple food prices in sub-Saharan Africa and South and Southeast Asia compared, 2011 and 2017 (in current US$)

	Sub-Saharan Africa		*South and Southeast Asia*		*SSA/SSEA*	*SSA/SSEA*
	US$/kg	*US$/kg*	*US$/kg*	*US$/kg*		
	2011	*2017*	*2011*	*2017*	*2011*	*2017*
Maize, unprocessed	0.78	0.62	0.72	0.69	*1.1*	*0.9*
Maize flour (white)	1.94	1.12	1.50	1.71	*1.3*	*0.7*
Rice (combi. of 4 types)	1.19	1.12	0.75	0.69	*1.6*	*1.6*
Wheat flour (not self-rising)	1.15	1.09	0.99	0.75	*1.2*	*1.5*
Dried noodles	2.13	2.03	0.91	0.94	*2.3*	*2.2*
Brown potatoes	1.15	0.94	0.69	0.70	*1.7*	*1.3*
Sweet potatoes	0.65	0.64	0.55	0.37	*1.2*	*1.7*
Chicken eggs (large)	1.23	1.03	0.82	0.86	*1.5*	*1.2*
Palm oil (litre)	3.12	2.13	1.83	1.45	*1.7*	*1.5*
Sunflower oil (litre)	3.40	2.62	2.92	2.35	*1.2*	*1.1*

Sources: All price data are from the World Bank ICP 2011 and 2017 rounds. Underlying datasheets available upon request. For details on the methods of price recording see ICP Report 2011 (World Bank, 2015).

Notes: Small countries such as Brunei Darussalam and Equatorial Guinea, where food prices are high as a result of massive oil revenues, are excluded. Small island states (e.g. the Maldives, Fiji, Mauritius, Cape Verde, Comoros, Seychelles) and city-states (e.g. Singapore, Djibouti), where food prices are high because of a lack of a sizeable local agricultural sector and/or a lack of scale economies in trans-oceanic freight and/or an exceptionally large tourist industry, are also excluded. The comparison thus involves the larger mainland countries of SSA and Madagascar versus the rapidly developing economies of SSEA. Since price data were not always available for every country in the region, there are minor differences in sampling. The inconsistencies in the sample do not affect the overall picture, but may cause small biases in the estimated average prices and the price ratios.

large (World Bank, 2023b). Notwithstanding the significant intra-regional income differences, however, Table 9.1 shows that the average prices of nearly all listed food items were higher in SSA than in SSEA. In 2011 and 2017, a kilogram of rice was c. 60% more expensive in SSA than in SSEA. While this may not be that surprising given the ecological and historical advantages that many Asian countries have in rice production, the price gap is very large. The same goes for wheat flour, chicken eggs, potatoes, palm oil and sunflower oil. As these items constitute a fairly representative sub-stratum of daily caloric intakes of urban lower-income groups in both regions, such price gaps are meaningful and concerning. Only for maize, the leading African staple crop, prices were lower in SSA in 2017 than they were in SSEA.

Large variation in and volatility of staple food prices in SSA is inextricably connected to weakly integrated markets (Porteous, 2019; Zant, 2018, 2021; Frankema et al., 2024). The African continent constitutes a vast landmass with several landlocked countries (16 in total), and in which many cities are located at considerable distance from the coast. In SSEA Laos is the only landlocked country and sea-based transportation has played a larger role. Historical studies have shown that in Asia, international and national trade networks have spurred food price convergence since the middle of the 19th century, especially in the case of rice (Studer, 2008; Huff & Caggiano, 2007; Marks, 2010). In Africa, staple food prices tend to be lower in countries where a substantial part of consumption is supplied by local production systems. For example, maize is relatively cheap in countries that are either self-sufficient or net exporters, such as Tanzania, Uganda, Ethiopia or Malawi (see Okou et al., 2022, p. 26), but (much) more expensive in net-importing counties such as Angola or Senegal. The same goes for rice, which is relatively cheap in major producing countries such as Madagascar and Guinea, but more expensive in

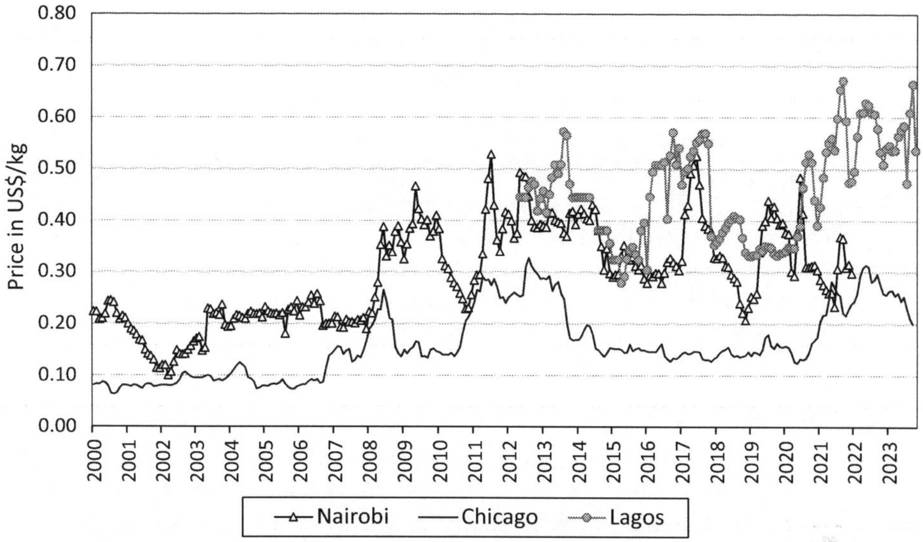

Figure 9.2 Monthly wholesale maize prices in Nairobi (Kenya), Lagos (Nigeria) and Chicago (USA), in US$/kg, 2000–2023.

Source: For Nairobi and Lagos from FAO (2023) and for Chicago from USDA (2023b).

countries where it is not locally produced. If markets are fully integrated, imported food does not necessarily have to be much more expensive compared to the areas where it is produced.[1]

Finally, it is worth considering that expensive food is not caused by short-term (international) price shocks. It has a more structural character. Figure 9.2 shows monthly wholesale prices of maize in Nairobi and Chicago to illustrate this pattern: prices of maize, the leading staple in East Africa, have been consistently higher than those in the USA (the world market leader) for a long time. Historical research suggests that average maize prices in East African countries converged to world market prices starting already in the late 1940s, and that these were consistently above the world market price since the 1970s and in some places long before (Frankema et al., 2024). Also for West Africa scattered evidence suggests that at least since the 1960s maize prices were mostly above the world market price (Hesp, 1985). All of this does not mean that maize prices were "high" everywhere, as in the middle of major maize-producing zones, prices can be far lower than in areas and cities located at considerable distance from the main producing areas. One example is Songea, a city located in the heart of Southern Tanzania's Ruvuma province, a major maize-growing area, where wholesale maize prices between 2010 and 2014 were below Chicago prices (FAO, 2023). Yet, overall, high urban food prices are a structural phenomenon and this problem has in most countries emerged during the second half of the 20th century.

3 What causes expensive food?

It has been estimated that if West African food baskets could be obtained at Indian prices, consumers would save between 19 and 33% of their incomes depending on the country they live in (Allen, 2017, p. 10). Expensive food not just compromises food access, but also reduces the means to diversify food intake and secure other basic needs such as schooling for children, health care, clothing, transportation or housing. If daily subsistence costs are high, employers who pay wages to urban employees must take these into account as well. Hence, wages tend to rise in response to increasing costs of living. High wages, however, imply reduced competitiveness of (potential) export sectors, and more specifically the labour-intensive industries where wage bills are an important part of total production costs. High wages also affect the domestic economy, as they lead to higher overall prices of non-tradables. The ICP data reveal that such price-wage effects exist for hairdressers, office cleaners and kitchen aids: controlled for income levels, these wages in SSA tend to exceed those in SSEA (see also Frankema & van Waijenburg, 2018). Similar differences can be observed in wage data for office cleaners or kitchen aids. The good news, however, is that if the problem of expensive food can be tackled, flywheel effects can speed up economic development and poverty reduction. But what causes expensive food in SSA? In this section I discuss five interconnected factors. How these factors play out in each particular country is beyond the scope of this chapter.

3.1 Low agricultural productivity

Low agricultural productivity is the ultimate cause of high food prices. If the conversion of inputs such as land, labour, seeds, fertilizer and water into a harvestable crop is inefficient, the relative costs per unit of output rise accordingly. Low agricultural productivity has, with few exceptions, been a longstanding feature of African agriculture. While value added per worker employed in agriculture usually tends to be somewhat lower than in manufactures or services, such gaps narrow as economies develop and transform (Timmer, 2009). However, in many SSA economies the inter-sectoral productivity gaps are much larger than elsewhere (Gollin et al., 2014; Suri & Udry, 2022).

The reasons for low agricultural productivity range from ecological to institutional. Broadly speaking, and despite pockets of highly favourable conditions, factors such as rainfall variability, soil quality and heterogeneity, and plant or animal disease incidences are less favourable to agricultural intensification in SSA than in many other parts of the world (Sachs & Warner, 1997; Hayami et al., 1998; Frankema, 2014; Barrett et al., 2017; Suri & Udry, 2022). In addition to ecological conditions, the literature has emphasized that African agriculture has long suffered from "urban bias" in economic policy, from a lack of context-specific research, from ill-conceived structural adjustment programmes, from deliberate rent-seeking and ethnic favouritism, hampering agricultural productivity growth in a myriad of ways (Lipton, 1977; Djurfeldt et al., 2005; Larson & Otsuka, 2013; Henley, 2015).

Low productivity, however, also means there is ample scope for output growth. Research by van Ittersum et al. (2016) reveals that African countries currently produce at c. 20–40% of their maximum yield potential under rainfed conditions (see also Tittonell & Giller 2013; Schut & van Ittersum, this volume). While yield gap estimates are primarily about land productivity, they also indicate there is ample potential for labour productivity growth associated with intensification of production.

Low agricultural productivity growth has translated into a growing reliance on food imports. With the exception of roots and tubers which are costly to transport and crops that degrade quickly (e.g. fruits, vegetables), substantial shares of the cereals, flours, dairy, oils and fats are imported from outside Africa. Net food imports into Africa began to rise in particular after 1950 and are now close to 20% of Africans' total caloric intake (Joshipura, 2024). Importing food also became more attractive as real-world market prices of staple foods have continued to decline after World War II. Figure 9.3 shows the long-run world market price index of an unweighted basket of maize, rice and wheat prices. These crops are at present about four times as cheap compared to other products and services as in the mid-19th century, when yields worldwide were much lower. After the shocks of the First and the Second World War and the oil crisis of

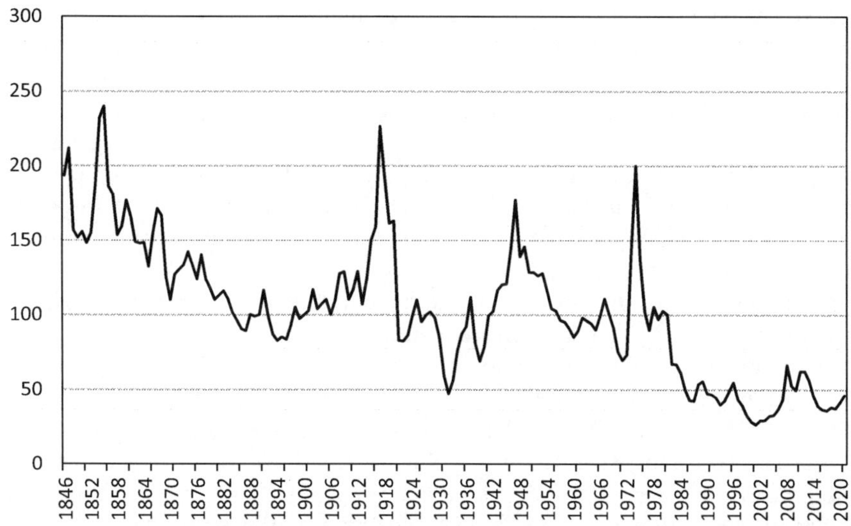

Figure 9.3 Real-world market price index for an unweighted basket of maize, rice and wheat, 1846–2021 (1900 = 100).

Source: Author's own based on the following commodity price data series: 1960–2021 (World Bank, 2023a); 1900–1960 (Grilli & Yang, 1988); 1846–1900 (Sauerbeck, n.d., various issues).

Note: Real prices estimated with a US Consumer Price Index from Officer and Williamson (2023).

1974, the prices have further declined and become more stable, even though the declining trend has halted in the early 2000s. The availability of cheap imports has, in turn, also constrained incentives and investment margins for productivity improvements (see Bulte et al., this volume).

Following the Balassa-Samuelson theorem, the prices of imported food (i.e. tradables) tend to converge towards world market prices and will thus be comparatively expensive for low-income countries. Governments can intervene and keep the price of imported foodstuffs low through overvalued exchange rates or large-scale subsidy schemes, but both approaches entail a burden on respectively the export sector (reduced competitiveness) and public finances. Overvalued exchange rates do not only raise incentives to rely on food imports at the expense of domestic farmers' produce, but can also become a trap when local currencies depreciate. Nigeria offers an example where windfall revenues from spiking oil exports in the 1970s were used to supply urban consumers with cheap food, and especially wheat imports from North America, sparking off an exponential growth of local mills, bakeries and distribution centres. This episode, which has been referred to as the Nigerian "wheat trap", turned the country into one of Africa's largest net food importers in a single decade. However, the collapse of the Naira in 1984 wreaked havoc among millions of urban unemployed who were confronted with skyrocketing bread prices (Andrae & Beckman, 1986). The Nigerian government also had to deal with the effects of a major international food price hike between 2006 and 2009, which almost doubled retail prices of wheat, pulling other staple food prices up as well (Olomola, 2014, pp. 277–280).

3.2 *High transport, storage and processing costs*

A second determinant of high food prices in SSA are the costs associated with *transportation, storage and processing* (TSP) of food. Transportation costs impact input prices (e.g. seeds, fertilizer) as well as output prices, and are especially critical for perishable products.

When TSP costs become prohibitive, subsistence production will prevail over commercial farming. In SSA one of the major compounds of low agricultural productivity is post-harvest losses due to low-quality transport and storage facilities and/or inefficiencies in industrial processing (Stathers et al., 2020). Such losses are ultimately factored into urban retail prices.

High TSP costs result from two types of causes. One is related to specific geographical and ecological conditions. Whereas the lion's share of locally produced crops in SSA have to be transported overland over considerable distances to reach an ocean port, transport networks (roads, rail) remain underdeveloped (Suri & Udry, 2022). In case cities are not directly located at the coast with a harbour, imported food will also have to be transported overland. Studies point out that improvements in road networks or the building of new bridges have a measurable downward effect on prices of food in connected urban markets (e.g. Zant, 2018, 2021).

A second factor of high TSP costs is to do with the way the markets for transportation and storage services and processing activities are organized, an issue I will elaborate in Sections 3.3–3.5. In short, this is about the costs that producers or intermediaries face ranging from taxes and fees to licences or expensive capital supplies (e.g. interest rates), as well as the margins they are able to reap as a result of unequal market power, for example when companies obtain a (near) monopoly or form a cartel. Many SSA economies suffer from high TSP costs due to a combination of these factors.

3.3 High transaction costs

Closely related, but conceptually distinct, transaction costs constitute a layer of the price structure that deals with the problem of risk and incomplete information. Transaction costs are a key concept in institutional economic theory, where they are defined as costs involved in making, monitoring and enforcing contracts, including the costs to collect information about the products and actors involved in the transaction. States can reduce transaction costs by designing a regulatory environment (i.e. institutions) in which actors (e.g. farmers, cooperatives, investors, traders, food factories, insurance companies, banks, retailers) can trust that the expectations they hold from a particular transaction will be fulfilled, and if not, that they are entitled to compensation. Farmers who supply their crop to an intermediary at a given price want to be ensured that they will receive their money in time to buy inputs for the next growing season. The intermediary would want assurance that the offered crop is of a minimum quality and that it will be delivered at the agreed date.

Risks in transactions include a breach of contract, mutual misunderstandings, exploitation of information asymmetries or, what economists call, hold-up and principal-agent problems. In SSA transaction costs are high as a result of weak institutional regulations which raise entrepreneurial risks. Institutions can reduce transaction costs by providing relevant information, by creating stability in financial and consumer markets and by building trust in judicial systems as a means of last resort in case of conflict over contracts. Risks also include unanticipated taxes levied by parastatal or criminal organizations, red tape required to pass-by road blocks or state borders, insecurities about the development of exchange rates or interest rates and climate shocks that affect harvests or stored food. The scale of markets and consolidation of production also impact on transaction costs: traders can operate contracts at lower costs when the market grows beyond minimum efficient scale (Dorward et al., 1998).

Insurances can deal with the risks that are not taken away by reliable market institutions. Yet, the costs of insurance are determined by another range of factors, including assessment costs. A key example is weather shocks. Due to climate change, rainfall levels and seasons are becoming more erratic in many parts of Africa, resulting in larger harvest insecurities. To shield farmers from climate-induced risk, several private-sector initiatives have been launched to reduce the costs of indemnity assessments through the use of sophisticated damage detection techniques, such as satellite images and geo-located rainfall indices. So far, these insurance schemes have had disappointing take-up (Ali et al., 2020; Kramer et al., 2022). Many smallholders prefer to avoid risk by limiting the investments in their commercial activities and take losses as they come. This constrains local food production and translates into higher and more volatile food prices. Effective insurance schemes can help lift agriculture out of such low productivity traps, but are not a silver bullet: it is one link in a long chain of factors shaping transaction costs.

3.4 *Unequal market power*

Agro-industrial value chains in SSA have a long history of systemic inequality in market power: what opportunities do companies have to establish a monopoly or monopsony through particular insiders' knowledge, first-mover advantages, capital intensities or political connections? Institutions may enhance competition by facilitating market entry and prohibiting disproportional market power, but can also enhance inequalities, for instance, via selective licences or lucrative concessions. Whereas smallholders are underorganized, companies engaged in the transportation, storage or industrial processing tend to have (much) more market power and can use it to raise their profit margins, especially when they receive political support. The classic example is the marketing boards that were established in many SSA countries from the 1930s through the 1980s. Whereas most of these government-controlled agencies were set up with the explicit aim to stabilize prices and farmer's income, they often distorted market incentives, drained resources away from agriculture, increased administration costs, introduced logistic inefficiencies and monopolized (major) parts of the value chain, including transportation services (Bates, 1981; Jayne & Jones, 1997; van der Laan et al., 1999; Barrett & Mutambatsere, 2008).

Teravaninthorn and Raballand (2009) argue that in theory the costs of physically moving commodities along four major transport corridors in different parts of Africa are not significantly different from the transportation costs in corridors of similar length in China. They argue that transport prices are much higher in Africa because of the many informal payments (i.e. red tape) that are required in order to pass borders and road blocks, as well as the large market power of trucking companies backed by governments who restrict the entry of new companies. They also show that when Rwanda deregulated its transport sector in 1994, ending a *de facto* monopoly of a parastatal trucking company (STIR), transport prices in this landlocked country fell by a staggering 75% in real terms (pp. 23–24).

3.5 *Food price volatility and subsidy programmes*

The sudden ban on rural-urban traffic in several African countries during the Covid pandemic (2020–2022) demonstrated how integrated markets reduce price volatility: prices of locally produced crops in the countryside dropped, as farmers could not sell their products in urban markets, while prices in cities shot up, and prices of *imported* food items skyrocketed in the countryside. When these bans were lifted, prices converged again.

To overcome exposure of low-income groups to price volatility, a considerable number of developing economies have implemented food subsidy schemes (Alderman et al., 2018). Besides providing a social safety net, such programmes are often also motivated by a political desire to curb social unrest and instability that may emerge in the wake of food price hikes (Gutner, 2002). Countries with large subsidy schemes include India, Indonesia, Sri Lanka and Mexico, as well as the North African countries mentioned above. Food subsidy schemes can take many different forms, with different impacts on the functioning of markets: through direct price controls or subsidies (heavy market distortion), in-kind food transfers, voucher systems or (un) conditional cash transfers (limited market distortion).

Food subsidy schemes tend to be expensive and are largely funded by domestic tax-payers. The extensive costs of such schemes are one reason why their adoption has been limited in SSA. Food subsidy schemes are also a political choice, as they redistribute resources from tax-payers to subsidy-recipients. If this involves an effective transfer from rich to poor, such programmes enhance solidarity. However, these subsidy schemes can spin out of control. In Egypt the rise in food subsidy spending led to fiscal deficits as high as 20% of GDP in the early 1980s, which forced the Mubarak regime to adopt drastic austerity measures (Abdalla & Al-Shawarby, 2018, p. 118). In Venezuela, food subsidies were long backed-up with oil revenues, but also involved the expropriation of large farmers and retailers, mounting fiscal deficits and excessive corruption (Howard-Hassmann, 2015). When state finances collapsed, the shops were empty and millions of poor Venezuelans were forced to flee the country to prevent starvation. Subsidy schemes can thus distort markets and not infrequently result in parallel (black) markets, unintended food shortages and/or fiscal imbalances. They also open up possibilities for abuse and corruption. These disadvantages will have to be weighed against the prevention of hunger and the provision of minimum welfare nets. In sum, whether there is a case for state interventions to stabilize food prices or enhance food access through other means (e.g. cash transfers) will ultimately depend on the quality of design and governance.

4 Conclusion

What can be done about the problem of expensive food in SSA, given that dependence on urban retail markets will only further grow in the coming decades? This survey translates into four, admittedly rather general, guidelines for policymakers.

First, price levels vary across SSA due to context-specific socio-economic and agricultural conditions, and also within countries, as a result of poorly integrated markets. There is no silver-bullet solution in policies aimed at curtailing or stabilizing food prices. The most universally applicable recommendation is to improve the integration of markets. Policies aimed at lowering transportation and transaction costs will, however, necessarily come at the expense of a sub-stratum of farmers and traders who are currently sustaining their livelihoods under (partial) protection from market competition.

Second, food import dependency is part of the reason why SSA food prices are high, but reducing food imports, if at all feasible, will not automatically lead to lower prices. In many cases, imported food is simply cheaper than domestic produce, and any policies geared towards domestic market protection will first raise food prices. Structurally lower food prices can only follow from major improvements in agricultural output and productivity, paired with the reduction of margins in the entire agri-food value chain. This requires a long-run policy perspective.

Third, excessive costs due to monopoly profits and rent-seeking may appear at various stages of the value chain. Effective policies to remove such rents or to dismantle monopolies will, given their context-specificity, have to be based on an in-depth analysis of country- or region-specific value chains. Ultimately, this is not just a matter of policy design, but primarily one of political economy: how to navigate, negotiate or compensate the financial interests of specific stakeholders that are interwoven with the maintenance of sub-optimal market institutions?

Fourth, and to end on a more optimistic note: lowering trade or transactions costs is bound to generate positive spill-overs in other parts of the value chain. There is a major virtuous cycle that can be set in motion, once food prices are declining as a result of targeted policy interventions.

Lower prices will improve the purchasing power of urban consumers, which will lead to enhanced demand and additional economies of scale and scope. There are countries south of the Sahara, such as Tanzania, that have managed to reduce their historical food import dependence and which have become net exporters of major staple crops. There are also countries, such as Côte d'Ivoire, that have managed to sustain a successful model of export-led growth in agriculture, the revenues of which can be used to import food without depleting foreign reserves. Such models can be extended. After all, the efficiency gains required for lower prices are not reached when every country in Africa strives to become self-sufficient. Instead, food produced for and by Africans should be grown where the production costs are lowest, and trade within the region helps to make sure these price advantages are reaped by all.

Note

1 In parts of Asia farmers also enjoy subsidized farm inputs (e.g. seed, energy, fertilizer), part of which trickles down in lower consumer prices (see for example the subsidies in Indian agriculture: Zafar et al., 2023).

References

Abdalla, M., & Al-Shawarby, S. (2018). The Tamween food subsidy system in Egypt. Evolution and recent implementation reforms. In H. Alderman, U. Gentilini, & Y. Ruslan (Eds.), *The 1.5 billion people question. Food, vouchers, or cash transfers?* (pp. 107–150). World Bank.

Andrae, G., & Beckman, B. (1986). The Nigerian wheat trap. In P. Lawrence (Ed.), *World Recession and the food crisis in Africa* (pp. 213–230). James Curry.

Alderman, H., Gentilini, U., & Yemtsov, R (Eds.). (2018). *The 1.5 billion people question. Food, vouchers, or cash transfers?* World Bank.

Ali, W., Abdulai, A., & Mishra, A.K. (2020). Recent advances in the analyses of demand for agricultural insurance in developing and emerging countries. *Annual Review of Resource Economics*, *12*(1), 411–430.

Allen, T., (2017). *The cost of high food prices in West Africa.* OECD West African Papers No. 8.

Asea, P.K., & Corden, W.M. (1994). Thirty years of the Balassa-Samuelson model. *Review of International Economics*, *2*(3), 191–325.

Barrett, C.B., Christiaensen, L., Sheahan, M., & Shimeles, A. (2017). On the structural transformation of rural Africa. *Journal of African Economies*, *26*(suppl_1), i11–i35.

Barrett, C.B., & Mutambatsere, E. (2008). Marketing boards. In S.N. Durlauf & L. Blume (Eds.), *The new Palgrave dictionary of economics* (2nd ed.; pp. 91–95). Palgrave Macmillan.

Bates, R.H. (1981). *Markets and states in tropical Africa. The political basis of agricultural policies.* University of California Press.

Djurfeldt, G., Holmén, H., Jirström, M., & Larsson, R. (2005). *The African food crisis: Lessons from the Asian green revolution.* CABI Publishing.

Dorward, A., Kydd, J., & Poulton, C. (1998). *Smallholder cash crop production under* market *liberalisation: A new institutional economics perspective.* CAB International.

FAO. (2023). Food monitoring and price analysis (FPMA) tool. https://fpma.fao.org/giews/fpmat4/#/dashboard/tool/discontinued

Frankema, E. (2014). Africa and the green revolution. A global historical perspective. *NJAS - Wageningen Journal of Life Sciences*, *70–71*, 17–24.

Frankema, E., de Haas, M., Joshipura, T., & Westland, T. (2024). *The political economy of maize markets in East Africa, 1900-2020. How a cheap staple became expensive.* Work in progress.

Frankema, E., & van Waijenburg, M. (2018). Africa rising? A historical perspective. *African Affairs*, *117*(469), 543–568.

Gelb, A., & Diofasi, A. (2015). *What determines purchasing-power parity exchange rates?* Centre for Global Development Working Paper 416.

Gollin, D., Lagakos, D., & Waugh, M.E. (2014). Agricultural productivity differences across countries. *American Economic Review*, *104*(5), 165–170.

Grilli, E., & Cheng Yang, M. (1988). Primary commodity prices, manufactured goods prices, and the terms of trade of developing countries: What the long run shows. *The World Bank Economic Review*, *2*(1), 1–47.

Gutner, T. (2002). The political economy of food subsidy reform: The case of Egypt. *Food Policy*, *27*(5), 455–476.

Hayami, Y., Platteau, J.-P., & Dasgupta, P. (1998). Resource endowments and agricultural development: Africa versus Asia. In Y. Hayami & M. Aoki (Eds.), *The institutional foundations of East Asian economic development: Proceedings of the IEA conference held in Tokyo, Japan* (pp. 357–412). St. Martin's Press in association with the International Economic Association.

Henley, D. (2015). *Asia-Africa development divergence. A question of intent*. ZedBooks.

Hesp, P. (1985). *Producer prices in tropical Africa. A review of official prices for agricultural products, 1960–1980*. African Studies Centre Research Reports 23/1985.

Howard-Hassmann, R.E. (2015). The right to food under Hugo Chavez. *Human Rights Quarterly*, *37*(4), 1024–1045.

Huff, G., & Caggiano, G. (2007). Globalization, immigration and Lewisian elastic labor in pre-World War II Southeast Asia. *The Journal of Economic History*, *67*(1), 33–68.

Jayne, T.S., & Jones, S. (1997). Food marketing and pricing policy in Eastern and Southern Africa: A survey. *World Development*, *25*(9), 1505–1527.

Joshipura, T. (2024). *An inquiry into the nature and historical roots of sub-Saharan Africa's 20th century food import dependence*. PhD-thesis' work in progress.

Kramer, B., Hazell, P., Alderman, H., Ceballos, F., Kumar, N., & Timu, A.G. (2022). Is *agricultural insurance fulfilling its promise for the developing w*orld? A review of recent evidence. *Annual Review of Resource Economics*, *14*(1), 291–311.

Larson, D.F., & Otsuka, K. (2013). Towards a green revolution in sub-Saharan Africa. In D. Larson & K. Otsuka (Eds.), *An African green revolution. Finding ways to boost productivity on small farms* (pp. 281–300). Springer Verlag.

Lipton, M. (1977). *Why poor people stay poor: A study of urban bias in world development*. Temple Smith.

Marks, D. (2010). Unity or diversity? On the integration and efficiency of rice markets in Indonesia, c. 1920–2006. *Explorations in Economic History*, *47*(3), 310–324.

Officer, L.H., & Williamson, S.H. (2023). *The annual consumer price index for the United States, 1774-present*. MeasuringWorth. https://www.measuringworth.com/uscpi/

Okou, C., Spray, J., & Unsal, D.F. (2022). *Staple food prices in sub-Saharan Africa. An Empirical Assessment*. IMF Working Papers 22/135.

Olomola, A.S. (2014). The political economy of food price policy in Nigeria. In P. Pinstrup-Andersen (Ed.), *Food price policy in an era of market instability: A political economy analysis* (pp. 275–295). Oxford University Press.

Porteous, O. (2019). High trade costs and their consequences: An estimated dynamic model of African agricultural storage and trade. *American Economic Journal: Applied Economics*, *11*(4), 327–366.

Sachs, J.D., & Warner, A.M. (1997). Sources of slow growth in African economies. *Journal of African Economies*, *6*(3), 335–376.

Sauerbeck, A. (n.d.) Wholesale prices 1845–1950. *Journal of the Royal Statistical Society*, 49–114 (various issues from annual publications 1886–1951).

Stathers, T., Holcroft, D., Kitinoja, L., Mvumi, B.M., English, A., Omotilewa, O., … & Torero, M. (2020). A scoping review of interventions for crop postharvest loss reduction in sub-Saharan Africa and South Asia. *Nature Sustainability*, *3*(10), 821–835.

Studer, R. (2008). India and the great divergence: Assessing the efficiency of grain markets in eighteenth- and nineteenth-century India. *Journal of Economic History*, *68*(2), 393–437.

Suri, T., & Udry, C. (2022). Agricultural technology in Africa. *Journal of Economic Perspectives*, *36*(1), 33–56.

Teravaninthorn, S., & Raballand, G. (2009). *Transport prices and costs in Africa. A review of the international corridors*. World Bank.
Timmer, C.P. (2009). *A world without agriculture: The structural transformation in historical perspective*. Aei Press.
Tittonell, P., & Giller, K.E. (2013). When yield gaps are poverty traps: The paradigm of ecological intensification in African smallholder agriculture. *Field Crops Research*, *143*, 76–90.
UN DESA. (2023). World urbanization prospects 2018. https://population.un.org/wup/Download/
USDA. (2023a). Economic research services. Chart gallery. https://www.ers.usda.gov/data-products/chart-gallery/gallery/chart-detail/?chartId=58372
USDA. (2023b). Economic research services. Feed grains: Yearbook table. https://www.ers.usda.gov/data-products/feed-grains-database/feed-grains-yearbook-tables.aspx
Van Ittersum, M.K., van Bussel, L.G., Wolf, J., Grassini, P., van Wart, J., Guilpart, N., … & Cassman, K.G. (2016). Can sub-Saharan Africa feed itself? *Proceedings of the National Academy of Sciences*, *113*(52), 14964–14969.
Van der Laan, L.H., Dikstra, T., & van Tilburg, A. (Eds.) (1999). *Agricultural marketing in Tropical Africa*. African Studies Centre Research Series 15. Routledge.
World Bank. (2015). *Purchasing power parities and the real size of world economies: A comprehensive report of the 2011 international comparison program*. World Bank.
World Bank. (2020). *Purchasing power parities and the size of world economies: Results from the 2017 international comparison program*. World Bank.
World Bank. (2022). Africa's pulse 25, April 2022. https://issuu.com/world.bank.publications/docs/9781464818714
World Bank. (2023a). Commodity price data (the pink sheet). https://www.worldbank.org/en/research/commodity-markets
World Bank. (2023b). World development indicators. https://databank.worldbank.org/source/world-development-indicators
Zafar, S., Aarif, M., & Tarique, M. (2023). Input subsidies, public investments and agricultural productivity in India. *Future Business Journal*, *9*(54), 1–12 (early view).
Zant, W. (2018). Trains, trade, and transaction costs: How does domestic trade by rail affect market prices of Malawi agricultural commodities? *The World Bank Economic Review*, *32*(2), 334–356.
Zant, W. (2021). Measuring trade cost reductions through a new bridge in Mozambique: Who benefits from transport infrastructure? *Journal of African Economies*, *31*(4), 384–408.

10 Addressing malnutrition in all its forms in Africa requires a radical paradigm shift

Alida Melse-Boonstra, Inge D. Brouwer, Folake Samuel, and Mofu Musonda

1 Current state of food and nutrition security in Africa

1.1 The devastating burden of malnutrition in Africa is not declining

Whereas maternal and child mortality and child growth deficits have steadily declined in Latin America and Southeast Asia in previous decades, little progress has been made in sub-Saharan Africa (FAO et al., 2022). On top of that, recent global crises such as the COVID-19 pandemic and conflicts have disproportionally affected Africa with increases in hunger and undernutrition (Osendarp et al., 2021). This reveals the fragility of African food and health systems to shocks. Seemingly in contradiction with the above, the prevalence of overweight and obesity is rising rapidly on the African continent as the first signposts for non-communicable diseases (NCDs), such as diabetes, high blood pressure, cardiovascular diseases, and cancer. Indeed, in 2019, 37% of all deaths in the African region were already attributable to NCDs (Wangou et al., 2023).

Low-quality diets are the major driver of the global burden of morbidity and mortality worldwide (Development Initiatives, 2021). The leading dietary risk factors are low intake of fruits and vegetables (responsible for 5–8% of premature mortality across regions) and whole grains (2–5%), high intake of red and processed meat (1–6%), and high prevalence of overweight and obesity (5–13%). Low-quality diets confront the African region as a whole with unresolved health issues due to undernutrition, which are increasingly superimposed by overweight, obesity, and diet-related NCDs and its related steep increases in health care expenses. In the long run, these multiple forms of malnutrition undermine human capital in terms of physical health, cognitive capacity, and (agricultural) productivity, unless preventive action is taken to turn the tide (Hawkes et al., 2020).

1.2 Current food systems do not support healthy diets and are unsustainable

Healthiness of diets depends on individual needs, which vary by age, gender, physiological stage (e.g. pregnancy), and physical activity, but in general healthy diets can be defined as being diverse and proportional in food groups. Unbalanced diets (including monotonous and undiversified diets) often supply sufficient calories derived from cereals and root crops but lack essential fats, amino acids, vitamins, and minerals to maintain health. Overconsumption of foods that contain too much or the wrong type of fat, sugar, and salt challenges the body in other ways. The shares of food and beverage markets grow most rapidly in low- and middle-income countries (LMICs), including African countries such as Kenya, Nigeria, and Cameroon. Increasingly, foods that do not fit in a healthy diet are the easiest choice for ever more African consumers in terms of availability and affordability. Moreover, global food production poses uneven

DOI: 10.4324/9781032649696-13

stress on environmental resources such as land and water, and contributes a large share (>30%) to greenhouse gas emissions.

Sustainable healthy diets (SHDs) can in short be defined as dietary patterns that prevent all forms of malnutrition, promote health, and have a low impact on the environment (FAO/WHO, 2019). SHDs are accessible, affordable, safe and equitable, and culturally acceptable. A healthy diet includes enough fruits, vegetables, nuts, seeds, whole grains, and legumes, and sufficient but not excessive amounts of starchy staples and animal-sourced foods (milk, eggs, poultry, and fish). SHDs need to be safe, and limited in food groups, or substances that could lead to health risks when eaten in excess, such as free sugars (including sugar-sweetened beverages), saturated fat, salt, red and processed meats, and ultra-processed foods (Ruel & Brouwer, 2021). SHDs vary between and within regions, dependent on the trade-offs between health and environmental burden depending on how food is produced (HLPE, 2017). For instance, in areas of sub-Saharan Africa where the prevalence of undernutrition and anaemia is very high, and where animal-sourced foods are currently consumed in relatively low amounts, increased consumption of meat, poultry, or fish may support health (Brouwer et al., 2021).

In 2019, the EAT-Lancet Commission on Healthy Diets from Sustainable Food Systems initiated a global discourse on the 'great transformation of food systems' to deliver healthy diets for a growing population within planetary boundaries by 2050 (Willett et al., 2019). Health and sustainability go hand in hand to quite some extent, with whole grains, fruits, vegetables, nuts, and legumes contributing positively, while red and processed meat contributing negatively to both health and environmental impacts. No region in the world is currently on track to meet the Sustainable Development Goals in limiting environmental burdens related to diets (GNR, 2021). In view of lower agricultural and industrial inputs, Africa and Asia among all regions comply best with keeping food production within planetary boundaries. However, environmental impact is expected to increase when moving towards healthy diets in these regions, and environmental sustainability will be of great concern if these regions develop food systems resembling those in higher-income countries.

1.3 One billion people in Africa cannot afford a healthy diet

It is estimated that approximately 40% of the world population cannot afford to buy a healthy diet due to the high cost of nutrient-rich non-staple foods (FAO et al., 2023). Surprisingly, the cost of a healthy diet in the USA (3.32 PPP (purchasing power parity) dollars per person per day in 2021) was almost the same for Africa (3.57 PPP dollars) (FAO, 2023; Frankema, this volume). The vast majority of people in Africa (~80%) cannot afford a healthy diet due to low income, in particular those living in East Africa (85%). For comparison, this proportion is 72% in South Asia, and only 1.2% in high-income countries (FAO, 2023).

Food choices are strongly driven by food prices. Staple foods such as wheat, maize, rice, and vegetable oil are generally cheaper per calorie compared to more nutrient-dense foods such as meat, dairy, vegetables, and fruits (Darmon & Drewnowski, 2015). Poor smallholder farmer households, who consume much of the food they produce, still purchase a significant part of their diet, but their food choice is largely limited by their budget: most purchases go to the cheaper staple grains, small amounts of condiments and vegetables, affordable sources of animal protein such as dried fish, and increasingly to cheap processed convenience foods. In sub-Saharan Africa, where more than 60% of farmers experience a single cropping season and a prolonged dry season, the availability of perishable but nutrient-dense foods like fruits, vegetables, and animal-source foods diminishes towards the end of the dry season. During this 'hunger season', food prices rise, the costs of a healthy diet increase, and dietary diversity decreases (de Jager et al., 2022). Local and global economic shocks also reveal how food prices drive nutrition and health, as shown by the case of Indonesia where an increase in child anaemia was seen after the economic crisis in the late 1990s (Block et al., 2004); by the surge in undernourishment in LMICs following the global bank crisis in 2008 (Brinkman et al., 2010); and more recently following the COVID-19 pandemic (Laborde et al., 2021). These shocks demonstrate the need for more resilience in food systems to warrant that healthy diets can be achieved and sustained.

In high-income countries, a reduction in consumption of animal-sourced foods will mutually benefit environmental sustainability and the wallet of consumers. However, in many LMICs it may be desirable to modestly increase the consumption of animal-source foods to meet nutritional requirements, leaving little room to improve affordability nor sustainability. Clearly, affordability is very critical when striving for food and nutrition security within the planetary boundaries.

2 Sustainable healthy diets and the four paths

In the last five decades, Africa's food demand has increased due to population growth, while rapid urbanization has increased the demand for convenience foods with a high content of fat, sugar, and salt. Both trends are predicted to continue in the decades to come. As outlined above, African diets do currently not meet health requirements within planetary boundaries and are vulnerable to (economic) shocks. In the face of population growth and climate change, there is a strong call for a transformation of African food systems to support human and planetary health. de Haas et al. (this volume) not only describe how countries' food security is linked to average income per capita, but also observe substantial heterogeneity in this relationship. They then compare and categorize African countries based on the productivity of their agricultural sector (proxied by cereal yields) and the extent of structural transformation (proxied by their agricultural labour share). Interacting these variables, they distinguish four paths, which shape the nature of food systems and the feasibility of transformation. But to what extent do agricultural gains and structural changes result in greater nutrition security? Can SHD for all Africans be achieved along these different paths in the coming decades? To understand this, we first address several common narratives and beliefs (or misconceptions) that hinder the progress of transforming food systems towards delivering SHD.

2.1 Economic growth will resolve malnutrition in the long term, right?

Except for the path of 'agricultural stagnation' (situations in which yields are low and stagnant, and agricultural employment predominates), all of the paths describe some form of agricultural and/or economic growth. Whereas this is a valuable goal, it is questionable that such growth will automatically solve the problem of malnutrition. Although increased income may translate to a more diversified and nutritious diet for some privileged population groups, it is not that straightforward. Long-term studies, for instance in India, have shown that childhood malnutrition does not necessarily improve when national income grows (Subramanyam et al., 2011). Disparities in economic gain within countries can also enlarge the nutrition and health gap between rich and poor. Moreover, the balance can easily tip over to the other side: several middle-income countries, such as Mexico, Brazil, and South Africa, exemplify how unguided food systems can become the main contributors to morbidity and mortality in society, by stimulating overconsumption of unhealthy foods (Popkin, 2015). This should not be seen as collateral damage of economic development and something that needs to be endured, or a temporal phase that will simply pass, because the damage will be profound and long-term. Therefore, economic growth in itself is no guarantee for better nutritional outcomes, and may even lead to vast increases in malnutrition-related health care costs.

2.2 *Closing yield gaps might provide enough food, but will it nourish people?*

Worldwide, agricultural investments in food systems for human consumption have mainly fuelled the production, productivity, value chains, and markets for just a few staple crops, notably wheat, maize, rice, cassava, and potatoes, with the goal to produce more calories and to stimulate economic growth. As outlined in the previous section, it is unlikely that economic growth by itself will resolve the problem of malnutrition. But there is a second issue at stake: there is also a mismatch between agricultural investments and nutritional requirements of consumers. Contemporary African diets are often unbalanced in the sense that too many calories are derived from cheap staple foods (>60% of total energy intake), leaving too little space for foods that deliver other essential nutrients. A continued narrow-minded focus on staple crops will maintain the *status quo* of unbalanced diets and poor nutritional health. Instead, more emphasis is needed on innovation, value creation, and financial incentives of nutrient-dense crops such as vegetables, fruits, dairy, nuts, and legumes. Another mismatch may exist between food crops and climate suitability, which may lead to increased environmental burden (Mahaut et al., 2022). Thus, from the perspective of both health and sustainability, a healthy diet-oriented food systems agenda regarding agricultural production is required. Since local food systems do not exist in isolation, such an agenda should not only address local production for internal markets, but also include dietary targets for trade and import policies.

2.3 *Consumers demand what is best for them, don't they?*

Food choice depends on a complex interplay between culture, knowledge, availability, price, quality, palatability, preference, convenience, perceived food safety, needs, and aspiration (Verain et al., 2021). Food choice is by no means a purely rational process (Aunger & Curtis, 2013). Consumers depend on what is on offer, but their choice will ultimately be based on a mixture of rational and emotional drivers with health concern as only one factor in this mix (Blake et al., 2021). Lack of awareness or ignorance of what the body needs to support present and future health may withhold consumers from making healthy dietary choices. But even when they do know what is good for them they may not act on it, and therefore guidance is required (Brouwer et al., 2021).

2.4 *Are the four paths equipped to nourish Africa by 2050?*

From the above, it is apparent that neither economic growth nor closure of yield gaps will result in a healthier population *per se*. Of the four paths, those that are more inclusive, notably the paths of agriculture-inclusive structural change and agriculture-led growth, are more likely to bring about equitable food and nutrition security, but are still unlikely to tackle malnutrition in all its forms. As can be seen in Table 10.1, there are no clear differences in nutritional and dietary outcomes between the four paths. This is likely because nutritional and dietary outcomes are the resultant of the way food systems are structured, and the degree by which nutrition policies have been implemented. It is therefore apparent that much more is needed to improve basic nutrition and health for Africa's growing population in spite of the path that is followed.

Table 10.1 Prevalence of key nutritional outcomes in example countries categorized by their situation along four paths of agricultural and economic change[1]

	Low birth weight[2] %	*Stunting*[3] *(children <5 years)* %	*Anaemia*[4] *(adult women)* %	*Food group diversity score*[5]	*All-5*[6] %	*Global dietary recommendation*			*Obesity*[10,11] *(adult women)* %	*Diabetes*[11] *(adult women)* %
						Overall score[7]	*NCD-Protect*[8]	*NCD-Risk*[9]		
Agriculture-inclusive structural change										
Ghana	14.2	17.5	35.4	4.5	20	10.5	3.0	1.5	23.6	9.1
Senegal	18.5	17.9	52.7	5.5	28	10.9	3.6	1.7	18.7	9.8
South Africa	14.2	21.4	30.5	5.5	23	9.1	3.3	3.2	46.7	16.7
Agriculture-extensive structural change										
Gabon	14.2	17.0	52.4	5.1	24	9.6	2.8	2.2	26.1	14.4
Namibia	15.5	22.7	25.2	NA	NA	NA	NA	NA	34.1	10.5
Nigeria	NA	31.5	55.1	4.5	23	9.7	2.7	2.0	20.4	8.1
Agricultural growth										
Ethiopia	NA	36.8	23.9	NA	NA	NA	NA	NA	11.6	7.2
Uganda	NA	25.4	32.8	5.0	32	11.3	3.6	1.3	14.0	7.2
Zambia	11.6	34.6	31.5	NA	NA	NA	NA	NA	18.6	8.8
Agricultural stagnation										
Burundi	15.1	50.9	38.5	NA	NA	NA	NA	NA	14.3	6.7
Niger	NA	44.4	49.5	3.9	15	10.6	2.7	1.2	14.1	7.7
Zimbabwe	12.6	23.5	28.9	4.8	24	10.9	3.5	1.5	31.2	9.0

Data sources: [1] For the methodology behind the classification of countries in one of the four 'paths': de Haas et al. (this volume). The classification is based on a cross-sectional stocktaking of the situation in 2021, as in their Figure 1.5. For the nutritional outcomes, the following sources have been used: 2022 Global Nutrition Report, https://globalnutritionreport.org/reports/2022-global-nutrition-report/; Global Diet Quality Project, https://www.dietquality.org/; and NCD Risk Factor Collaboration, https://ncdrisc.org/index.html. [2] Birth weight < 2,500 g, reported for the year 2015. [3] Height-for-age Z score < −2, reported for the year 2018. [4] All women, either pregnant (haemoglobin concentration < 110 g/L) or non-pregnant (<120 g/L), reported for the year 2019. [5] The Food Group Diversity Score is the number of food groups consumed the previous 24 hours, out of ten food groups: (1) grains, white roots and tubers, and plantains, (2) pulses, (3) nuts and seeds, (4) dairy, (5) meat, poultry, and fish, (6) eggs, (7) dark green leafy vegetables, (8) other vitamin A-rich fruits and vegetables, (9) other vegetables, and (10) other fruits. A higher average population score (ranging from 0 to 10) indicates inclusion of more food groups in the diet. [6] Proportion of the population who consumed all five food groups typically recommended for daily consumption in FBDGs around the world: at least one vegetable; fruit; pulse, nut, or seed; animal-source food; and starchy staple in the previous day or night. [7] The Global Dietary Recommendations (GDR) score indicates meeting WHO recommendations of food consumption from nine health-protective food groups (NCD-Protect) and eight food groups to limit or avoid (NCD-Risk) during the previous day or night. The score ranges from 0 to 18 and is expressed as an average score. [8] The NCD-Protect score is an average score for consumption of foods that are protective against NCDs (such as fruits, vegetables, whole grains, pulses, nuts and seeds, and fibre) during the previous 24 hours, ranging from zero to nine.[9] The NCD-Risk score is an average score for consumption of foods that pose a risk for NCDs (free sugar, salt, total and saturated fat, and red and processed meat) during the previous 24 hour, ranging from zero to nine. [10] Body mass index (BMI) > 30 kg/m^2; [11] Data are projections for 2025. NA: data not available.

3 Shaping future-proof African food systems: placing consumers in the centre

To eradicate malnutrition in all its forms, it is imperative to bring an SHD perspective into the discourse. This perspective requires us to reverse our thinking: the starting point is first and foremost to understand what consumers require to nourish their bodies so that they can live healthy, productive, and purposeful lives, while respecting planetary boundaries. This should then dictate what, how, and where the required food should be produced, manufactured, and traded to bring SHD in reach for all Africans, while providing a decent income for those involved. Consumers also need to be empowered to demand what they require to live healthy lives in healthy environments. Agriculture can then be developed as a viable sector within these constraints.

Addressing malnutrition in all its forms in Africa thus requires a radical re-focus of local food systems, such that all actors in the system strive to attain SHD for all individuals in society. This contradicts the current (global) situation where production and purchase of calories and immediate economic benefit are prioritized over long-term health and well-being. As also outlined by de Haas et al. (this volume), transformation of agri-food systems is a complex and multi-faceted process, being a necessary but not sufficient component of a much broader food systems transformation where the many actors who produce, process, distribute, and market foods to consumers interact intensively with social and ecological systems. In the remainder of this chapter, we describe what we believe is needed to reach the ultimate goal to not just feed, but to nourish Africa's population with SHD in 2050. Placing SHD in the centre of food system transformation will support such a paradigm.

3.1 What do consumers need? Defining diet quality within the local context

Nutritional requirements are defined in terms of the amounts of essential nutrients that are required to achieve and maintain good health, specified for different population groups, such as women, men, and children. However, these requirements have little value towards consumers, since their reference frame is foods, not nutrients. Therefore, nutritional requirements need to be translated to food-based recommendations within local contexts. Formulation of national food-based dietary guidelines (FBDGs) can bridge the gap between global recommendations and context-specific realities. FBDGs comprise food intake recommendations that guide consumers towards healthier dietary habits, while taking cultural, socioeconomic, and environmental aspects into account. To date, 11 African countries have formulated national FBDGs.[1] Nevertheless, these guidelines are generally not widely known, and, to be effective, still require to be fully adopted into policies and public health communication as well as in agricultural sector and food industry actions. Environmental sustainability can be integrated into FBDGs, as exemplified by several high-income countries. FBDGs can be used to educate consumers on SHD, as well as all actors in the food system, from producers to vendors and policy makers, to improve the quality and sustainability of the food on offer.

3.2 Nutrition-sensitive agriculture

To cater for SHD for consumers, it is crucial to define a healthy diet-targeted food systems agenda that involves all actors in the agricultural sector. The question of what, how, and where to produce, manufacture, and offer foods will then be guided by human and planetary health, be it for the local, regional, or global market. This implies not to increase investments in staple cereals, as usually occurs, but to put investments and innovation in nutrient-dense commodities such as fruits, vegetables, legumes, and nuts. Innovation may come from sustainable agriculture and

Table 10.2 Food system typologies based on a composite of four indicators

	Value added per worker	*Dietary energy share derived from cereals, roots, and tubers*	*Number of supermarkets per 100,000 people*	*Degree of urbanization*	*Examples of African countries*
	US$	*Energy %*	*N*	*%*	
Rural and traditional	653–1,444	61–70	0.5	26–42	Malawi, Ethiopia, Mali
Informal and expanding	1,698–3,886	50–61	1.6	37–58	Ghana, Nigeria, Angola
Emerging and diversifying	3,999–7,527	41–49	4.4	50–70	South Africa, Algeria
Modernizing and formalizing	11,261–20,060	33–43	9.6	66–84	–
Industrialized and consolidated	29,726–82,625	27–33	17.3	77–92	–

Source: Marshall et al. (2021).

mixed production systems with long-term benefits, such as restoration of soil health and reversal of biodiversity loss. It may also imply that traditional knowledge and practices to grow and process long forgotten crops are revisited in the light of climate adaptation. Demand and supply will need to be aligned, guided, and measured by what constitutes SHD. All actors in the supply chain can actively contribute to this by making product portfolios more compliant with SHDs. Greater benefits are achieved when such nutrition-sensitive agriculture includes behaviour change communication targeting nutrition and health outcomes, interventions for safe water, sanitation and hygiene, food fortification, and when women's empowerment is addressed (Ruel et al., 2018).

3.3 Transformation of pluriform food systems

Food systems can be characterized by their productivity, share of dietary energy derived from starchy staples, number of supermarkets per 100,000 inhabitants, and degree of urbanization (Table 10.2). Africa harbours three out of the five types of food systems, and any country within Africa is home to different types. Food systems are not uniform between and within regions, and are constantly evolving. When measuring these food system typologies against features of SHD, more whole grains and pulses and less processed convenience foods are consumed in 'rural and informal' systems (types 1–2), while consumption of fruits and vegetables and packaged foods is higher in 'emerging and industrializing' food system types (types 3–5). None of the food system types currently adhere to SHD. When thinking through the transformational processes that food systems require, strategies to ensure SHD for consumers will need to be aligned with stages of food system development, both nationally and sub-nationally. To inform transformational processes, we need a deeper understanding of how consumers make food choices within their context-specific food environment.

3.4 The food environment as entry point for behavioural change

Central to food system transformation is the question how people's food choices can be steered towards SHD within their local context. Enabling food environments are required that educate and entice consumers to make choices in line with SHD. Food environments are the physical spaces where consumers choose their food (Turner et al., 2018). These span from kitchen gardens to farm fields, from roadside stalls and kiosks to wet markets and supermarkets, from schools and universities to office canteens and sports facilities, from kindergartens to elderly homes, from rural health posts to hospitals, from street food vendors to restaurants, and from bill boards to TV commercials (Downs et al., 2021). Provision of information and awareness creation about SHD, using national FBDG as a tool through all available communication channels, can be a first important step (HLPE, 2017). Whereas price is an important consideration in food choice, availability and acceptability are just as important. True social and behavioural change can only be expected when emotional and subconscious motives are targeted consistently. It will therefore be key to consider psychological traits, emotions, and social norms of consumers when creating demand for more sustainable and healthier food options.

In view of their heterogeneity, 'one size fits all' measures to transform food systems are likely to fail. Nevertheless, lessons can be learned from other environments that can be tried and tested elsewhere (Turner et al., 2018). There are several good examples of how consumers can be nudged towards healthier choices by reframing social norms, vendor retail innovations, product labelling, taxing of undesired foods, and making healthy food the default choice. Such social and behaviour change strategies can support innovations in the food environment to steer consumers into the right direction (Fretes, Marshall and Leroy, 2024). Within the African context, some well-designed school feeding programmes have led to more food security, increased demand for local produce, and better knowledge and attitudes regarding dietary practices. More recently, several innovative strategies in local food environments have been developed successfully, such as 'baby restaurants' in Madagascar, where mothers can come with their infant to receive a balanced meal, and door-to-door selling of healthy porridges by mobile food vendors (MQSUN+, 2018), ready-to-cook vending of vegetables-on-wheels in Nigeria (Snoek et al., 2022), increased vegetable offer in meals by street-food vendors in Benin, and milk vending machines that allow choice of quantity (Ayuya et al., 2020), or pre-cooked dehydrated beans in small packages, both of which are more affordable, in Kenya. More evidence on what works and what can be scaled within the context of food environments in African food systems is needed to guide decision and policy makers.

3.5 Challenges in shaping future-proof African food systems

Food systems are characterized by multiple intersecting inequalities that undermine their resilience. Therefore, socioeconomic, gender, health, and environmental inequalities need to be addressed when transforming African food systems. Change will undoubtedly create tensions between the interests of different actors in the food system, which is probably best managed through social protection of the poorest and most vulnerable people. Food system transformation will inevitably meet resistance when trade-offs extend to more powerful players on the local, national, regional, and global market, or to other sectors, such as the environmental, health, transportation, and energy sectors. However, there may also be unexpected synergies that can serve multiple purposes and can be strengthened. Timely identification of threats and opportunities and guidance how to navigate them will enhance the implementation of effective interventions towards food system change, and for consumers to achieve SHD.

4 Multi-sectoral collaboration is crucial to achieve SHD for all

No single stakeholder group within the food system will be able to bring about a food system transformation like the one described above. Bold leadership and governance will be needed to unite consumers, farmers, retailers, investors, researchers, policy makers, and many others, in order to build a shift in Africa towards SHD, supported by policies that prioritize and support the alignment of dietary and food system change. This requires strong political commitment and governing bodies at (sub) national levels, and engagement with stakeholders from public and private sectors, and international organizations (FAO, 2022). Governments do have a toolbox to control agri-food production, e.g. trade and market interventions, price incentives and fiscal subsidies to producers and consumers, and general services support, which can be employed to influence the availability, affordability, and accessibility of the required foods.

Of note, market price controls such as minimum prices for staple foods like wheat, maize, rice, and sugar may negatively affect the production of more desirable crops in terms of nutrient density, such as fruits, vegetables, and legumes. Currently, support of agricultural production mostly favours cereals, sugar, and meat, while fruits and vegetables are even discouraged in some LMICs (FAO, 2022). Trade and tariff policies can also be barriers for nutritious foods in domestic markets and can gravely undermine the availability and affordability of SHD. Governments need to increase expenditures that collectively support food and agriculture to overcome gaps in sustainable production of nutritious foods (for instance by availing resources and green technologies), and to provide a decent income to improve affordability of healthy diets (FAO, 2022). At the same time, overproduction and overconsumption of emission-intensive or otherwise problematic commodities may be dis-incentivized.

Clearly, food system transformation to achieve SHD is not something that can be done overnight. It will require strong political will, careful planning, and patient lobbying. It will require a coherent plan, grounded in national policies, and implemented and governed by institutions that are equipped with adequate resources, capacity, and authority. Stand-alone interventions that only target one specific component of the food system may even be counterproductive. Food system transformation will require all actors to deliver their part, guided by a common vision. Multi-sectoral collaborations, consisting of consumers and public and private sectors, will be crucial to reach that goal and to mitigate trade-offs. Essential elements include those that can accelerate transformation such as development of national action plans as a follow-up of the UN Food System Summit 2021 and its bi-annual 'Stocktaking Moments', to build trust and to enable social licencing (Figure 10.1).

5 Conclusion

In the first chapter of this book, four paths of change are highlighted to describe the context of African agri-food systems and how these may change towards 2050. We conclude that neither of these paths will lead Africa towards SHD for its inhabitants, unless an SHD perspective is adopted. Taking the dietary needs of consumers as the starting point in reshaping food systems to deliver SHD represents a long-term investment in a strong workforce and manageable health care expenses. A continued narrow-minded focus on productivity, calories, yields, and inputs of a few staple and cash crops is unlikely to serve human and planetary well-being and equity in Africa. What is needed is leadership and a common long-term vision that guides re-shaping of food systems to nourish the continent.

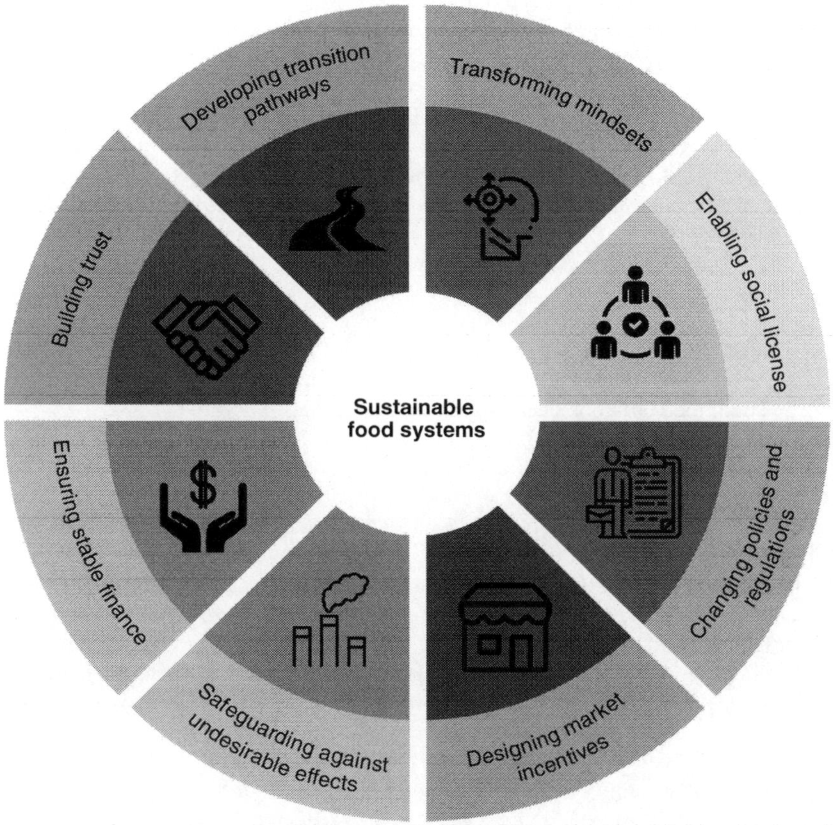

Figure 10.1 Essential elements for accelerating the systemic transformation of food systems.
Source: Herrero et al. (2020).

Note

1 See https://www.fao.org/nutrition/education/food-based-dietary-guidelines.

References

Ayuya, O.I., Ireri, D.M., Kithinji, J., Ndambi, A., Kilelu, C., Bebe, B.O., Ndung'u, P., & van der Lee, J. (2020). *Milk dispensing machines in Kenya's dairy industry: Trends and scenario analysis.* 3R Kenya research report 012/Wageningen Livestock Research, Report 1253.

Aunger, R., & Curtis, V. (2013). The anatomy of motivation: An evolutionary-ecological approach. *Biological Theory*, *8*, 49–63.

Blake, C.E., Frongillo, E.A., Warren Andrea, M., Constantinides, S.V., Rampalli, K.K., & Bhandari, S. (2021). Elaborating the science of food choice for rapidly changing food systems in low- and middle-income countries. *Global Food Security*, *28*, 100503.

Block, S.A., Kiess, L., Webb, P., Kosen, S., Moench-Pfanner, R., Bloem, M.W., & Timmer, C.P. (2004). Macro shocks and micro outcomes: Child nutrition during Indonesia's crisis. *Economics & Human Biology*, *2*(1), 21–44.

Brinkman, H.J., de Pee, S., Sanogo, I., Subran, L., & Bloem, M.W. (2010). High food prices and the global financial crisis have reduced access to nutritious food and worsened nutritional status and health. *Journal of Nutrition*, *140*(1), 153S–161S.

Brouwer, I.D., van Liere, M.J., de Brauw, A., Dominguez-Salas, P., Herforth, A., Kennedy, G., … & Ruel, M. (2021). Reverse thinking: Taking a healthy diet perspective towards food systems transformations. *Food Security*, *13*, 1497–1523.

Darmon, N., & Drewnowski, A. (2015). Contribution of food prices and diet cost to socioeconomic disparities in diet quality and health: A systematic review and analysis. *Nutrition Reviews*, *73*(10), 643–60.

De Jager, I., van de Ven, G.W.J., Giller, K.E., & Brouwer, I.D. (2022). Seasonality and nutrition-sensitive farming in rural Northern Ghana. *Food Security*, *15*, 381–394.

Development Initiatives. (2021). *Global nutrition report: The state of global nutrition*. Development Initiatives.

Downs, S.M., Ahmed, S., Fanzo, J., & Herforth, A. (2021). Food environment typology: Advancing an expanded definition, framework, and methodological approach for improved characterization of wild, cultivated, and built food environments toward sustainable diets. *Foods*, *9*(4), 532.

FAO & WHO. (2019). *Sustainable healthy diets – Guiding principles*. Food and Agricultural Organization of the United Nations and World Health Organization.

FAO, IFAD, UNICEF, WFP, & WHO. (2022). *The state of food security and nutrition in the world 2022. Repurposing food and agricultural policies to make healthy diets more affordable*. Food and Agriculture Organization.

FAO, IFAD, UNICEF, WFP, & WHO. (2023). *The state of food security and nutrition in the world 2023. Urbanization, agrifood systems transformation and healthy diets across the rural–urban continuum*. Food and Agricultural Organization.

Fretes, G., Marshall, Q., & Leroy, J.L. (2024). Food environments. Improving their healthfulness. In *Global food policy report: Transforming food systems after COVID-19*. International Food Policy Research Institute.

Hawkes, C., Ruel, M.T., Salm, L., Sinclair, B., & Branca, F. (2020). Double-duty actions: Seizing programme and policy opportunities to address malnutrition in all its forms. *The Lancet*, *395*, 142–155.

Herrero, M., Thornton, P.K., Mason-D'Croz, D., Palmer, J., Benton, T.G., Bodirsky, B.L., … & West, P.C. (2020). Innovation can accelerate the transition towards a sustainable food system. *Nature Food*, *1*, 266–272.

High Level Panel of Experts (HLPE). (2017). *Nutrition and food systems. A report by the high level panel of experts on food security and nutrition of the committee on world food security*.

Laborde, D., Herforth, A., Headey, D., & de Pee, S. (2021). COVID-19 pandemic leads to greater depth of unaffordability of healthy and nutrient-adequate diets in low- and middle-income countries. *Nature Food*, *2*, 473–475.

Mahaut, L., Pironon, S., Barnaguad, J.Y., Bretagnolle, F., Khoury, C.K., Mehrabi, Z., Milla, R., Phillips, C., Rieseberg, L.H., Violle, C., & Renard, D. (2022) Matches and mismatches between the global distribution of major food crops and climate suitability. *Proceedings of the Royal Society B*, *289*, 20221542. https://doi.org/10.1098/rspb.2022.1542

Marshall, Q., Fanzo, J., Barrett, C.B., Jones, A.D., Herforth, A., & McLaren, R. (2021). Building a global food systems typology: A new tool for reducing complexity in food systems analysis. *Frontiers in Sustainable Food Systems*, *5*, 746512.

Maximising the Quality of Scaling Up Nutrition Plus (MQSUN+). (2018). *Where business and nutrition meet: Review of approaches and evidence on private sector engagement in nutrition*. Washington, DC.

Osendarp, S., Kweku Akuoku, J., Black, R.E., Headey, D., Ruel, M., Scott, N., … & Heidkamp, R. (2021). The COVID-19 crisis will exacerbate maternal and child undernutrition and child mortality in low- and middle-income countries. *Nature Food*, *2*, 476–484.

Popkin, B.M. (2015). Nutrition transition and the global diabetes epidemic. *Current Diabetes Reports*, *15*(9), 64.

Ruel, M., & Brouwer, I.D. (2021). Nutrition: Transforming food systems to achieve healthy diets for all. *Global food policy report: Transforming food systems after COVID-19*. International Food Policy Research Institute.

Ruel, M.T., Quisumbinga, A.R., & Balagamwala, M. (2018). Nutrition-sensitive agriculture: What have we learned so far? *Global Food Security*, *17*, 128–153.

Snoek, H.M., Raaijmakers, I., Lawal, O.M., & Reinders, M.J. (2022). An explorative study with convenience vegetables in urban Nigeria—The veg-on-wheels intervention. *PLoS One*, *17*(9), e0273309.

Subramanyam, M.A., Kawachi, I., Berkman, L.F., & Subramanian, S.V. (2011). Is economic growth associated with reduction in child undernutrition in India? *PLoS Medicine*, *8*(3), e1000424.

Turner, C., Aggarwal, A., Walls, H., Herforth, A., Drewnowski, A., Coates, J., Kalamatianou, S., & Kadiyala, S. (2018). Concepts and critical perspectives for food environment research: A global framework with implications for action in low- and middle-income countries. *Global Food Security*, *18*, 93–101.

Verain, M.C.D., Snoek, H.M., Onwezen, M.C., Reinders, M.J., & Bouwman, E.P. (2021). Sustainable food choice motives: The development and cross-country validation of the sustainable food choice questionnaire (SUS-FCQ). *Food Quality and Preference*, *93*, 104267.

Wangou, M.M., Bouesso, B.R.O., Nobert, L., Bataliack, S.M., Karamagi, H.C., & Makubalo, L.E. (2023). *What are the leading causes of death in the African region?* Integrated African Health Observatory, World Health Organization.

Willett, W., Rockström, J., Loken, B., Springmann, M., Lang, T., Vermeulen, S., …, Murray, C.J.L. (2019). Food in the anthropocene: The EAT-lancet commission on healthy diets from sustainable food systems. *The Lancet*, *393*(10170), 447–492.

Part 3

Threats

11 Political instability and food security: the long view towards 2050

Han van Dijk and Nakar Syntyche Djindil

1 Introduction

Political instability and conflict reduce the capacity of people to access healthy diets, sustain an active and healthy life, and ensure food security through productive activities. Political instability negatively affects agriculture, health, and education facilities, weakens social cohesion, and limits families' access to resources and basic needs for sustaining food security including production and value chains.

Chronic malnutrition, not only as a result of conflict, was estimated at 795 million people in 2015 worldwide, with 161 million children under five years old being stunted (Loewenberg, 2015). Most of the malnourished people live in areas affected by both violent conflict and ecological fragility leading to recurrent natural disasters (FAO, 2019).

According to the World Bank, 30 out of 48 sub-Saharan African countries have experienced at least one situation of fragility, conflict, or violence since 1998.[1] In addition to their serious repercussions on the population's well-being, conflicts hamper the ability to reduce food insecurity. By causing loss of life and material damage, conflicts destabilize economic activity, disrupt food supplies, and increase the risk of famine. At present, many of these violent conflicts are transformed into protracted crises[2] and exacerbate the consequences of natural disasters, trigger crises and chronic hunger, and cause the breakdown of livelihoods and infrastructure.

In addition, the nature of present-day conflicts has changed. They are hardly ever neat and clear-cut wars between clearly defined parties. Often, conflict and insecurity stretch out over long periods of time in diverse configurations, coalitions between rebels, regular armies, organized crime, militias, and extremists (Kaldor, 2013; see also Bøås, 2015). These conflicts are not confined to national states but spread over multiple countries and may even attain regional dimensions such as the Lord's Resistance Army (LRA) conflict originating in North Uganda, and subsequently spreading into South Sudan and Central African Republic, the Boko Haram insurgency starting in northeast Nigeria and then affecting Niger, Cameroon, and Chad. The Liptako-Gourma region straddling Niger, Burkina Faso, and Mali is another example of a trans-border controlled by Jihadists. In these cases, rural areas (OECD, 2022) are most vulnerable to food security and marginal from a human development perspective.

Countries emerging from conflict face enormous challenges in reaching stability and peace (Themnér & Wallensteen, 2012; Malejaq, 2016). Usually, political instability exacerbates local patterns of vulnerability and increases socioeconomic inequities, injustice, and oppression. As a result, endemic conflict zones experience severe long-term food insecurity, due to the weaknesses of food system coordination and innovation systems, and the education deficit of their youth.

Researchers and policy circles often are focused on the short-term consequences of conflict, centring on emergency aid and refugee care. Yet, little is known about the impact on agriculture

DOI: 10.4324/9781032649696-15

and food production in the long run (Adelaja & George, 2019). There has been even less attention to the long-term consequences of conflict for food security. Retrospective studies of long-term protracted crises are rare (Blattman & Miguel, 2010). Looking towards 2050, conflict may have severe long-term consequences for income and economic growth, because it might reduce productive capacity, interrupt socialization and school careers, and impair the physical and mental development of children and youth. Despite the possible magnitude of the problem, we have little information and micro-level data on the mechanisms leading to long-term food insecurity and the magnitude of the impacts (Blattman & Miguel, 2010).

This chapter attempts to provide insight in the nexus between political instability, protracted crises, and food security. We present a case study from Chad, a context in which long-term protracted conflict has profound impact on food production, social cohesion, and the future production capacity of a generation who grew up in this situation (see also Djindil, 2021).

We start with a note on terminology and the complexities of pinning down the nature of political instability and protracted crises. The short- and long-term mechanisms leading to long-term food security are then discussed, focusing on social cohesion, natural resource management institutions, and the physical development of children growing up in protracted conflict. We conclude by discussing possible ways forwards and setting a research agenda to build more knowledge.

2 Political Instability and Protracted Crises

The emergence of protracted crises in Africa has mainly been attributed to the weakness of African states. Many African states struggle to exert sufficient authority over their territories, resulting in the so-called "ungoverned spaces" (Clunan & Trinkunas, 2010). Violent conflicts are no longer regular interstate wars fought by regular armies. Most wars in Africa are what has been labelled "new wars" fought in "varying combinations of networks of state and non-state actors – regular armed forces, private security contractors, mercenaries, jihadists, warlords, paramilitaries, etc." (Kaldor, 2013: p.2). These wars are related to hybrid forms of state formation (Boege et al., 2009) where non-state actors fill the holes in the state's monopoly on violence. These holes in the security system are filled by violent actors that provide non-state governance and alternative forms of security (Boege et al., 2009; Clunan & Trinkunas, 2010).

Often, these security holes appear in ecologically and politically marginal food-insecure rural areas and are accompanied by other emergencies such as droughts caused by erratic rainfall (Anderson et al., 2021). In the Sahel and the Horn of Africa, insurgencies originate in remote dryland areas with low population densities where poverty and food insecurity are rampant and government presence and services are poor.

So, the affected populations are often extremely vulnerable. Even in relatively low-intensity conflicts (i.e. where the number of casualties is relatively low) such as in the Central Sahel, many people are at risk for long-term food insecurity. UNICEF (2023) estimates that 10 million children are at risk in this region.

In many cases regular armed forces collapse and cede control over a large part of national territory to insurgents or use proxy forces such as self-defence militia and vigilantes (Wolff, 2011). However, some regular African armies have been operating on a regional basis (Wolff, 2011), such as the Uganda People's Defence Force operated in South Sudan to neutralize the rebels of the LRA (Schomerus, 2012), or the Rwandan army that intervened in Eastern Democratic Republic of the Congo to protect Tutsi-related groups (Auteserre, 2008) and Chadian forces that operated in Niger, Cameroon and Nigeria in counterinsurgency operations against the Muslim extremists of Boko Haram (Albert, 2017). As a result, pockets of political instability and violence are often regional and concentrated in border areas (Wolff, 2011). These armies are not as regular and disciplined as they should be. In some instances, human rights abuses, racketeering and extortion by these armies are reported, adding to the burden of violence rather than relieving it (see, e.g., Schomerus, 2012; Auteserre, 2008).

Other violent actors are harder to pin down, as they manifest themselves in very diverse ways, and may also transform over time. In Mali, Muslim extremist movements started as small remnants of the Groupe Islamique Armé (GIA) terrorists on the run from Algeria under the pressure of counter-insurgency operations. These small isolated groups transformed into more home-grown rural insurgencies affiliated with Al-Qaida, as they were able to recruit among local rural discontented youth (Thurston, 2020; Van Dijk & de Bruijn, 2022). At the time of writing, foreign elements related to Islamic State (IS) are slowly taking over these rural insurgencies with more radical and violent insurgency strategies (Cold-Ravnkilde & Ba, 2022). These groups use a variety of tactics and means to ensure the flow of resources, weapons and recruits for the continuation of their operations (Bøås, 2015)

Other armed groups that are organized for self-protection, such as vigilante groups, tend to degrade into violent behaviour, including human rights abuses and other crimes against the people they are supposed to protect (Meagher, 2012; Hagberg et al., 2019; Poudiougou, 2023).

3 Political conflict and long-term food insecurity

This raises the question to what extent acute and chronic food insecurity is not the result but a driver of conflict, given that insurgencies in rural peripheries can also be understood as revolts against the government and the international community, who failed to address poverty, absence of public services, and governance challenges (Hendrix & Brinkman, 2013).

Globally, the risk of food insecurity contributing to conflict is higher in some countries depending on the level of oppression and the numbers of suffering populations. Within the 19 countries the Food and Agriculture Organization (FAO) classifies as being in a protracted food crisis, many are also affected by conflict and violence (Holleman et al., 2017). In conflict-affected countries from sub-Saharan Africa, the number of undernourished people increased by 23.4 million between 2015 and 2018 – and at a faster rate compared to countries not exposed to conflict (FAO, 2019).

Given the complexity of the conflict-food insecurity-development nexus (see Martin-Shields & Stojetz, 2019; Kemmerling et al., 2022) and the fragility of many of these areas it is hard to predict whether the number of people, countries, and regions affected will expand or contract towards 2050. What is the scope for rebuilding peace and stabilizing states and institutions in countries affected by protracted crises? Given that underlying causes for insurgencies are present in many situations, do we foresee a further proliferation of protracted conflict? To what extent will external interventions, focusing on food system transition and rebuilding resilient livelihoods, successfully redress the impact of conflict and reduce the vulnerability of the populations? What will be the influence of the population growth in these conflict areas on the pressure on natural resources and competition for land?

It is impossible to predict how current protracted crises will develop from here to 2050, given the geopolitical situation and how food system interventions and rehabilitation policies will mitigate and address the consequences of these conflicts. What are the consequences of protracted crises for human resources, i.e. life chances and capacities of children growing up with hunger, with no access to education and adequate healthcare? The African population will double by 2050 (van Wissen, this volume). In the absence of industrial development (see Pieters, this volume), the question emerges: how people will be able to eke out an existence in rural areas?

4 Short-term consequences of violent conflicts for food security

The most immediate and obvious disruption of food security is when major conflicts cause massive numbers of displaced refugees, such as the Darfur conflict in West Sudan, and the LRA insurgence in Northern Uganda. This sudden displacement disrupts food production activities over large areas. Areas where refugees and internally displaced persons settle become dependent on emergency aid and international assistance. The influx of large numbers of people increases pressure on local food markets causing price hikes and shortages (Alix-Garcia & Saah, 2009).

In many cases, people remain in their home area, because they either lack the means to leave, are unsure where to go, or simply decide to ride out the conflict. As a result, these victims often remain out of reach for humanitarian assistance and face severe food insecurity. In addition, some have to carry the burden of feeding insurgents who are present in the area and imposed all kind of unwanted taxes and demands for food. This threat may also come from the side of the regular army (see, e.g., de Bruijn & van Dijk, 2007).

Such was the impact of the Boko Haram uprising in Northern Nigeria, and the long-term crisis in South Sudan (Loewenberg, 2015, 2017; Lokosang et al., 2016). The denial of access or the deliberate disruption of transport and food supply may also be strategies of warfare and are labelled 'famine crimes' (de Waal, 2018).

Second, in many cases, common resources like forests, shrublands, and rangelands essential for sustaining livelihoods become less accessible. These areas are often used as hiding places by insurgent groups or patrolled by the army and other armed groups. Sources of income and sustenance, such as wood, fruits, resins, and pasture, are crucial, especially for women, marginal farmers, and pastoralists. They supplement diets and provide additional income. However, local people may face cattle rustling, sexual harassment, and other forms of violence when venturing outside the relative safety of villages and small towns. For pastoralists, rangelands controlled by hostile actors become inaccessible, disrupting normal livestock grazing patterns. This can lead to loss of livestock because of cattle rustling and reduced production, posing an existential threat to many pastoralists in Nigeria (Ibrahim, 2023).

Third, inevitably, there is an increase of female-headed households in conflict zones. Men are more likely to be killed, displaced, or forced to flee to avoid recruitment by armed groups. Some migrate to other areas in search of additional income or join rebels. As a result, female-headed households are exposed to all kinds of risk and become extremely vulnerable. They face heightened food insecurity due to labour shortages and are more susceptible to sexual and economic exploitation.

Fourth, an important factor contributing to food insecurity is the destruction of crucial infrastructure, such as wells, food processing, roads, and storage facilities. Water, electricity, and energy facilities may also be destroyed and exacerbating the difficulty of rural households to earn an income. Markets depend on secure supply lines and reliable institutions. With increasing risk, traders avoid markets and increase prices, compounding food insecurity.

5 Long-term impact of conflict and political instability

The short-term consequences of conflicts can be overcome once peace is restored and people can rebuild their lives. Yet often emergency cases become decades-long protracted crises with many people becoming permanently dependent on international assistance such as in the case of the Darfur conflict and Somalian and South-Sudanese refugees in Kenya and Uganda (see, e.g., Jansen, 2018). It is impossible to estimate the number of people affected by long-term food insecurity in their youth and the impact pathways consequences in later life.

While the link between food insecurity and instability is intrinsically understood in policy and academic circles, there is less attention to protracted crises and the nature of conflicts. Often researchers rely on anecdotal evidence to capture these situations.

Analysis of food insecurity in countries that have experienced decades of violent conflict is essentially historical, and the experience should be considered to understand the present. Notably, the onset of peace must not be confused with the end of food insecurity. Few post-conflict countries have achieved credible stability and peace (Themnér & Wallensteen, 2012; Malejaq, 2016) to sustain the local economy and food production. Often, the reduction or mitigation of violence, the intervention of international diplomacy, and the establishment of democracy as a governance system mask structural violence which continues to harm the living conditions of inhabitants, especially those who live in rural areas and do not have the power of decision and action, such as unemployed wandering youth who remain opposed and question the legitimacy of these regimes (De Bruijn & Both, 2017). Most of them perpetually live under the pressure of insecurity and do not have access to healthcare systems and education.

5.1 Human resources

The most important long-term impact of protracted crises is on the health and development of children affecting their long-term capacity to be productive adults. Close to 80% of the World's 155 million stunted children and about 60% of undernourished people live in conflict-affected countries (FAO et al., 2017, p.29). Especially in remote rural areas, chronic malnutrition can lead to high rates of child mortality. Those who survive are damaged for life.

According to some studies, an individual's height is a key indicator of well-being. In childhood, it helps to track growth rate and identify any potential growth disorders or deficiencies, and adequate nutrition. The growth in height of an individual depends highly on the genetic potential for growth, and also on environmental conditions, particularly nutrition and health, which is why height is the best outcome indicator of a child's well-being early in life (see Melse-Boonstra et al., this volume). As argued by Golden (2009), children can only reach their growth potential if they have adequate nutrition and a healthy environment, and receive appropriate healthcare. The first 1,000 days are critical for the continuity of this physiological process, followed by episodes of rapid growth during infancy and puberty (see, e.g., Cusick & Georgieff, 2016). Adverse conditions manifest themselves in stunting and depress life chances in the form of education and income-earning opportunities later in life (Hoddinot et al., 2013). Even experiencing sudden and severe malnutrition *in utero* has life-long effects as was shown in a study of adults born during the Hunger Winter in 1944–1945 in the Netherlands on epigenetic changes and morbidity and blood pressure in later life (Heijmans et al., 2008).

Prolonged exposure to violence and instability can lead to mental health issues, hunger, and damage to healthcare and education infrastructure leading to a decline in education outcomes and increased mortality rates. The loss of culture may disturb intergenerational knowledge transfer.

5.2 Education

There is wide evidence that education (or the lack thereof) is an important predictor of future income and food security (Justino et al., 2013). Short- and long-term conflict disrupts education. School careers of young children are broken and often not compensated later. Research outside Africa (East Timor) suggests that episodes of violence have severe negative consequences for school careers and a significant loss of human capital (Justino et al., 2013). In Africa, a problem is that remote rural areas where most of the conflicts occur are also areas where education is often absent. In these cases, conflict may wipe out the last remnants of child education, as is for example the case in areas in the Sahel, where Jihadist movements close down schools because they are considered haram. The Boko Haram (literally "Western education is forbidden") insurgency explicitly turns itself against Western education (Adelaja & George, 2019).

5.3 Long-term impact on institutions, governance of natural resources, and land tenure security

The literature on institutional change provides us with contrasting cases. People in a conflict situation may be able to fall back on all kinds of help and solidarity relations may be reinforced as they are the only basis to cope, as was observed in Somalia (Sexsmith, 2009). However, the opposite may happen when people choose different sides and mistrust enters communities (Sexsmith, 2009).

In a situation of conflict many people are displaced and leave their homes, land, and belongings behind. More often, these vacant places are occupied by other people (most often from outside), and actors may take advantage of the situation and grab land. In Northern Uganda the government increased its presence in the context of the LRA rebellion to gain influence over land and disown communities for the benefit of outside investors (Kobusingye et al., 2017). In South Sudan, post-conflict land reform increased rather than reducing the degree of conflict, which undermined both agricultural production and land tenure security (Justin and van Dijk, 2017). In Burundi, dispute settlement between returning refugees and those who occupied their land was still ongoing even a decade after the Arusha Peace Agreement (Tchatchoua-Djomo and van Dijk, 2022). In Chad, military and ex-rebels have become large cattle owners, and send large herds into agricultural areas with armed herders, which has damaged the relationship between farmers and herders and accentuated conflicts in rural areas.

6 The case of Chad

Chad's case study is a good example to give insights into the long-term consequences of conflict for food insecurity. After more than three decades of political instability (civil war, violent conflicts) (see Azevedo & Nnadozie, 2019), power was taken over by Idriss Deby on December 1, 1990 and the country entered a period of relative stability, despite numerous rebellions against his regime and despite the harsh living conditions of the majority of the Chadians population (Tubiana & Debos, 2017). In 2021, Deby died in a clash with rebels and his son Mahamat Kaka Deby became the head of the military government.

Under the regime of Idriss Deby major investments were made by NGOs, EU, and UN agencies to maintain security and peace in the Sahelian region and reinforce rural development including food security (Pegg, 2006). The World Bank provided loans to start up the production of oil in 2003. Despite all Chad remains an extremely fragile country with a Fragility State Index score of 104.6 in 2023 (100–110) (FFP, 2023). Classified as 120th out of 125 countries on the Global Hunger Index (GHI, 2023), with a hunger index of 35.1 and a stunting rate of 47.7%, Chad remains among the poorest and undeveloped in the world. At the sub-national level, the Guéra region in Central Chad is a good example. It is a marginal region vulnerable to recurrent droughts and where the first mass peasant revolt erupted in 1965 against the post-independence government creating a generalized insecurity and fear in the country. Since then, the Guéra region has been embroiled in continuous episodes of violence and conflict combined with drought over several decades until the present day (de Bruijn & van Dijk, 2007; de Bruijn et al., 2004). The inhabitants lost most of their assets that underpinned their livelihoods and could not go back to their normal life (Djindil, 2021).

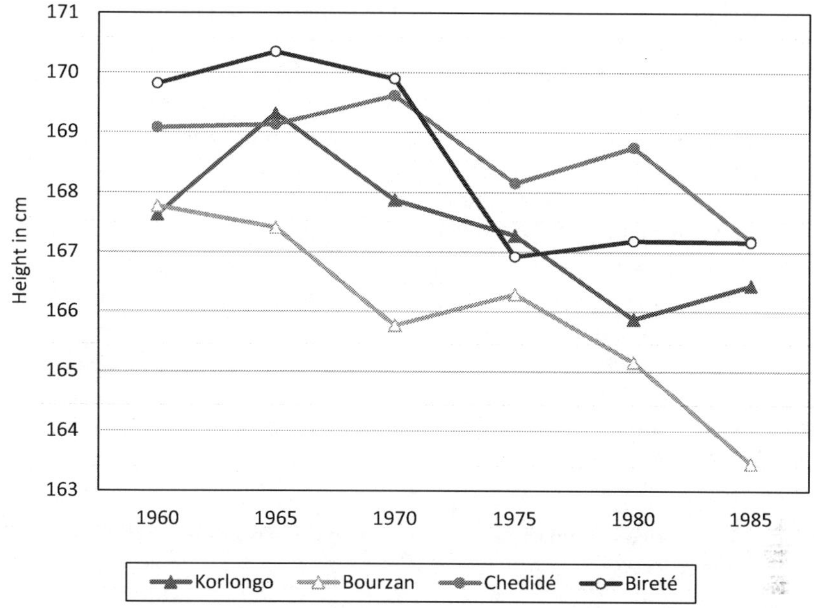

Figure 11.1 Adult height by cohort and village of individuals born between 1960 and 1985.
Source: Djindil (2021).

To assess the long-term impact we collected anthropometric data in four villages that were exposed to violent conflict and mostly dependent on agriculture and local resources. Based on our observations, semi-structured interviews, and focus group discussions over several years, villages were differently exposed to violence and this reflected in their livelihoods and income (de Bruijn & van Dijk, 2007; De Bruijn et al., 2004; Djindil, 2021).

We used adult height as the main outcome indicator of the combined impact of long-term ecological stress and violent conflict. As described above, adult height is a widely accepted indicator for life-chances and life expectancy (Deaton, 2007). The sampled population was subdivided into different cohorts according to their year of birth. Figure 11.1 presents the average adult height by cohort and village. In all four villages, mean height values varied across cohorts and time. In the 1960s cohort, composed of individuals born before 1960, a period when villages faced the same food problems, the difference in adult height among the villages was not significant ($p = 0.352$) (Djindil, 2021).

In the 1965s cohort, just after independence, and before the outbreak of conflicts, the trends show an increase in adult height with a significant difference between villages ($p = 0.008$). Under the combined effect of political and ecological stresses, all the four villages suffered the same fate manifested in a decrease in adult height (with no significant differences, $p = 0.288$). The years marked by violence continued into the 1980s but the degree vary from one village to another, led to a significant difference in the decrease in average adult height between the four villages ($p > 0.001$). For example, Korlongo fared slightly better in this period, with a 1 cm increase in height compared with the other three villages. According to elder persons, Korlongo was less affected by the violence under the regime of Hissène Habré (1982–1990) because most of the people had enough property to satisfy the demands of the rebels and many had already abandoned their belief in the Margays (the animist religion of the Guéra region) in favour of the

Muslim religion. In Bideté and Bourzan, the population did not have large resources and was subjected to violence and looting by rebels as well as abuse by national military forces. Chédidé did not experience rebel violence during this period, but the population was exploited by state agents who imposed several taxes (Djindil, 2021).

The average height deficit in the 1975 and 1985 cohorts shows that surviving children carry the scars of growth retardation into adulthood and this may affect their productivity and the households' economy in the future. Children exposed to these hazards miss the opportunity to benefit from optimal growth and a conducive social environment for their physical and psychological development with serious impact on their productivity, and overall well-being, as exemplified by Nigeria's civil war (Akresh, Bhalotra, Leone, & Osili, 2012).

This case study showed a trend linking height growth to conflict episodes in Chad. Selective chance allowed some rural children to reach adulthood with their full growth potential. In rural Chad, physical strength is a key factor for agricultural productivity. Poor nutrition and health resulting from violent conflicts systematically reinforces a poverty cycle that exacerbates food insecurity and social vulnerability.

7 Discussion: more knowledge needed on the long-term impact of protracted crises

As we have shown the consequences of exposure to violence and ecological stress over a long period are manifested in reductions in adult height and productivity with impacts on the capacities of individuals and families to be food secure now and in the future. Village-level differences in experiences of violence were shown to be correlated with income-earning capacity, adult height and directly translated in food insecurity, given the general level of poverty in the villages and the percentages of stunted children which were inversely related to income (Djindil, 2021).

Since data collection, national indicators in Chad relating to political fragility, the Human Development Index (HDI), and malnutrition rates have not improved significantly despite more political stability. None of the villages studied has a school or access to healthcare in its immediate surroundings. In short, Chad is still conditioned by its history of protracted crises. A longitudinal study with precise data on birth years, growth monitoring, integrating health indicators, and infection histories will provide more conclusive results.

Our findings have important implications for other ongoing protracted crises across Africa and the prospects of food security in 2050. The impacts of violent conflict are felt beyond their immediate consequences. The damage inflicted on children in their early years may never be fully repaired in the course of their life. Children born today will bear the consequences towards 2050 and beyond. Such damage is present not only in the physical development and health of adults who experienced trauma, but also in their psychological condition, their ability to relate to and trust other individuals in society, and access to the networks and resources needed to secure their food security. To this we must add the unknown costs of psychological trauma and stress related to violent events and their impact on the future functioning of individuals in society.

The societal damage in terms of the fallout caused by malfunctioning institutions, and the effort to rebuild trust and restore social order, has yet to be researched systematically. Given the increased number of conflicts in Africa, their duration, and the weak prospects for peacebuilding in the medium and long term, it is clear that more serious and systematic attention to the implications of long-term political instability for food (in)security in Africa towards 2050 is required, to identify more focused and efficient policy approaches.

A better understanding of the link between conflicts and long-term productivity deficits and population growth is needed to provide insights into the relation between protracted crises and long-term food insecurity and lessons for better-coordinated actions for food security by 2050. Following the Food Systems Summit of 2021, the UN established the Humanitarian-Development-Peace (HDP) Nexus Coalition to support member states affected by food crises and conflict in operationalizing and implementing their National Food Systems Transformation Pathways to promote food security.[3] However, this initiative is still far from being an operational and mainstream institution and a smarter and a long-term strategy is needed for promoting food security in conflict areas.

Notes

1 Fragility and poverty in sub-Saharan Africa: two sides of the same coin (worldbank.org), viewed September 5, 2024.
2 https://inee.org/eie-glossary/protracted-crisis, consulted January 20, 2024.
3 https://hdpnexus.fightfoodcrises.net/.viewed September 5, 2024.

References

Adelaja, A., & George, J. (2019). Effects of conflict on agriculture: Evidence from the Boko Haram insurgency. *World Development*, *117*, 184–195. https://doi.org/10.1016/j.worlddev.2019.01.010

Albert, I.O. (2017). Rethinking the functionality of the multinational task force in managing the Boko Haram crisis in the lake Chad Basin. *Africa Development*, *XLII*(3), 199–135.

Alix-Garcia, J., & Saah, D. (2009). The effect of refugee inflows on host communities: Evidence from Tanzania. *World Bank Economic Review*, *24*(1), 148–170. https://doi.org/10.1093/wber/lhp014

Akresh, R., Bhalotra, S., Leone, M., & Osili, U.O. (2012). War and stature: Growing up during the Nigerian civil war. *The American Economic Review*, *102*(3), Papers and proceedings of the one hundred twenty fourth annual meeting of the American economic association (May 2012), pp. 273–277. https://www.jstor.org/stable/23245541

Anderson, W., Taylor, Ch., McDermid, S., Ilboudo-Nébié, E., Seager, R., Schlenker, W., … & Markey, K. (2021). Violent conflict exacerbated drought-related food insecurity between 2009 and 2019 in sub-Saharan Africa. *Nature Food*, *2*, 603–615. https://www.nature.com/articles/s43016-021-00327-4

Auteserre, S. (2008). The trouble with Congo: How local disputes fuel regional conflict. *Foreign Affairs*, *87*(3), 94–110.

Azevedo, M.J., & Nnadozie, E.U. (2019). *Chad. A nation in search of its future.* Routledge.

Blattman, C., & Miguel, E. (2010). Civil war. *Journal of Economic Literature*, *48*(1), 3–57. https://doi.org/10.1257/jel.48.1.3

Bøås, M. (2015). Crime, coping and resistance in the Mali-Sahel periphery. *African Security*, *8*(4), 299–319. https://doi.org/10.1080/19392206.2015.1100506

Boege, V., Brown, M.A., & Clements, K.P. (2009). Hybrid political orders, not fragile states. *Peace Review: A Journal of Social Justice*, *21*(1), 13–21.

Clunan A.L., & Trinkunas, H.A. (2010). Conceptualizing ungoverned spaces: Territorial statehood, contested authority and softened sovereignty. In A.L. Clunan & H.A. Trinkunas (Eds.), *Ungoverned spaces: Alternatives to state authority in an era of softened sovereignty* (pp. 17–33). Stanford University Press.

Cold-Ravnkilde, S.M., & Ba, B. (2022). Jihadist ideological conflict and local governance in Mali. *Studies in Conflict and Terrorism*. https://doi.org/10.1080/1057610X.2022.2058360

Cusick, S.E., & Georgieff, M.K. (2016). The role of nutrition in brain development: The golden opportunity of the "first thousand days". *Journal of Pediatrics*, *175*, 16–21.

De Bruijn, M.E., Dijk, H., & Djindil, N.S. (2004). *Central Chad revisited. The long-term impact of drought and war in the Guéra*. African Studies Centre.

De Bruijn, M.E., & Both, J. (2017). Youth between state and rebel (dis)orders: Contesting legitimacy from below in sub-Sahara Africa. *Small Wars and Insurgencies*, *28*(4–5), 779–798.

De Bruijn, M.E., & van Dijk, H. (2007). The multiple experiences of civil war in the Guera region of Chad, 1965–1990. *Sociologus*, *57*(1), 61–98.

De Waal, A. (2018). The end of Famine? Prospects for the elimination of mass starvation by political action, *Political Geography*, *62*, 184–195.
Deaton, A. (2007). Height, health and development. *PNAS*, *104*(33), 13232–13237. www.pnas.org/cgi/doi/10.1073/pnas.0611500104
Djindil, N.S. (2021). *Insécurité Alimentaire et stress politique au Guéra, Tchad 1965–2006: Étude de l'impact à long terme des violences sur la production agricole, l'organisation sociale et l'accès aux services de santé*. Leiden, Leiden University, PhD thesis. https://scholarlypublications.universiteitleiden.nl/handle/1887/3244045
FAO. (2019). Monitoring food security in countries with conflict situations A joint FAO/WFP update for the members of the United Nations Security Council, Issue no 6. August 2019, Rome FAO and WFP.
FAO, IFAD, UNICEF, WFP, & WHO. (2017). *The state of food security and nutrition in the world 2017: Building resilience for peace and food security.* FAO.
FFP. (2023). Fragile state index: Annual report 2023, Washington, Fund for Peace. https://fragilestatesindex.org/country-data/
GHI. (2023). *2023 global hunger index: The power of youth in shaping food systems*. Welthungerhilfe.
Golden, M.H. (2009). Proposed recommended nutrient densities for moderately malnourished children. *Food and Nutrition Bulletin*, *30*(3), S267–S342.
Hagberg, S., Kibora, L.O., Barry, S., Cissao, Y., Gnessi, S., Kaboré, A., Koné, B., & Zongo, M. (2019). *Sécurité par le bas: Perceptions and perspectives citoyennes des défis de sécurité au Burkina Faso.* Uppsala University, Uppsala Papers in African Studies 5.
Heijmans, B.T., Tobi, E.W., Stein, A.D., Putter, H., Blauw, G.J., Susser, E.S., Slagboom, P.E., & Lumey, L.H. (2008). Persistent epigenetic differences associated with prenatal exposure to famine in humans. *PNAS*, *105*(44), 17046–17049. https://doi.org/10.1073/pnas.0806560105
Hendrix, C.S., & Brinkman, H-.J. (2013). Food insecurity and conflict dynamics; causal linkages and complex feedbacks. *Stability. International Journal of Security and Development*, *2*(2), 1–18. https://doi.org/10.5334/sta.bm
Hoddinot, J., Behrman, J.R., Maluccio, J.A., Melgar, P., Quisimbing, A., Ramirez-Zea, M., ... & Martorell, R. (2013). Adult consequences of growth failure in early childhood. *American Journal of Clinical Nutrition*, *98*, 1170–1178.
Holleman, C., Jackson, J., Sánchez, M.V., & Vos, R. (2017). *Sowing the seeds of peace for food security. Disentangling the nexus between conflict food security and peace*. FAO Agricultural Development Economics Study 2. Rome, FAO.
Ibrahim, S. (2023). Livelihood transition and economic well-being in remote areas under the threat of cattle rustling in Nigeria. *Geojournal*, *88*, 1–16. https://doi.org/10.1007/s10708-022-10583-x
Jansen, B.J. (2018). *Kakuma refugee camp: Humanitarian urbanism in Kenya's accidental city*. Zed Books.
Justin, P.H., & van Dijk, H. (2017). Land reform and conflict in South Sudan: Evidence from Yei River county. *Afrika Spectrum*, *52*(2), 3–28. https://nbn-resolving.org/urn:nbn:de:gbv:18-4-1073
Justino, P., Leone, M., & Salardi, P. (2013). Short-and long-term impact of violence on education: The case of Timor Leste. *The World Bank Economic Review*, *28*(2), 320–353. https://doi.org/10.1093/wber/lht007
Kaldor, M. (2013). In defence of new wars. *Stability. International Journal of Security and Development*, *2*(1), 4. https://doi.org/10.5334/sta.at
Kemmerling, B., Schetter, C., & Wirkus, L. (2022). The logics of war and food security. *Global Food Security*, *33*, 100634. https://doi.org/10.1016/j.gfs.2022.100634
Kobusingye, D.N., van Leeuwen, M., & van Dijk, H. (2017). The multifaceted relationship between land and conflict: The case of the Apaa evictions in Amury district, Northern Uganda. *Journal of Modern African Studies*, *55*(3), 455–477. https://doi.org/10.1017/S0022278X17000106
Loewenberg, S. (2015). Conflicts worsen global hunger crisis. *The Lancet*, *386*, 1719–1721.

Loewenberg, S. (2017). Famine fears in northeast Nigeria as Boko Haram fight rages. *The Lancet*, *389*, 352.

Lokosang, L., Ramroop, S., & Zewotir, T. (2016). The effect of weakened resilience on food insecurity in protracted crisis: The case of South Sudan. *Agriculture & Food Security*, *5*, 2. https://doi.org/10.1186/s40066-016-0051-y

Malejaq, R. (2016). Warlords, intervention, and state consolidation: A typology of political orders in weak and failed states. *Security Studies*, *25*, 85–110. https://doi.org/10.1080/09636412.2016.1134191

Martin-Shields, Ch.P., & Stojetz, W. (2019). Food security and conflict: Empirical challenges and future opportunities for research and policy making on food security and conflict. *World Development*, *199*, 150–164. https://doi.org/10.1016/j.worlddev.2018.07.011

Meagher, K. (2012). The strength of weak states? Non-state security forces and hybrid governance in Africa. *Development and Change*, *43*(5), 1073–1101.

OECD. (2022). *Conflicts are increasingly becoming rural*. OECD. https://oecd-development-matters.org/2022/07/26/conflicts-are-becoming-increasingly-rural-in-north-and-west-africa/ consulted April 1, 2023.

Pegg, S. (2006). Can policy intervention beat the resource curse? Evidence from the Chad-Cameroon pipeline project. *African Affairs*, *105*(418), 1–25. https://doi.org/10.1093/afraf/adi090

Poudiougou, I. (2023). Appropriation foncière, migrations agricoles at conflits armés en pays Dogon (Mali) (PhD thesis). Leiden University, Università de Torino. https://scholarlypublications.universiteitleiden.nl/handle/1887/3643682

Schomerus, M. (2012). "They forget what they came for" Uganda's army in Sudan. *Journal of East African Studies*, *6*(1);124–153. https://doi.org/10.1080/17531055.2012.664707

Sexsmith, K. (2009). Violent conflict and social transformation: An institutionalist approach to the role of informal economic networks. *European Journal of Development Research*, *21*(1), 81–94. https://doi.org/10.1057/ejdr.2008.8

Tchatchoua-Djomo, R., & van Dijk, H. (2022). Ambiguous outcomes of returnees' land dispute and restitution in war-torn Burundi resolution. *Land*, *11*, 191. https://doi.org/10.3390/land11020191

Themnér, L., & Wallensteen, P. (2012). Armed conflicts, 1946–2011. *Journal of Peace Research*, *49*(4), 565–575. https://doi.org/10.1177/0022343312452421

Thurston, A. (2020). *Jihadists of North Africa and the Sahel: Local politics and rebel groups*. Cambridge University Press.

Tubiana, J., & Debos, M. (2017). *Deby's Chad: Political manipulation at home, military intervention abroad, challenging times ahead*. United States Institute for Peace.

UNICEF. (2023). *Extreme jeopardy. Ten million children in the central Sahel need humanitarian assistance amid spiralling conflict and punishing climate*. UNICEF.

Van Dijk, H., & de Bruijn, M.E. (2022). Religious movements in the drylands: Ethnicity, jihadism, and violent extremism. In A. Kronenburg García, T. Haller, H. van Dijk, C. Samimi, & J. Warner (Eds.), *Drylands facing change: Interventions, investments and identities* (pp. 155–173). Routledge. https://doi.org/10.4324/9781003174486-12

Wolff, S. (2011). The regional dimensions of state failure. *Review of International Studies*, *37*, 951–972. https://doi.org/10.1017/S0260210510000951

12 Climate change: impacts and adaptation in smallholder farming systems in sub-Saharan Africa

Katrien Descheemaeker, Patricia Masikati, and Dilys Sefakor MacCarthy

1 Introduction

Convincing scientific evidence shows that climate change is caused by human activities, is already happening now and is seriously impacting agri-food systems globally. The impacts are likely to worsen in future. Especially if global action to curb greenhouse gas emissions fails to limit global warming to 1.5°C or 2°C, adaptation to climate change will become increasingly complicated. The situation is particularly worrying on the African continent for several reasons. First, regions located at the lower latitudes will be hardest hit by increasing temperatures, while the semi-arid and sub-humid regions are likely to witness trends of drying and increased variability in rainfall and seasonal patterns. This will lead to vast areas across the African continent becoming less or even not suitable for agriculture. Second, the African agricultural sector is dominated by smallholders, who are particularly vulnerable to climate change. A third concern results from the fact that currently African countries typically lack sufficient resources to invest in adaptation and coping measures to improve the resilience of their agricultural systems. As a result, the prospects for food security in the coming decades look rather grim.

In this chapter we unpack the challenge posed by climate change on agricultural systems, by giving an overview of its expected effects on weather patterns, and the consequences for crop and livestock productivity. We explore the reasons for the high vulnerability of African smallholders, at the farm level and beyond. From the vast body of literature on adaptation, we distil insights not only on the options and opportunities, but also on the limits and constraints in adaptation, including the risk of maladaptation. This leaves room only to conclude that to reach food security under a changing climate, more drastic and systemic changes are needed. Based on examples from Ghana and Zimbabwe, we reflect on what is required to enable a transformation of agricultural systems that are climate-resilient and sustainable.

2 Projected changes in the climate across the African continent

Global warming has already led to observed changes in the climate across Africa, with notable increases in mean temperature and more frequent heatwaves, whereas trends in rainfall vary across regions (Trisos et al., 2022).

Future climate projections depend on the greenhouse gas emission scenario, the climate model and the period considered. Notwithstanding the variability in predictions, mean air temperature is expected to rise and temperature extremes will become more frequent across the entire continent (Trisos et al., 2022). For precipitation, the projected changes differ among regions. Mean annual rainfall is expected to increase over East Africa, Central Africa and the eastern Sahel region, whereas the western Sahel region, southwestern southern Africa and coastal north Africa will

DOI: 10.4324/9781032649696-16

likely become drier. Heavy precipitation events and associated floods are expected to become more frequent and intense across most of the African continent. Likely shifts in the agricultural season, due to a later onset and/or a later or earlier end of the rainy season, are also reported.

3 Expected impacts on crop and livestock production

Climate change will affect food production in different ways that also interact with each other and with various components of the agricultural system. For example, whereas the increased incidence of heat stress has direct effects on the productivity of crops and animals, it also affects agricultural labour, provided both by people and draught animals (Kjellstrom et al., 2016), thus also leading to indirect effects on agricultural productivity.

3.1 How climate change affects crop growth

Increased temperatures lead to accelerated crop development and changes in the photosynthesis process. Whether this has a positive or negative impact on crop yield depends on the type of crop and the base temperature, as in colder conditions such as at high elevations, warming will increase crop yield. More frequent high temperature extremes will result in crop damage due to heat stress. With respect to water, drought stress and flooding are the prominent causes damaging crops. Increased CO_2 concentrations can partly compensate these negative impacts by stimulating photosynthesis, biomass accumulation and water use efficiency. These effects are stronger in C3 crops (wheat, rice, legumes, tree crops) compared to C4 crops (maize, sorghum, millet, sugarcane). In addition, higher CO_2 concentrations may lead to decreased nutrient and protein contents in crops. Climate change also affects the prevalence of weeds, pests and diseases and the biology of pollinators, which, in turn, have their impact on crop production. Another indirect effect originates from changes in hydrology that affect water availability for irrigation.

3.2 Impacts across crops and regions

The expected impacts on crop yields are generally negative, but vary per crop and by region (Trisos et al., 2022). For example, for maize, median yields in western Africa are expected to decline by 9–41% compared with 2005 yields for projected global warming levels of 1.5°C and 4°C respectively and without adaptation. In eastern and southern Africa, the declines are more severe and amount to roughly 50% for 4°C warming, albeit with large variability in the predictions, reflecting the variability in elevation. Besides yield impacts, changes in suitability for cultivating specific crop species will entice farmers to adjust cropping patterns (Wichern et al., 2019), shift from cropping to livestock rearing (Jones and Thornton, 2009) or abandon farming altogether. For example in the Sahel, the area suitable for maize is expected to decline by 40% (Birkmann et al., 2022). As millet and sorghum are seen as more climate-robust cereals than maize, transitioning back to these more traditional crops could increase the resilience of cropping systems (Trisos et al., 2022).

3.3 Impacts on livestock

Animals are better adapted to dry and variable climates than crops, yet they are directly impacted by climate change as increased temperatures affect animal physiology, fertility, susceptibility to diseases and feed intake, with repercussions on overall productivity (Godde et al., 2021). Especially in lower- and middle-income countries in the lower latitudes, where the risk of extreme heat stress is increasing, outdoor livestock keeping is likely to become unviable (Thornton et al., 2021). Livestock are also indirectly affected through climate change impacts on the quantity and quality of farm-produced fodder and grazed feed resources (Descheemaeker et al., 2018), changes in pest and disease incidence and changes in the availability and quality of water resources. In addition, the increased probability of droughts increases the risk of losing entire herds. While pastoralism is regarded as a form of adaptation to temporal and spatial feed variability and scarcity, it is under increasing pressure from climate change, which, together with other drivers, exacerbates herder-farmer conflicts (van Dijk & Djindil, this volume).

4 Vulnerability and adaptive capacity of the smallholder sector

Vulnerability results from the interplay between exposure, sensitivity and adaptive capacity. We focus here on the vulnerability of the African agricultural sector, which is dominated by smallholders.

Most of African agriculture is rain-fed, and the development of irrigation infrastructure is slower than on other continents (Veldwisch et al., this volume). As a result, crop yields are sensitive to changes in the amount and distribution of rainfall. In areas with large rainfall variability, such as the semi-arid regions, the risks associated with unpredictable rainfall discourage farmers to invest in agricultural inputs and adaptive management. Climate change is expected to exacerbate rainfall variability, with changes in the start and length of the growing season and more dry spells. Without access to water for (supplemental) irrigation as an adaptation measure, agricultural production by smallholder will be negatively affected.

Vast areas across the African continent face soil fertility constraints. Together with poor access to and the unaffordability of agricultural inputs, this leads to large yield gaps and stagnant development in yield gains (Schut & van Ittersum, this volume). Moreover, crops experiencing nutrient stress are more severely affected by increasing temperatures and benefit less from the CO_2 fertilization effect (Rosenzweig et al., 2014).

In smallholder systems, human labour is an essential resource for many farming operations and to maintain farm flexibility in the face of increased climate variability. Yet, the hiring out of own labour by poor households to earn money jeopardizes this flexibility. Also, with increased temperatures and humidity, and changes in the prevalence of human disease vectors, climate change may compromise human health, with negative repercussions on labour productivity (Kjellstrom et al., 2016). Furthermore, agricultural labour will become even more strenuous with increased heat and simply impossible in certain areas.

Over recent decades and across the continent, crop diversity is declining. For example, due to its relatively high yield, ease of processing and good marketability, maize became the dominant cereal. So far, the demographic and market drivers that push systems towards the most profitable or calorie-dense crop have been overriding the interest in crops like sorghum and millet that are considered to be more drought-resistant. The loss of agro-biodiversity results in the lost ability to exploit species complementarity in resource use and susceptibility to pests and diseases. As such, uniform systems are more susceptible to the changes and shocks associated with climate change.

(Agro-)pastoralists rely on the mobility of their herd to exploit spatial and temporal variation in the availability of water and grazing resources. As such, they would seem better armed against climate change than sedentary, purely crop-oriented farmers. However, since colonial times, the pastoralist lifestyle has been increasingly endangered due to settlement policies and changes towards more individual land tenure, the expansion of cultivated land and land use conflicts hampering herd mobility (van Dijk & Djindil, this volume). These ongoing trends further increase the already high vulnerability of herders to climate change impacting the availability of water resources (for livestock drinking) and the quality and quantity of the biomass for grazing.

Compared with farmers in industrialized countries, African smallholders have poor access to information on weather and prices, and to knowledge on agronomic management. This prevents taking timely decisions and adapting farm management to the changing weather conditions. Likewise, poor extension services and limited access to credit, assistance and insurance preclude the investment in modern agricultural inputs and adaptive farm management that could cushion farmers against climate change. Additionally, increased weather variability will further discourage farmers to invest and use inputs (Hansen et al., 2019). As a consequence, the prevalence of low-risk subsistence activities with their associated low productivity and profit may have a stronger impact than the direct effects of climate change.

Smallholder farmers have remarkably few preventative and curative strategies at their disposal to cope with climate variability and change (Wichern et al., 2023). Moreover, some existing strategies further aggravate households' vulnerability. In particular, common practices like selling assets (e.g. livestock) or eating less in response to a shock have strong impacts on income and food security, which are felt over prolonged periods, and bear the risk of pushing households into a poverty trap.

5 Adaptation potential, limits and constraints

Given Africa's exposure to climate change effects, the sensitivity of the agricultural systems to these changes and the severe vulnerability of smallholders, adaptation to climate change is of vital importance.

Intergovernmental Panel on Climate Change (IPCC) (2022) advocates for an integrated adaptation approach that combines technical options with climate information services, policies, strategic financial investments and enabling institutions, each integrating indigenous and local knowledge systems. The challenges with respect to developing supporting services are enormous in many current African contexts. For example, the provision of weather information to smallholders is hampered by farmers' capacity to interpret the information and by data unavailability (Carr et al., 2020), related to poor weather observation infrastructure. Another frequently touted product is weather insurance (Jensen and Barrett, 2017). However, besides their limited uptake, insurance products may mostly benefit wealthier farmers and lead to further homogenization of cropping systems, thus counteracting resilience.

Zooming in to technical agricultural adaptation, numerous options target the crop, the livestock or the entire farm system (Trisos et al., 2022) and can be classified according to the main strategy of risk management, diversification and sustainable intensification (Descheemaeker et al., 2016). Risk management aims to reduce the variance in crop yield or farm production, and examples include adjusting planting time, improving post-harvest storage and seasonal herd migration. Diversification builds resilience by combining crops, livestock species or different livelihood activities that smoothen the overall effect of a shock. Smallholders can implement combinations of locally adapted crops, cultivars and livestock species, agroforestry, intercropping and rotations. Sustainable intensification aims to increase the productivity of the system, without causing negative side-effects on other sustainability dimensions. Examples relevant for the African context include integrated soil fertility management, water harvesting and small-scale irrigation, and small-scale mechanization. A lot is expected from breeding varieties that are tolerant to abiotic and biotic stresses. Unfortunately, most large-scale breeding efforts are targeted to major staple crops like maize, wheat and rice, counteracting the needed diversification of some African cropping systems.

The multitude of available options may give the impression that adapting to climate change is a matter of simply choosing and implementing the right option. However, several factors caution for optimism on the adaptation potential. First, the concept of adaptation limits helps to detect when and where crop productivity cannot be secured through adaptive actions or the costs of adaptation and residual damage are too high. Even with current global warming levels, the adaptation limits for many crops in tropical Africa are being reached, and this will become more stringent and more widespread across the continent in future (Trisos et al., 2022). Second, many constraints and barriers prevent smallholders from adopting promising adaptation options (Descheemaeker et al., 2016). These limitations largely overlap with the factors explaining smallholders' vulnerability (see Section 4). Third, smallholder farmers are typically risk-averse, which discourages investments under high climate variability (Hansen et al., 2019). Another factor that discourages the investment in inputs, is that climate change reduces the positive effect on yields of using agricultural inputs, such as fertilizer (Traoré et al., 2017). Finally, whereas the effect of an adaptation option may be sizeable at the field level, the impact on farm-level income or food self-sufficiency evaporates due to small farm sizes (Giller & Andersson, this volume). Households living on small farms may thus not be motivated to implement improved practices, especially when farming is not their primary livelihood activity.

Adapting to climate change requires sound decision-making by a variety of actors, but when decisions are based on incorrect or incomplete information, maladaptation, i.e. actions that increase the risk of adverse outcomes, may be the result (IPCC, 2022). Different types of maladaptation can be distinguished (Campbell, 2022). A first type emerges when short-term adaptive measures lower the long-term adaptive capacity. For example, when groundwater is over-used for irrigation, the lowering of the groundwater table may compromise irrigation in the future. The second type of maladaptation results from shifting the vulnerability from one group or area to another, leading to increased inequity and overall impairment of sustainable development. This may happen if, for example, insurance products are preferentially directed to more powerful stakeholders who can pay the premiums. The third maladaptation category erodes sustainable development because of increased emissions, or deteriorating conditions in other sustainability domains. The diversification of smallholder livelihood activities into charcoal production to generate income is a typical example. Whereas this diversification buffers agricultural income losses resulting from climate shocks, it leads to increased emissions and deforestation, thus removing the climate buffering function of trees.

In conclusion, while many options are available, on-farm adaptations are insufficient to lift smallholders out of poverty and increase food production such that the smallholder sector could feed the fast-growing human population (Trisos et al., 2022). Up to 2°C warming, negative impacts can be somewhat compensated for, but from mid-century onwards and especially for the more pessimistic warming scenarios, the costs of adaptation will be hard to bear. It is expected that this will be most acute in the warmer regions at low latitudes, which often coincide with low-income conditions, where financial and other support mechanisms are lacking. Unfortunately, the combination of these conditions is quite common on the African continent, indicating the need for more drastic or transformative changes.

6 Need for transformative change: climate-resilient development in two case study countries

Future climate change and corresponding changes in agricultural systems are likely to exceed thresholds beyond which incremental adjustments will not suffice to sustain functions (IPCC, 2022) and more drastic changes are required. IPCC (2022, p. 823) defines a transformation of the agricultural system as "a redistribution of at least a third in the primary factors of production (land, labour, capital) and/or the outputs and outcomes of production". To structure the process of transformation, Climate-Resilient Development (CRD) pathways (Schipper et al., 2022) can guide climate adaptation and mitigation actions in food system transitions. The CRD pathways are defined as "trajectories in time reflecting a particular sequence of actions and consequences against a background of autonomous developments leading to a specific future situation". However, the operationalization of CRD is hampered by limited indicators, a lack of robust, integrated analyses and little evidence on approaches to include local perspectives, norms and values.

For the African context in particular, there is a lack of clarity about what transformation could concretely look like and few examples shed light on stakeholder-informed processes of pathway development. Here we describe the process and outcomes of the Agricultural Model Intercomparison and Improvement Project (AgMIP) Regional Integrated Assessment (RIA) approach (Antle et al., 2018), as it was applied in two case studies, namely semi-arid northern Ghana and a dryland zone in western Zimbabwe. AgMIP-RIA integrates climate, crop, livestock and economic data and models for impact assessment in heterogeneous farming systems (Homann-Kee Tui et al., 2021), in combination with an iterative, multi-stakeholder engagement process.

Agriculture in Ghana is highly vulnerable to climate change and extremes since it relies mainly on rainfall. The vulnerability increases from the South, where forest vegetation predominates with two growing seasons in a year, to the North, which has savannah vegetation with only one growing season in a year. Northern Ghana is characterized by a semi-arid agro-ecology, with rain-fed agriculture accounting for about 70% of livelihoods. The current agricultural productivity is far below its potential, with average maize yield at 0.9 t/ha (Freduah et al., 2019). The dominant smallholder farming system is characterized by soils that are generally shallow and low in fertility, erratic rainfall patterns during one growing season and a low use of external inputs. Households are extremely vulnerable to climate shocks, as poverty rates are above 70% and food shortage periods amount up to six months. Agriculture is an important sector to the economy of Ghana, household livelihoods and food security. Despite the recent transition of the economy to services (46%) and industry (33%), agriculture still contributes significantly (21%) to the country's GDP. Nationally, it employs about half of the workforce out of which about 71% are rural populations. Ghana is recognized to be vulnerable to climate change according the Global Risk Index 2020 which ranked it 109 out of 181 countries.

The situation is very similar in semi-arid Zimbabwe, as, also here, rainfall is variable and soil fertility is poor. Mixed crop-livestock systems predominate with crop residues used as dry-season feed, and livestock providing draft power and manure to crop production (Homann-Kee Tui et al., 2015). Current crop yields are low, with maize yielding on average 0.7 t/ha and other crops like sorghum, millet and groundnut even less. Also livestock productivity is poor and constrained by high mortality rates. With about three quarters of the rural population below the poverty line and food self-sufficiency achieved for only three to ten months per year, rural households are extremely vulnerable to the adverse effects of climate change. Zimbabwe has been experiencing frequent and severe droughts and is among the world's most drought-prone countries. According to the Global Climate Risk Index 2021, Zimbabwe was among the top three most affected countries by extreme weather events. The agricultural sector contributes approximately 17% to national GDP and produces 60% of industrial raw material and 60–70% of the population derive their livelihoods from agriculture. Because of the predominantly rain-fed nature of agriculture, climate shocks strongly affect national food production and the national economy. For example, during the 2019/2020 period crop production was well below average and the country was faced with a national food deficit of about 40% while the economic performance went down by 9%.

The expected impacts of climate change on the current systems in both Ghana and Zimbabwe were negative for cereals (up to 20% of current yields) and positive for legumes (up to 10%). The cereal yield decline was mainly due to the effects of higher temperature on crop phenology, resulting in a shorter cycle and less time to take up resources for crop development. Legume crops also witnessed a shortening of the crop cycle but because they benefited from CO_2 fertilization, legume yields were not negatively impacted. The effects on crop yields were more pronounced on better-quality soils and for those farms that applied more nutrients. Livestock productivity was negatively affected in the drier climate change scenarios due to reduced on-farm crop residue production and rangeland productivity. Farms with larger livestock stocking densities were more sensitive to feed gaps and hence more strongly affected by a drying of the climate. The relatively better-off farms with larger herds faced greater losses in farm net returns compared to the poorer households who are impoverished already in the current system with poverty rates amounting to 95%. Although the poorest households are less sensitive to climate change, they are very vulnerable because they operate at extremely low levels of income and food self-sufficiency. Hence, in current systems, climate change does not represent the most acute problem, but poor-quality soils, low input access and use, and poor land, capital and labour assets.

In collaboration with a diverse set of key stakeholders and experts from provincial and national levels, contrasting representative agricultural pathways (RAPs) for 2050 and adaptation packages were co-developed (Valdivia et al., 2021). The pathways encompassed contrasting but internally coherent sets of policies and drivers, and socio-economic and biophysical conditions. Contrasts were sought between business as usual, sustainable development and rapid economic growth driven by fossil-fuel development. For both countries, the adaptation package that was co-designed with stakeholders comprised heat- and drought-tolerant crop varieties and improved livestock feeding practices.

The integrated quantitative analysis of future systems revealed that agricultural development, consisting of larger farm areas, more land dedicated to legume crops, intensified input use, and improved market access and functioning, led to improvements in agricultural productivity, economic returns and food self-sufficiency and dietary quality. The effects of climate change were more pronounced than in the current system, because current limitations, in particular with respect to crop nutrient stress, were alleviated. The adaptation package was effective in terms of offsetting some of these negative climate change impacts and increases income and food security of the farming households. All pathways left future farmers considerably better off than today, even under climate change. However, a clear differentiation between pathways emerged: a transition towards sustainable development based on crop diversification, soil rehabilitation, crop-livestock integration and better feed technologies was more profitable and more equitable than rapid economic growth. Stakeholders agreed that the drastic changes in the pathways would require the strong support of institutions and policies, which are currently lacking. One notable example of this comes from the Zimbabwean case where diversification towards crops that are better adapted to a changing climate than maize is currently hampered by national policies that are biased towards maize as a staple crop. Hence, increased support for legume, sorghum, millet, and livestock value chains is needed to counteract the current poor returns on investment which lead to underutilized agricultural land and limited use of mineral fertilizer, improved seed and feed technologies. In Ghana, maize also receives more attention from government policy support and international donors and research institutes, compared to other crops that could be more climate-resilient. While the Ghanaian agricultural systems are more diversified compared to Zimbabwe, there is still a bias towards cereals, even though legume crops may benefit from climate change because of CO_2 fertilization besides providing many other benefits. Subsidies on fertilizers and seeds are usually restricted to the cereals, thereby hampering the effective development of the value chains of the other crops which are increasingly becoming more relevant.

The analysis of future development pathways revealed that even though farming households would be better off under the transformed agricultural systems, not all households would be lifted out of poverty. Hence, for the poorest households in rural communities, interventions to build climate resilience need to be coupled with social protection mechanisms (Ulrichs et al., 2019). These can include cash transfers, farmer-managed storehouses and asset (e.g. boreholes) development, which support short-term coping mechanisms while at the same time building longer-term resilience. Both in Ghana and Zimbabwe, drought early warning and forecasting systems are being developed to act as a basis for fast action in case of weather-related risks hitting vulnerable populations.

In conclusion, the AgMIP-RIA transdisciplinary approach was geared towards cross-fertilization of stakeholder engagement processes and integrated modelling analysis, which created a space for co-learning by researchers and the users of research information. The process generated actionable, contextualized and evidence-based information for climate change adaptation in agricultural development pathways that take into account diversity across farming households.

7 Conclusions

With warmer temperatures, changes in rainfall patterns and more frequent extreme events climate change is being felt already across the African continent. These changes will become more pronounced in the coming decades, even if global warming could be limited to 1.5°C or 2°C. The impacts of climate change on African agri-food systems will be profound. With an agricultural sector that is dominated by smallholder farmers, who are extremely vulnerable to climate change, food production will be negatively impacted. Most of the available on-farm adaptation options will be insufficient to offset the negative impacts of climate change on crop and livestock productivity, and increasing land areas will become unsuitable for agriculture. Hence, the goal to increase agricultural productivity in order to obtain food security for the fast-growing population is becoming more challenging to achieve. As growing areas of land become not only unsuitable for agriculture but also more and more inhabitable, migration to urban areas or other countries will increase, and food imports are likely to play an important role in meeting growing food demands.

Overall, scientific evidence provides a clear warning that the climate change impacts expected for African food systems, in combination with non-climatic drivers and risks, are likely to result in chronic poverty and food insecurity for millions of Africans. Likewise, the causes of vulnerability and limited adaptive capacity of smallholders clearly indicate that drastic changes in the African agri-food systems are urgently required. Such transformation requires concerted action at different levels and in different sectors. In consultation with various stakeholders climate-resilient development pathways can be explored to identify policies and institutions to eliminate structural inequalities, build the financial and human capacity of farmers, enable the diversification of rural income-generating activities and improve climate information services. A role for science in this is to provide stakeholders with evidence-based, context-specific information about the likely effects of climate change and adaptation options and the trade-offs and co-benefits in other sustainability dimensions.

Given the challenges ahead, it is difficult, yet essential, to stay optimistic and grab the rapidly narrowing window of opportunity for concerted climate action that would enable the sustainable and climate-resilient transformation of African agri-food systems.

References

Antle, J.M., Homann-Kee Tui, S., Descheemaeker, K., Masikati, P., & Valdivia, R.O. (2018). Using AgMIP regional integrated assessment methods to evaluate vulnerability, resilience and adaptive capacity for climate smart agricultural systems. In *Climate smart agriculture: Building resilience to climate change. Natural resource management and policy* (vol. 52, pp. 307–333). Springer.

Birkmann, J., Liwenga, E., Pandey, R., Boyd, E., Djalante, R., Gemenne, F., … & Wrathall, D. (2022). Poverty, livelihoods and sustainable development. In H.-O. Pörtner, D.C. Roberts, M. Tignor, E.S. Poloczanska, K. Mintenbeck, A. Alegría, … & B. Rama (Eds.), *Climate change 2022: Impacts, adaptation and vulnerability. Contribution of working group II to the sixth assessment report of the intergovernmental panel on climate change* (pp. 1171–1274). Cambridge University Press. https://doi.org/10.1017/9781009325844.010

Campbell, B. (2022). *Climate change impacts and adaptation options in the agrifood system – A summary of recent IPCC sixth assessment report findings.* FAO. https://doi.org/10.4060/cc0425en

Carr, E.R., Goble, R., Rosko, H.M., Vaughan, C., & Hansen, J. (2020). Identifying climate information services users and their needs in sub-Saharan Africa: A review and learning agenda. *Climate and Development, 12*, 23–41. https://doi.org/10.1080/17565529.2019.1596061

Descheemaeker, K., Oosting, S.J., Homann-Kee-Tui, S., Masikati, P., Falconnier, G.N., & Giller, K.E. (2016). Climate change adaptation and mitigation in smallholder crop–livestock systems in sub-Saharan Africa: A call for integrated impact assessments. *Regional Environmental Change, 16*, 2331–2343.

Descheemaeker, K., Zijlstra, M., Masikati, P., Crespo, O., & Homann-Kee Tui, S. (2018). Effects of climate change and adaptation on the livestock component of mixed farming systems: A modelling study from semi-arid Zimbabwe. *Agricultural Systems, 159*, 282–295.

Freduah, B. S., MacCarthy, D. S., Adam, M., Ly, M., Ruane, A. C., Timpong-Jones, E. C., … & Adiku, S. G. K. (2019). Sensitivity of Maize yield in smallholder systems to climate scenarios in semi-arid regions of West Africa: Accounting for variability in farm management practices. *Agronomy, 9*(10), 639. https://doi.org/10.3390/agronomy9100639

Godde, C.M., Mason-D'Croz, D., Mayberry, D.E., Thornton, P.K., & Herrero, M. (2021). Impacts of climate change on the livestock food supply chain; a review of the evidence. *Global Food Security, 28*, 100488. https://doi.org/10.1016/j.gfs.2020.100488

Hansen, J., Hellin, J., Rosenstock, T., Fisher, E., Cairns, J., Stirling, C., … & Campbell, B. (2019). Climate risk management and rural poverty reduction. *Agricultural Systems, 172*, 28–46. https://doi.org/10.1016/j.agsy.2018.01.019

Homann-Kee Tui, S., Descheemaeker, K., Valdivia, R.O., Masikati, P., Sisito, G., Moyo, E.N., … & Rosenzweig, C. (2021). Climate change impacts and adaptation for dryland farming systems in Zimbabwe: A stakeholder-driven integrated multi-model assessment. *Climatic Change, 168*, 10.

Homann-Kee Tui, S., Valbuena, D., Patricia Masikati, P., Descheemaeker, K., Nyamangara, J., Claessens, L., … & Nkomboni, D. (2015). Economic trade-offs of biomass use in crop-livestock systems: Exploring more sustainable options in semi-arid Zimbabwe. *Agricultural Systems, 134*, 48–60.

IPCC. (2022). In H.-O. Pörtner, D.C. Roberts, M. Tignor, E.S. Poloczanska, K. Mintenbeck, A. Alegría, … & B. Rama (Eds.), *Climate change 2022: Impacts, adaptation and vulnerability. Contribution of working group II to the sixth assessment report of the intergovernmental panel on climate change* (3056 pp). Cambridge University Press. https://doi.org/10.1017/9781009325844

Jensen, N., & Barrett, C., 2017. Agricultural index insurance for development. *Applied Economic Perspectives and Policy, 39*, 199–219. https://doi.org/10.1093/aepp/ppw022

Jones, P.G., & Thornton, P.K. (2009). Croppers to livestock keepers: Livelihood transitions to 2050 in Africa due to climate change. *Environmental Science & Policy, 12*, 427–437.

Kjellstrom, T., Briggs, D., Freyberg, C., Lemke, B., Otto, M., & Hyatt, O. (2016). Heat, human performance, and occupational health: A key issue for the assessment of global climate change impacts. *Annual Review of Public Health, 37*, 97–112. https://doi.org/10.1146/annurev-publhealth-032315-021740

Rosenzweig, C., Elliott, J., Deryng, D., Ruane, A.C., Müller, C., Arneth, A., … & Khabarov, N. (2014). Assessing agricultural risks of climate change in the 21st century in a global gridded crop model intercomparison. *Proceedings of the National Academy of Sciences*, *111*, 3268–3273. https://doi.org/10.1073/pnas.1222463110

Schipper, E.L.F., Revi, A., Preston, B.L., Carr, E.R., Eriksen, S.H., Fernandez-Carril, L.R., … & Singh, P.K. (2022). Climate resilient development pathways. In H.-O. Pörtner, D.C. Roberts, M. Tignor, E.S. Poloczanska, K. Mintenbeck, A. Alegría, … & B. Rama (Eds.), *Climate change 2022: Impacts, adaptation and vulnerability. Contribution of working group II to the sixth assessment report of the intergovernmental panel on climate change* (pp. 2655–2807). Cambridge University Press. https://doi.org/10.1017/9781009325844.027

Thornton, P., Nelson, G., Mayberry, D., & Herrero, M. (2021). Increases in extreme heat stress in domesticated livestock species during the twenty-first century. *Global Change Biology*, *27*, 5762–5772. https://doi.org/10.1111/gcb.15825

Traoré, B., Descheemaeker, K., van Wijk, M.T., Corbeels, M., Supit, I., & Giller, K.E. (2017). Modelling cereal crops to assess future climate risk for family food self-sufficiency in southern Mali. *Field Crops Research*, *201*, 133–145.

Trisos, C.H., Adelekan, I.O., Totin, E., Ayanlade, A., Efitre, J., Gemeda, A., … & Zakieldeen, S. (2022). Africa. In H.-O. Pörtner, D.C. Roberts, M. Tignor, E.S. Poloczanska, K. Mintenbeck, A. Alegría, … & B. Rama (Eds.), *Climate change 2022: Impacts, adaptation and vulnerability. Contribution of working group II to the sixth assessment report of the intergovernmental panel on climate change* (pp. 1285–1455). Cambridge University Press. https://doi.org/10.1017/9781009325844.011

Ulrichs, M., Slater, R., & Costella, C. 2019. Building resilience to climate risks through social protection: From individualised models to systemic transformation. *Disasters*, *43*(Suppl 3), S368–S387. https://doi.org/10.1111/disa.12339

Valdivia, R., Homann-Kee Tui, S., Antle, J., Subash, N., Singh, H., Nedumaram, S., … & Dickson, C. (2021). Representative agricultural pathways: A multi-scale foresight process to support transformation and resilience of farming systems. In C. Rosenzweig, C.Z. Mutter, & E.M. Contreras (Eds.), *Handbook of climate change and agroecosystems: Climate change and farming system planning in Africa and South Asia: AgMIP stakeholder-driven research* (pp. 47–102). World Scientific Publishing.

Wichern, J., Descheemaeker, K., Giller K.E., Ebanyat, P., Taulya, G., & van Wijk, M.T. (2019). Vulnerability and adaptation options to climate change for rural livelihoods – A country-wide analysis for Uganda. *Agricultural Systems*, *176*, 102663.

Wichern, J., Hammond, J., van Wijk, M.T., Giller, K.E., & Descheemaeker, K. (2023). Production variability and adaptation strategies of Ugandan smallholders in the face of climate variability and market shocks. *Climate Risk Management*, *40*, 100490.

13 Reconciling food security and forest conservation in Africa: challenges and opportunities

Marieke Sassen

1 Introduction

1.1 The role of forest in food and nutr ition security

Forests, including woodlands,[1] play a key role in supporting food security in Africa and worldwide through various ecological, economic, and social mechanisms. They support globally vital ecosystem services, including climate regulation and ensuring water supply for food production (Sheil, this volume). More locally, forests support services that benefit food production such as water regulation, pollination, natural pest control, and soil fertility (Balima et al., 2020; Bele et al., 2015). Forests are both a source of land for crop and livestock production and a direct source of food and nutrition (Figure 13.1). They provide habitat for wild edible plants, animals, and fungi that are an essential part of food and nutrition security in forested areas of Africa (see comprehensive reviews in Arnold et al., 2011; Sunderland et al., 2013; Vinceti et al., 2013), including for urban populations (Bojang, 2014 and associated papers in the issue). Forest resources are especially important during "lean periods" in farming and for vulnerable groups during disasters such as droughts and war. Wild fruits, vegetables, and meat supply micronutrients not provided by staple foods in many rural communities (Rasolofoson et al., 2018). Moreover, non-timber forest products and timber provide sources of income that indirectly benefit food security, including for investing in future food production (i.e. purchase inputs) (e.g. Teketay, 2004). Forests and woodlands are an important source of fuelwood, on which most people in Africa depend for cooking, and in some cases, heating (FAO, 2017; Kebede et al., 2010).

Africa holds the largest tropical moist forest after the Amazon in the Congo Basin forests, with 180 million ha of primary forest (Mayaux & Malingreau, 2001) and extensive areas of wooded savanna (Searchinger et al., 2015). Yet, the need to provide food and energy for a growing and urbanising world population is leading to changes in land use with important consequences for remaining forested areas and the services they provide, especially in the tropics (Lambin & Meyfroidt, 2011; Laurance et al., 2014). Agricultural expansion is considered the main driver of forest loss across Africa (Pendrill et al., 2019), and between 2010 and 2020, it lost nearly 40 million ha of forest, the largest net loss globally (FAO, 2020).

1.2 Food production and forest loss in Africa

Historically, West and East Africa have been most impacted by deforestation (Aleman et al., 2018). After Europeans harvested most of the commercially valuable timber during the early 1900s, infrastructure development and the post-Second World War economic boom triggered a rapid expansion of commercial crops, such as cocoa, coffee, rubber, oil palm, and bananas

DOI: 10.4324/9781032649696-17

Figure 13.1 Forest-food linkages: forest for crop and grazing land, non-timber forest products (e.g. mushrooms, fish, medicine), water regulation, fuelwood, and tools (*photo credits*: Marieke Sassen and Meilinda Wan).

mostly grown by African (smallholder) farmers (de Haas & Travieso, 2022). As a result, most of the once-densely forested landscapes in West and East Africa are now dominated by forest-agriculture mosaics dotted with a few remnant protected forests (Aleman et al., 2018; Norris et al., 2010). Only Liberia and Sierra Leone retain larger areas of forest outside protected areas. In East Africa, forest loss has been especially severe in the densely populated highlands (Naughton-Treves et al., 2005; Ojoatre et al., 2023; Soini, 2006) that attracted early settlement due to fertile soils and bi-modal rainfall allowing two cropping seasons in a year (Giller & Andersson, this volume). These areas are now under pressure due to amongst others decreasing farm sizes, land degradation, and climate change (Amede & Lemenih, 2019).

Central African forests were extensively populated and cultivated before European colonisation (Van Gemerden et al., 2003). After independence, the discovery of oil, conflict, and poor maintenance of infrastructure shifted government attention away from agriculture (Lambin & Geist, 2003) and since then deforestation in this region has remained relatively low (Megevand & Mosnier, 2013). Yet, recent studies find that the pressure from agricultural expansion on central African forests is mounting (Ordway et al., 2017) (see also Section 2.2). The Democratic Republic of Congo (DRC) recently ranked second, after Brazil, for the highest net deforestation rate (FAO, 2020).

Forest loss in the wooded savannas of Africa, such as the West African Guinea Savanna and the Miombo woodlands of South-Central Africa, is less well studied and understood. The overall reported trend is one of forest and tree cover loss due to agriculture and woodfuel extraction (Buchadas et al., 2023; Kalema et al., 2015; Ouedraogo et al., 2011), though locally they also seem to show a high dynamism of gains and losses over time (Bodart et al., 2013; Mitchard &

Flintrop, 2013). In some areas active regreening efforts are increasing tree cover, such as in the West African Sahel, or through natural processes in the Central African savannas (Aleman et al., 2018).

The demand for food not only leads to outright forest clearance but it can also drive forest degradation. Though this is often harder to detect than outright deforestation, most tropical forests have been and are influenced by human activity, including for food (under-canopy crop planting, and wild foods and fuelwood collection) (Van Gemerden et al., 2003). For example, overexploitation of wild plant foods can lead to severe depletion of the species in the wild (e.g. *Gnetum africana* in Cameroon, a vine leaf vegetable (Ingram et al., 2012)). Overhunting for meat is leading to the so-called "empty forest syndrome", particularly in African moist forests (Benítez-López et al., 2019), with important consequences for local livelihoods (Nasi et al., 2011) and for forest dynamics (Effiom et al., 2013). Very little is known about the role and impacts of hunting on wildlife in drier forest areas (van Velden et al., 2018).

1.3 Questions for the future

Sub-Saharan Africa has made significant commitments to address food and nutrition insecurity through agricultural development (African Union, 2014). These commitments have important implications for the relationship between food production and the environment (Diallo & Wouterse, 2023). The formulated objectives are firmly based on the Borlaug hypothesis, where technological change supporting higher yields will reduce demand for land and help drive a so-called "Forest Transition". The Forest Transition theory posits that economic development, followed by increased wealth and rural out-migration, will lead to a concentration of agriculture on the most productive land and the abandonment of less productive areas followed by forest recovery on these lands (Mather, 1992). It is represented by a curve showing first rapid net deforestation followed by slow forest recovery. Yet, so far implementation and progress on food security objectives have varied across the continent (Ulimwengu et al., 2023). Indeed, to date, increases in food production in Africa are still achieved mainly through expansion rather than by yield increases (de Haas et al., this volume) and largely at the expense of forests or woodlands. Expected increases in demand for food and other agricultural products (Schut & van Ittersum, this volume) are therefore likely to drive further agricultural expansion.

In the context of an increasing global drive (including through the Global Biodiversity Framework, the Sustainable Development Goals, the European Union, and other country laws against imported deforestation but also private sector commitments) to address the impacts of agriculture on forests, the need to meet growing demands for food triggers important questions, such as:

- How does food production affect forests in different contexts? What is the influence of drivers such as population, markets, global commodity trade and climate change?
- Will increasing yields on land outside forests lead to less deforestation and even increased forest cover over time or will it drive deforestation? And at what scales?
- What production systems or agricultural landscapes provide the best outcomes for food production and ecosystem services provided by forests and trees?

Answers to these questions likely depend on starting situations and context. Sub-Saharan Africa encompasses an enormous array of environmental, climatic, social, demographic, economic, and political contexts. These interact with global forces to affect past and future impacts of agricultural development on forests and the role forests can play in supporting food and nutrition security.

To inform discussions, it can be helpful to consider different archetypes of forest vs. agriculture interactions in Africa, for example:

- Densely populated areas with high historical deforestation, often moist tropical forest areas with high soil fertility such as the Afromontane forests (most notably the East African highlands of Ethiopia, Uganda, Kenya, Tanzania, Rwanda, Burundi, and DRC)
- Intermediate to high population density areas where most forests have already been converted to agriculture such as the cocoa belt in the Guinean forest area of West Africa (Guinea to Cameroon, via Cote d'Ivoire and Ghana)
- Sparsely populated, but with high population growth rates, dense Guineo-Congolian forests of the Congo Basin (Southern Cameroon, Gabon, Republic of Congo, DRC, Central African Republic, and Equatorial Guinea)
- Sparsely populated (densely) wooded savannas such as the West African Guinea Savanna woodlands (Guinea to Cameroon), and the Miombo woodlands across South-Central Africa (Angola to Tanzania including Southern DRC)

This chapter seeks to give an overview of relevant evidence and scientific debates to stimulate further discussion and research on the (future) interactions between forests and food. It is not a comprehensive review on the questions raised above. It first addresses the debate on the underlying drivers of deforestation associated with meeting demands for food. It then explores different avenues proposed for increasing food production in Africa – from (expanding) high-yielding agriculture outside forests to combining food production with forest values. Finally, I consider the implications for a potential forest transition in Africa.

2 Drivers of deforestation in Africa: the role of food production

2.1 Many poor people needing food

In Africa, poor and land-hungry smallholder farmers are considered the main threat to forests (Carr et al., 2005; Norris et al., 2010; Shapiro et al., 2023). At more local levels though, the picture is often more nuanced.

In densely populated areas, especially highly fertile ones such as some East African highlands, exacerbated competition over land and resources for food leading to agricultural encroachment into neighbouring (protected) forests certainly continues to be an important source of forest loss and degradation (Lung & Schaab, 2010; Uusivuori et al., 2002). Yet, high labour availability in combination with favourable agroecological conditions and effective forest protection (e.g. coffee areas of Kilimanjaro and Mt Elgon), can also help reduce pressure on forests as people invest on existing land, following the theory of Boserup (1965) (Meyfroidt et al., 2014; Sassen et al., 2013). Conversely, in areas with fewer people, one might expect lower pressure on forest. Yet in the Guineo-Congolian forest for example, low densities of smallholder farmers are the main agents of deforestation. In such areas though, farmers often expand into forests because they lack the resources or incentives to increase productivity and they prefer to save on labour and inputs than on land (Angelsen & Kaimowitz, 2001; Chamberlin et al., 2014).

Poor people who depend on forests for food and other resources are considered a major driver of forest loss and degradation in Africa (VanWey et al., 2005). Generally, poor people are drawn towards forest areas not because they provide attractive economic opportunities (e.g. near protected areas), as is sometimes suggested, but because they provide a safety net and because of a lack of employment opportunities elsewhere (Giller et al., 2013; Naughton-Treves et al., 2011). Forest edges tend to be the least accessible areas and the first in line when it comes to crop predation and the risk of eviction by authorities (from protected forests). According to the forest transition theory, wealth and economic development in general, including urbanisation, should create opportunities for alternative livelihoods outside agriculture, e.g. through trade, and help alleviate pressure on forests from agriculture. Yet, in many forested places, wealthier people have been shown to be more able and motivated to expand agriculture into forests or unsustainably extract resources (see chapters (on Africa) in Angelsen & Kaimowitz, 2001) (but see also Section 2.2). Moreover, urban areas are major markets for fuel, bushmeat, and other forest products (DeFries et al., 2010; FAO, 2017).

2.2 Markets, prices, and the growing influence of global trade

Access to markets and crop prices are seen as a major underlying driver of forest loss or degradation, often mediating the effect of population. Changes in access to market and prices can be reflected relatively rapidly in spatial patterns of land use (VanWey et al., 2005), though price fluctuations for seasonal crops are more often and more rapidly associated with deforestation than those for perennials such as coffee and cocoa (see cases in Angelsen et al., 1999; Soini, 2006). High prices and access to markets can drive agricultural expansion into forests, but they can also support investment in existing land or away from forests, if access to forest land is restricted (Sassen et al., 2013; Williams et al., 2021). However, when access to markets is good and prices are high, farmers tend to concentrate on a few dominant crops, leading to less diverse systems and potentially less resilience to shocks, especially for smallholder famers (see also VanWey et al., 2005). Market and price shocks can have dramatic consequences for nature in agricultural landscapes. For example in the mid-Zambezi valley where, after independence, a relatively reliable income from cotton served to restrict expansion of the area under agriculture, even under rapid population growth. Then the cotton sector collapsed and the past 15 years have seen a shift to more extensive cereal production and grazing with a major loss of natural habitat (Baudron et al., 2022). These examples illustrate the interplay between remunerative agriculture and land expansion: increased economic potential of farming can both slow and drive agricultural expansion – what has come to be known as Jevons' Paradox (York & McGee, 2016).

Whilst small-scale farming to meet domestic demands currently drives most agricultural expansion and deforestation in Africa, global markets increasingly influence local forest-agriculture interactions (Pendrill et al., 2022). In the past decade, the main agricultural commodities associated with deforestation in Africa have been cocoa, palm oil, and rubber in West Africa, including Cameroon (Ordway et al., 2017), and cashew and cotton in the savannas (Baudron et al., 2022; Masolele et al., 2024). For many of the globally traded commodities, there is little suitable land available outside remaining forests in most West and Central African countries (Ordway et al., 2017).

International and regional trade is not only affecting agricultural commodities and their impacts on forests: wild plants and meat from West and Central Africa are now also part of a global trade. At low (i.e. subsistence) levels of extraction, wild food collection can be sustainable, but commercial exploitation often leads to over-exploitation in the absence of effective regulation. In the case of wild meat, logging and mining, which increased access to previously remote forest areas and their wildlife, have exacerbated hunting pressure to often unsustainable levels (Nasi et al., 2011).

The studies referenced in Sections 2.1 and 2.2, and many others, illustrate how population density and poverty as the main drivers of deforestation in Africa cannot be generalised over time and at more local levels. Many other factors such as local agroecological conditions, land use history, forest governance, the availability of alternative livelihoods, and local markets and prices interact for different outcomes (Barraclough & Ghimire, 1996; Sassen et al., 2013). This interaction is then further affected by national and global economic contexts and forces (economic policies, macro-economics, trends in other sectors, etc.) to drive different outcomes for food production and forests at different scales. There is a rich literature on these topics, which cannot be treated here (e.g. Gbetnkom, 2009; Geist & Lambin, 2002; Leach, 2008; Meyfroidt et al., 2014).

3 Food security and maintaining forests and their services

The previous section illustrates how generalisations about population, poverty, and the growing influence of markets are not helpful to address negative interactions, and that local contexts influence potential challenges and opportunities for better outcomes. This section considers where food production might expand outside forests and the potential for combining forest and agriculture. Finally it considers the implications of different strategies to achieve a forest transition where both objectives are met.

3.1 Agricultural land in Africa: room for expansion outside forests?

Africa is widely considered to hold the largest area of so-called "unused" or "underutilised" lands that could potentially be brought into agricultural use or where agriculture could be made more productive. Estimates vary from 200 to 700 million hectares even when based on the same data sources (Alexandratos & Bruinsma, 2012; Morris et al., 2009). Different studies use different criteria for existing agricultural use, land productivity, access to markets, and availability of land for agriculture (e.g. excluding built-up areas, excluding or including protected areas and/or forests). According to Chamberlin et al. (2014), up to two-thirds of the region's surplus land lies under forests (including woodlands) and is found mainly in central, southern, and eastern Africa.

In Central Africa, governments have ambitious objectives for increasing agricultural production (Kamath et al., 2024 and references therein), based on the premise of large pools of land with potential for (profitable) expansion of food production (Chamberlin et al., 2014). Yet (foreign) investment into large-scale agriculture has so far been limited in most countries except Cameroon and, to some extent, in Gabon (Ordway et al., 2017). Africa's wetter and wooded savannas were the subject of much international interest especially around 2009–2015, as they are seen to have high potential for agricultural development with lower environmental costs compared with the moist tropical forests. They have also been referred to as "Africa's sleeping giant" (Morris et al., 2009). To support agricultural development in these savanna areas, the so-called "development corridors" are used as a strategy, seeking to concentrate public and private investment in designated (large) areas to drive economic growth. Some corridors explicitly focus on agricultural development, such as the Southern Agricultural Growth Corridor of Tanzania (SAGCOT). Yet, to date, policies and interventions to boost agricultural production and productivity in Africa have not had the impacts hoped for due to a lack of investment and other political and socio-economic factors (Ollenburger et al., 2016; Ulimwengu et al., 2023).

Overall, employment opportunities outside agriculture tend to be limited and dramatic increases in productivity are unlikely in the short to medium term. As a result, relatively low-yielding smallholder food production is likely to continue to drive most of the forest change in the region for the foreseeable future, especially in the DRC, which has the highest population growth rates in the region (van Wissen, this volume). Economic returns from agriculture are strongly limited by access to markets, especially for smallholders. The more easily accessible edges of moist forests and the wooded savannas will therefore likely be cleared first, in some places driven by mining activities, though this is poorly documented (Chamberlin et al., 2014).

More generally, much of the accessible land combining good soil fertility and sufficient rainfall in Africa is already farmed, and further expansion would likely extend into land now considered marginal for agriculture or protected forests. Lands that are marginal for (rainfed) agriculture are often remote and have sparse human population (Andersson et al., 2017). They tend to harbour significant biodiversity and carbon stocks, providing global environmental benefits (Searchinger et al., 2015; Oosting et al., this volume; van Dijk & Djindil, this volume), and supporting the livelihood strategies of the poor, such as extensive or pastoralist livestock grazing, hunting, and gathering. Converting such lands to agriculture may also affect local and regional water systems and soils, and lead to increased greenhouse gas (GHG) emissions (Wilson & Scholes, 2020; Oosting et al., this volume). Increasing food production in the so-called underutilised lands of Africa may therefore have high economic, social, and environmental costs (e.g. for biodiversity as illustrated in Figure 13.2) (van Soesbergen et al., 2017; WWF, 2015).

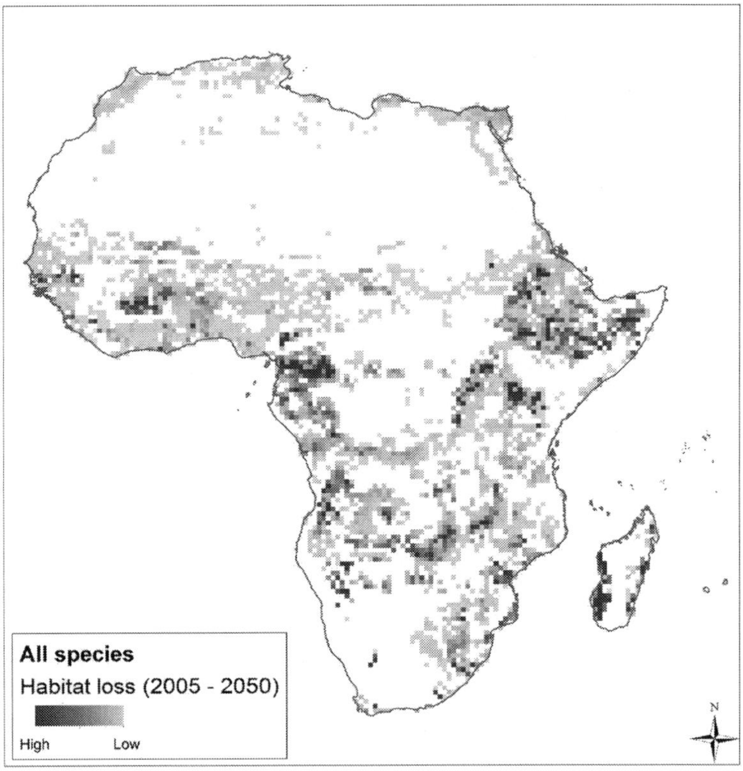

Figure 13.2 Projected habitat loss for all species as a result of projected changes in land use between 2005 and 2050 in Africa under a "strong regional integration but reactive governance" scenario with climate change as a given, which is considered a plausible scenario for the region.

Source: UN Environment-World Conservation Monitoring Centre in collaboration with the Climate Change Agriculture and Food Security research programme of the Consultative Group for International Agricultural Research (CGIAR Research Program on Climate Change, Agriculture and Food Security) (Vervoort et al., 2013) and Kassel University (Schaldach et al., 2011) as input to World Wildlife Foundation (WWF, 2015).

It is clear that the planning of future agricultural development in Africa will need to carefully balance trade-offs with maintaining or restoring forests, including in a context if increasing global commitments to reduce the role of food production in forest loss.

3.2 Trees on farm: combining forest and agriculture?

Trees on farms, such as in agroforestry systems, can help maintain or restore the environment for agriculture (and in some cases can be seen as a remnant of shifting cultivation practices where regenerating forest vegetation helped restore soil fertility) (Niether et al., 2020). Trees on farms can also provide numerous products such as food, fuel, construction materials, fodder, and mulch and support risk management (e.g. trees as a source of income in case of crop failure). Agroforestry systems that integrate nutrient-dense crops such as fruits, nuts, and vegetables can support objectives to improve or maintain healthy diets though this requires greater attention from investments and innovation (see also Melse-Boonstra, this volume). More diverse systems are also more likely to provide higher resilience to shocks in terms of food security and reduce reliance on neighbouring forests, especially in smallholder contexts (see Giller & Andersson, this volume). Depending on their level of complexity, such systems can be highly intensive in terms of labour and output per hectare but have relatively low dependence on external outputs (Vaast & Somarriba, 2014). High diversity and density of associated trees can lead to both trade-offs and benefits with regard to pests and diseases (Niether et al., 2020).

More than 40% of existing agricultural land globally has a least 10% tree cover (Zomer et al., 2014), though local patterns of trees on farms vary with agroecological, economic, and other contexts. In many places, access to traditional tree-based resources has decreased over time because of exclusion of local uses from protected areas or because of forest loss or degradation. There people often started to protect, plant, and manage the resources they found important on their own land (Rudel et al., 2005). Expanding markets for tree products (fuelwood and others) have accelerated this. In many places this meant that the density of planted trees has increased over time (Arnold & Dewees, 1997; Fairhead & Leach, 1996; Tiifen et al., 1993) whilst natural forests were lost. There is also evidence that smaller farms further away from forests include relatively more trees, as people seek to diversify (Sassen et al., 2015). In Kenya it was found that older households keep more trees as woodlots, because they need less labour than crops, but younger households tend to replace trees with cash crops as these are more lucrative. Having trees on land rather than crops can also free up labour so that people can work off-farm and still get a return from the land when desired (Arnold & Dewees, 1997). Trees, and especially highly valuable timbers, are a form of capital accumulation that can be harvested when cash is needed, including for food. In other places, such as the cocoa-growing areas of West Africa, farmers have historically been incentivised to reduce associated trees from their agroforestry systems in a drive to intensify production. There both forests and trees on agricultural land were lost.

In recent years, interest in agroforestry has regained momentum globally. This includes the interest to better understand the ecological, agronomic, socio-economic, and institutional factors that affect their productive performance and adoption by farmers. The conditions and incentives that encourage farmers to integrate more trees into their land-use system are often poorly understood, context-specific, and challenging the success of agroforestry programmes. Keeping trees on agricultural land makes sense when land and capital are available, when the trees provide higher (perceived) benefits than alternative uses of land and capital, and when land and tree tenure is secure. Then, people also have non-economic (e.g. cultural) reasons for keeping (certain) trees on their land. Further detail goes beyond the scope of this chapter, but for further reading we refer to Minang et al. (2021), who cover almost all aspects underpinning the potential for different tree crops to support sustainable development in Africa.

3.3 Land sharing or sparing for food and forests in Africa?

To increase food production whilst decreasing its impact on forest and other ecologically important areas, agricultural policies, nature organisations, and conservation scientists have traditionally advocated for a strict segregation between farming and nature, with yield maximisation as a major strategy to spare land for nature. More recently, the global agroecological movement and nature organisations tend to promote low-input, extensive "biodiversity friendly" agriculture. Yet, by needing larger areas to produce the same amount of food, such strategies can lead to further loss of forests and exacerbate rather than mitigate conflicts between farming and forest conservation (Baudron et al., 2021). "Intermediate" strategies include the targeting of already degraded forest areas for agricultural expansion, or increasing tree cover within agricultural landscapes, but outside agricultural fields.

The advantages and disadvantages of land sparing or sharing approaches in forested landscapes vary across biophysical and socio-economic contexts, such as the biophysical potential and the costs of production for maximising yields vs. biodiversity-friendly agriculture, the extent and type of forest present, the ability of local biodiversity to adapt to human disturbance, the resilience of agricultural systems to shocks, and the dependence on forests of local livelihoods (see also Grau et al., 2013). Land sparing, land sharing, and other strategies have important consequences for the ecological base on which food production depends. They may lead to both synergies and trade-offs with food and nutrition security objectives at different scales and over time.

In West Africa, where most forests have already largely been lost, remnants should be protected from conversion whilst efforts should be made to increase yields on existing agricultural land (land sparing). To reduce pressure on forests from the need for fuelwood and other tree products, tree cover and diversity in the surrounding agricultural landscapes should be increased where possible, for example, by incentivising agroforestry systems where appropriate (land sharing) (Vaast & Somarriba, 2014). From an ecological perspective it would make sense to focus such efforts in agricultural areas near forest remnants to support habitat connectivity, whilst more input-intensive, simpler food (and wood) production systems would be located further away from forests. Food production expansion in areas with large expanses of remaining forest should manage trade-offs with biodiversity and ecosystem services that support food, nutrition, and other objectives (Sassen et al., 2022). There both land sparing and sharing approaches can play a role, depending on local context and objectives (see also Section 3.1).

Land sparing is often seen as a strategy to support the forest transition, but as we have seen in previous sections and other chapters, maximising yields (i.e. "closing yield gaps") is unrealistic on the vast majority of land and for most farmers in Africa in the short to medium term. In most places, they will continue to expand into accessible forest areas to meet their needs for food and other resources. In areas of low population density, a stable small-scale crop-fallow system where, after one or more seasons, cleared land is left to regenerate to forest and a new patch of forest is cleared can likely be sustainable. This is however at risk of leading to forest degradation as population density increases (cf. DRC). In places where highly profitable food production is possible, this runs the risk of driving expansion into neighbouring forests if forest protection is not effective. What can be seen in different places though is that, in areas of land scarcity, shrinking forests and loss of access to forest resources lead to an increase in tree planting on agricultural land (see Section 3.2).

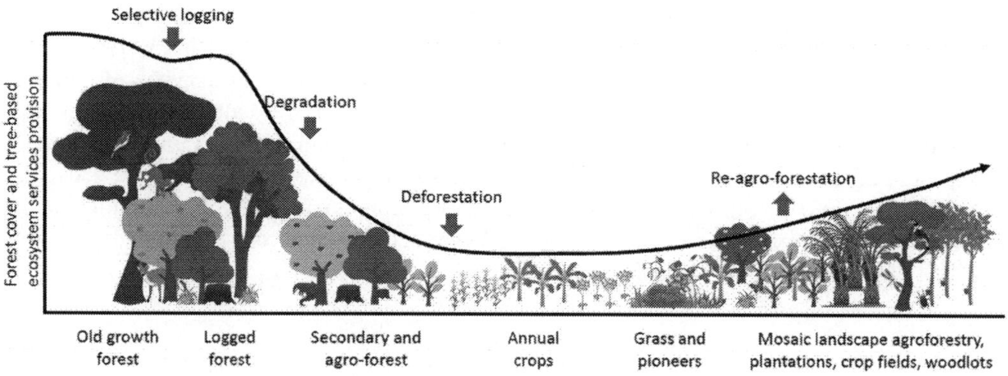

Figure 13.3 The forest transition curve: an example for moist forest areas in Africa.

Positioning countries or regions along the forest transition curve gives an indication of potential future trajectories for forest cover in relation to agriculture development, and potential interventions. For example, the moist forests of Central Africa and the Miombo woodlands are situated on the left of the curve, with strong risk for forest cover loss (see Section 3.1). Avoiding this may require providing the region with alternative sources of income from the crucial ecosystem services these forests provide to the world. The East African highlands and West African Guinean moist forests in countries like Ghana and Cote d'Ivoire are in the centre part of the transition curve, with remaining forests as protected "islands" in a sea of agriculture. An increase in tree planting in agricultural areas combined with forest restoration programmes may drive these areas towards a mosaic landscape of agroforestry, crop fields, tree plantations, and protected forests. Such "recovered" (agro-)forests will not provide the same ecosystem services as the original forests (Figure 13.3), but may be the most realistic end of the transition for much of Africa at least in the medium term (Figure 13.2).

Finally, climate change is impacting agri-food systems in Africa too (Descheemaeker et al., this volume) and will affect existing demographic and socio-economic forces. For crops and farms that cannot adapt to changing climatic conditions, these will either be abandoned or be shifted towards areas that are more suitable. These areas are highly likely to be forested. In the highlands, for example, this may mean a shift uphill to higher elevations, threatening the last remaining islands of tropical montane forests in Africa. In areas that become drier, crops may shift to wetter places, such as cocoa shifting from West to Central Africa. In other places, tree-based systems may be replaced by more drought-tolerant annual crops such as maize or sorghum, reducing tree cover. These effects will be combined with those of climate change on forests (e.g. Réjou-Méchain et al., 2021).

4 Conclusions

Forests play a key role in supporting food security in Africa. Yet, they are under increasing pressure from degradation and loss. Historical deforestation and future deforestation risks associated with the need to meet demands for food and nutrition vary across the region. The Congo Basin is an area of major risk whilst it also plays a crucial role for the global climate. Opportunities to achieve better outcomes for forests while meeting these demands vary across the continent and require different strategies. Depending on agroclimatic contexts, and position along the forest transition curve, forests in Africa may benefit most from a combination of land sparing and sharing approaches. These include improving the protection of remaining forests in the face of agricultural expansion and maintaining or increasing tree cover in existing agricultural landscapes whilst investing in increased productivity informed by local agroecological and socio-economic contexts. In some places the need to meet (growing) food demands will require balancing trade-offs between agricultural expansion and maintaining forest areas at local, national, and global scales. This chapter illustrates the strong and complex relationship between forests and food and nutrition security in Africa, and the context specificity of the challenges and opportunities to reconcile these two objectives, for which there is no "one-size-fits-all" solution.

Note

1 Forest is defined as land spanning more than 0.5 hectares with trees higher than 5 metres and a canopy cover of more than 10%, or trees able to reach these thresholds in situ, excluding land that is predominantly under agricultural or urban land use (Food and Agriculture Organization, FAO).

References

African Union. (2014). Malabo declaration on accelerated agricultural growth and transformation for shared prosperity and improved livelihoods doc. Assembly/Au/2 (Xxiii).

Aleman, J.C., Jarzyna, M.A., & Staver, A.C. (2018). Forest extent and deforestation in tropical Africa since 1900. *Nature Ecology & Evolution*, *2*(1), 26–33.

Alexandratos, N., & Bruinsma, J. (2012). *World agriculture towards 2030/2050: The 2012 revision*. ESA working paper No. 12–03. Food and Agriculture Organization of the United Nations. Rome, Italy.

Amede, T., & Lemenih, M. (2019). The highland mixed farming system of Africa: Diversifying livelihoods in fragile ecosystems. In J. Dixon, D.P. Garrity, J.- M. Boffa, T.O. Williams, T. Amede, C. Auricht, R. Lott, & G. Mburathi (Eds.), *Farming systems and food security in Africa. Priorities for science and policy under climate change* (pp. 248–281). Routledge.

Andersson, J., de Garine-Wichatitsky, M., Cumming, D., Dzingirai, V., & Giller, K. E. (2017). *Transfrontier conservation areas: People living on the edge*. Taylor & Francis.

Angelsen, A., & Kaimowitz, D. (2001). *Agricultural technologies and tropical deforestation*. CABi.

Angelsen, A., Katemansimba Shitindi, E.F., & Aarrestad, J. (1999). Why do farmers expand their land into forests? Theories and evidence from Tanzania. *Environment and Development Economics*, *4*(3), 313–331.

Arnold, J.E.M., & Dewees, P.A. (1997). *Farms, trees and farmers: Responses to agricultural intensification*. Earthscan.

Arnold, M., Powell, B., Shanley, P., & Sunderland, T.C. (2011). Forests, biodiversity and food security. *International Forestry Review*, *13*(3), 259–264.

Balima, L.H., Nacoulma, B.M.I., Bayen, P., Kouamé, F.N.G., & Thiombiano, A. (2020). Agricultural land use reduces plant biodiversity and carbon storage in tropical West African savanna ecosystems: Implications for sustainability. *Global Ecology and Conservation*, *21*, e00875.

Barraclough, S.L., & Ghimire, K.B. (1996). Deforestation in Tanzania: Beyond simplistic generalizations. *Ecologist*, *26*(3), 104–109.

Baudron, F., Govaerts, B., Verhulst, N., McDonald, A., & Gérard, B. (2021). Sparing or sharing land? Views from agricultural scientists. *Biological Conservation*, *259*, 109167.

Baudron, F., Guerrini, L., Chimimba, E., Chimusimbe, E., & Giller, K.E. (2022). Commodity crops in biodiversity-rich production landscapes: Friends or foes? The example of cotton in the Mid Zambezi Valley, Zimbabwe. *Biological Conservation*, *267*, 109496.

Bele, M.Y., Sonwa, D.J., & Tiani, A.-M. (2015). Adapting the Congo Basin forests management to climate change: Linkages among biodiversity, forest loss, and human well-being. *Forest Policy and Economics*, *50*, 1–10.

Benítez-López, A., Santini, L., Schipper, A.M., Busana, M., & Huijbregts, M.A. (2019). Intact but empty forests? Patterns of hunting-induced mammal defaunation in the tropics. *PLoS Biology*, *17*(5), e3000247.

Bodart, C., Brink, A.B., Donnay, F., Lupi, A., Mayaux, P., & Achard, F. (2013). Continental estimates of forest cover and forest cover changes in the dry ecosystems of Africa between 1990 and 2000. *Journal of Biogeography*, *40*(6), 1036–1047.

Bojang, F. (2014). Sustainable natural resources management in Africa's urban food and nutrition equation. *Nature & Faune*, *28*(2), 1–77.

Boserup, E. (1965). *The conditions of agricultural growth: The economics of agrarian change under population pressure*. Aldine Publishing Company.

Buchadas, A., Jung, M., Bustamante, M., Fernández-Llamazares, Á., Garnett, S.T., Nanni, A.S., Ribeiro, N., Meyfroidt, P., & Kuemmerle, T. (2023). Tropical dry woodland loss occurs disproportionately in areas of highest conservation value. *Global Change Biology*, *29*(17), 4880–4897.

Carr, D.L., Suter, L., & Barbier, A. (2005). Population dynamics and tropical deforestation: State of the debate and conceptual challenges. *Population and Environment*, *27*(1), 89–113.

Chamberlin, J., Jayne, T.S., & Headey, D. (2014). Scarcity amidst abundance? Reassessing the potential for cropland expansion in Africa. *Food Policy*, *48*, 51–65.

de Haas, M., & Travieso, E. (2022). Cash-crop migration systems in East and West Africa: Rise, endurance, decline. In M. de Haas & E. Frankema (Eds.), *Migration in Africa* (pp. 231–255). Routledge.

DeFries, R.S., Rudel, T., Uriarte, M., & Hansen, M. (2010). Deforestation driven by urban population growth and agricultural trade in the twenty-first century. *Nature Geoscience*, *3*(3), 178–181.

Diallo, M., & Wouterse, F. (2023). Agricultural development promises more growth and less poverty in Africa: Modelling the potential impact of implementing the Comprehensive Africa Agriculture Development Programme in six countries. *Development Policy Review*, *41*(3), e12669.

Effiom, E.O., Nuñez-Iturri, G., Smith, H.G., Ottosson, U., & Olsson, O. (2013). Bushmeat hunting changes regeneration of African rainforests. *Proceedings of the Royal Society B: Biological Sciences*, *280*(1759), 20130246.

Fairhead, J., & Leach, M. (1996). *Misreading the African landscape: Society and ecology in a forest-savannah mosaic*. Cambridge University Press.

FAO. (2017). *Incentivizing sustainable wood energy in sub-Saharan Africa: A way forward for policy*. Food and Agriculture Organization of the United Nations. Rome, Italy.

FAO. (2020). *Global forest resources assessment, 2020: Main report*. Food and Agriculture Organization of the United Nations. Rome, Italy. https://doi.org/https://doi.org/10.4060/ca9825en

Gbetnkom, D. (2009). Forest depletion and food security of poor rural populations in Africa: Evidence from Cameroon. *Journal of African Economies*, *18*(2), 261–286.

Geist, H.J., & Lambin, E.F. (2002). Proximate causes and underlying driving forces of tropical deforestation. *Bioscience*, *52*(2), 143–150.

Giller, K.E., Baudron, F., Matema, S., Milgroom, J., Chrispen, M., Guerbois, C., & Twine, W. (2013). Population and livelihoods on the edge. In J. Andersson, M. de Garine-Wichatitsky, D. Cumming, V. Dzingirai, & K. E. Giller (Eds.), *Transfrontier conservation areas* (pp. 62–84). Routledge.

Grau, R., Kuemmerle, T., & Macchi, L. (2013). Beyond 'land sparing versus land sharing': Environmental heterogeneity, globalization and the balance between agricultural production and nature conservation. *Current Opinion in Environmental Sustainability*, *5*(5), 477–483.

Ingram, V., Ndumbe, L.N., & Ewane, M.E. (2012). Small scale, high value: Gnetum africanum and buchholzianum value chains in Cameroon. *Small-Scale Forestry*, *11*(4), 539–556.

Kalema, V.N., Witkowski, E.T., Erasmus, B.F., & Mwavu, E.N. (2015). The impacts of changes in land use on woodlands in an equatorial African savanna. *Land Degradation & Development*, *26*(7), 632–641.

Kamath, V., Sassen, M., Arnell, A., van Soesbergen, A., & Bunn, C. (2024). Identifying areas where biodiversity is at risk from potential cocoa expansion in the Congo Basin. *Agriculture, Ecosystems & Environment*, *376*, 109216.

Kebede, E., Kagochi, J., & Jolly, C.M. (2010). Energy consumption and economic development in Sub-Sahara Africa. *Energy Economics*, *32*(3), 532–537.

Lambin, E.F., & Geist, H.J. (2003). Regional differences in tropical deforestation. *Environment: Science and Policy for Sustainable Development, 45*(6), 22–36.

Lambin, E.F., & Meyfroidt, P. (2011). Global land use change, economic globalization, and the looming land scarcity. *Proceedings of the National Academy of Sciences, 108*(9), 3465–3472.

Laurance, W.F., Sayer, J., & Cassman, K.G. (2014). Agricultural expansion and its impacts on tropical nature. *Trends in Ecology & Evolution, 29*(2), 107–116.

Leach, M. (2008). Pathways to Sustainability in the forest? Misunderstood dynamics and the negotiation of knowledge, power, and policy. *Environment and Planning A, 40*(8), 1783–1795.

Lung, T., & Schaab, G. (2010). A comparative assessment of land cover dynamics of three protected forest areas in tropical eastern Africa. *Environmental Monitoring and Assessment, 161*(1), 531–548.

Masolele, R.N., Marcos, D., De Sy, V., Abu, I.-O., Verbesselt, J., Reiche, J., & Herold, M. (2024). Mapping the diversity of land uses following deforestation across Africa. *Scientific Reports, 14*(1), 1681.

Mather, A.S. (1992). The forest transition. *Area, 24*, 367–379.

Mayaux, P., & Malingreau, J.P. (2001). Le couvert forestier d'Afrique centrale: Un nouvel état des lieux. *Bulletin des Séances académiques de la Société Royale d'Outre-Mer*, 46(2000-4), 475–486.

Megevand, C., & Mosnier, A. (2013). *Deforestation trends in the Congo Basin: Reconciling economic growth and forest protection*. World Bank Publications.

Meyfroidt, P., Carlson, K.M., Fagan, M.E., Gutiérrez-Vélez, V.H., Macedo, M.N., Curran, L.M., … & Lambin, E.F. (2014). Multiple pathways of commodity crop expansion in tropical forest landscapes. *Environmental Research Letters, 9*(7), 074012.

Minang, P.A., Duguma, L.A., & van Noordwijk, M. (2021). *Tree Commodities and Resilient Green Economies in Africa*. World Agroforestry. Available at https://www.cifor-icraf.org/gtci/publication/

Mitchard, E.T., & Flintrop, C.M. (2013). Woody encroachment and forest degradation in sub-Saharan Africa's woodlands and savannas 1982–2006. *Philosophical Transactions of the Royal Society B: Biological Sciences, 368*(1625), 20120406.

Morris, M.L., Binswanger-Mikhize, H.P., & Byerlee, D. (2009). *Awakening Africa's sleeping giant: Prospects for commercial agriculture in the Guinea Savannah Zone and beyond*. World Bank Publications.

Nasi, R., Taber, A., & Van Vliet, N. (2011). Empty forests, empty stomachs? Bushmeat and livelihoods in the Congo and Amazon Basins. *International Forestry Review, 13*(3), 355–368.

Naughton-Treves, L., Alix-Garcia, J., & Chapman, C.A. (2011). Lessons about parks and poverty from a decade of forest loss and economic growth around Kibale National Park, Uganda. *Proceedings of the National Academy of Sciences of the United States of America, 108*(34), 13919–13924.

Naughton-Treves, L., Holland, M.B., & Brandon, K. (2005). The role of protected areas in conserving biodiversity and sustaining local livelihoods. *Annual Review of Environment and Resources, 30*, 219–252.

Niether, W., Jacobi, J., Blaser, W.J., Andres, C., & Armengot, L. (2020). Cocoa agroforestry systems versus monocultures: A multi-dimensional meta-analysis. *Environmental Research Letters, 15*(10), 104085.

Norris, K., Asase, A., Collen, A., Gockowski, J., Mason, J., Phalan, B., & Wade, A. (2010). Biodiversity in a forest-agriculture mosaic: The changing face of West African rainforests. *Biological Conservation, 143*(10), 2341–2350.

Ojoatre, S., Zhang, C., Yesuf, G., & Rufino, M.C. (2023). Mapping deforestation and recovery of tropical montane forests of East Africa. *Frontiers in Environmental Science, 11*, 1084764.

Ollenburger, M.H., Descheemaeker, K., Crane, T.A., Sanogo, O.M., & Giller, K.E. (2016). Waking the sleeping giant: Agricultural intensification, extensification or stagnation in Mali's Guinea Savannah. *Agricultural Systems, 148*, 58–70.

Ordway, E.M., Asner, G.P., & Lambin, E.F. (2017). Deforestation risk due to commodity crop expansion in sub-Saharan Africa. *Environmental Research Letters, 12*(4), 044015.

Ouedraogo, I., Savadogo, P., Tigabu, M., Cole, R., Oden, P.C., & Ouadba, J.-M. (2011). Trajectory analysis of forest cover change in the tropical dry forest of Burkina Faso, West Africa. *Landscape Research, 36*(3), 303–320.

Pendrill, F., Gardner, T.A., Meyfroidt, P., Persson, U.M., Adams, J., Azevedo, T., … & De Sy, V. (2022). Disentangling the numbers behind agriculture-driven tropical deforestation. *Science, 377*(6611), eabm9267.

Pendrill, F., Persson, U.M., Godar, J., Kastner, T., Moran, D., Schmidt, S., & Wood, R. (2019). Agricultural and forestry trade drives large share of tropical deforestation emissions. *Global Environmental Change, 56*, 1–10.

Rasolofoson, R.A., Hanauer, M.M., Pappinen, A., Fisher, B., & Ricketts, T.H. (2018). Impacts of forests on children's diet in rural areas across 27 developing countries. Science Advances, *4*(8), eaat2853.

Réjou-Méchain, M., Mortier, F., Bastin, J.-F., Cornu, G., Barbier, N., Bayol, N., … & Deblauwe, V. (2021). Unveiling African rainforest composition and vulnerability to global change. *Nature, 593*(7857), 90–94.

Rudel, T.K., Coomes, O.T., Moran, E., Achard, F., Angelsen, A., Xu, J., & Lambin, E. (2005). Forest transitions: Towards a global understanding of land use change. *Global Environmental Change, 15*(1), 23–31.

Sassen, M., Sheil, D., & Giller, K.E. (2015). Fuelwood collection and its impacts on a protected tropical mountain forest in Uganda. *Forest Ecology and Management, 354*, 56–67.

Sassen, M., Sheil, D., Giller, K.E., & ter Braak, C.J. (2013). Complex contexts and dynamic drivers: Understanding four decades of forest loss and recovery in an East African protected area. *Biological Conservation, 159*, 257–268.

Sassen, M., van Soesbergen, A., Arnell, A.P., & Scott, E. (2022). Patterns of (future) environmental risks from cocoa expansion and intensification in West Africa call for context specific responses. *Land Use Policy, 119*, 106142.

Schaldach, R., Alcamo, J., Koch, J., Kölking, C., Lapola, D.M., Schüngel, J., & Priess, J.A. (2011). An integrated approach to modelling land-use change on continental and global scales. *Environmental Modelling & Software, 26*(8), 1041–1051.

Searchinger, T.D., Estes, L., Thornton, P.K., Beringer, T., Notenbaert, A., Rubenstein, D., … & Herrero, M. (2015). High carbon and biodiversity costs from converting Africa's wet savannahs to cropland. *Nature Climate Change, 5*(5), 481–486.

Shapiro, A., d'Annunzio, R., Desclée, B., Jungers, Q., Kondjo, H.K., Iyanga, J.M., … & Milandou, C. (2023). Small scale agriculture continues to drive deforestation and degradation in fragmented forests in the Congo Basin (2015–2020). *Land Use Policy, 134*, 106922.

Soini, E. (2006). *Livelihood, land use and environment interactions in the highlands of East Africa* (PhD thesis). Helsinki, Finland. Available at https://hdl.handle.net/10138/21188

Sunderland, T., Powell, B., Ickowitz, A., Foli, S., Pinedo-Vasquez, M., Nasi, R., & Padoch, C. (2013). Food security and nutrition. The role of forests. Center for International Forestry Research (CIFOR), Bogor, Indonesia.

Teketay, D. (2004). Causes and consequences of dryland forest degradation in Sub-Saharan Africa. *Walia, 2004*(24), 2–19.

Tiifen, C., Mortimore, M., & Gichuki, F.N. (1993). *More people, less erosion: Environmental recovery in Kenya*. John Wiley & Sons, Ltd.

Ulimwengu, J.M., Kwofie, E.M., & Collins, J. (2023). *African food systems transformation and the post-Malabo agenda*. ReSAKSS 2023 Annual Trends and Outlook Report. AKADEMIYA2063 and International Food Policy Research Institute (IFPRI).

Uusivuori, J., Lehto, E., & Palo, M. (2002). Population, income and ecological conditions as determinants of forest area variation in the tropics. *Global Environmental Change, 12*, 313–323.

Vaast, P., & Somarriba, E. (2014). Trade-offs between crop intensification and ecosystem services: The role of agroforestry in cocoa cultivation. *Agroforestry Systems, 88*(6), 947–956.

Van Gemerden, B.S., Olff, H., Parren, M.P., & Bongers, F. (2003). The pristine rain forest? Remnants of historical human impacts on current tree species composition and diversity. *Journal of Biogeography, 30*(9), 1381–1390.

van Velden, J., Wilson, K., & Biggs, D. (2018). The evidence for the bushmeat crisis in African savannas: A systematic quantitative literature review. *Biological Conservation, 221*, 345–356.

van Soesbergen, A., Arnell, A. P., Sassen, M., Stuch, B., Schaldach, R., Göpel, J., Vervoort, J., Mason-D'Croz, D., Islam, S., & Palazzo, A. (2017). Exploring future agricultural development and biodiversity in Uganda, Rwanda and Burundi: A spatially explicit scenario-based assessment. *Regional Environmental Change, 17*, 1409–1420.

VanWey, L.K., Ostrom, E., & Meretsky, V. (2005). Theories underlying the study of human-environment interactions. In E.F. Moran & E. Ostrom (Eds.), *Seeing the forest and the trees: Human-environment interactions in forest ecosystems* (p. 23). Mit Press.

Vervoort, J.M., Palazzo, A., Mason-D'Croz, D., Ericksen, P.J., Thornton, P.K., Kristjanson, P.M., … & Jost, C. (2013). *The future of food security, environments and livelihoods in Eastern Africa: Four socio-economic scenarios*. CCAFS Working Paper No. 63CGIAR Research Program on Climate Change Agriculture and Food Security (CCAFS). Copenhagen, Denmark.

Vinceti, B., Termote, C., Ickowitz, A., Powell, B., Kehlenbeck, K., & Hunter, D. (2013). The contribution of forests and trees to sustainable diets. *Sustainability*, *5*(11), 4797–4824.

Williams, D.R., Clark, M., Buchanan, G.M., Ficetola, G.F., Rondinini, C., & Tilman, D. (2021). Proactive conservation to prevent habitat losses to agricultural expansion. *Nature Sustainability*, *4*(4), 314–322.

Wilson, S.A., & Scholes, R.J. (2020). The climate impact of land use change in the miombo region of south central Africa. *Journal of Integrative Environmental Sciences*, *17*(3), 187–203.

WWF. (2015). *African ecological futures*. African Development Bank and WWF. Available at https://wwfint.awsassets.panda.org/downloads/wwf___afdb_african_futures_report__eng____final__1_.pdf

York, R., & McGee, J.A. (2016). Understanding the Jevons paradox. *Environmental Sociology*, *2*(1), 77–87.

Zomer, R.J., Trabucco, A., Coe, R., Place, F., Van Noordwijk, M., & Xu, J. (2014). *Trees on farms: An update and reanalysis of agroforestry's global extent and socio-ecological characteristics*. Working Paper 179. World Agroforestry Centre (ICRAF) Southeast Asia Regional Program. Bogor, Indonesia.

14 Food security for Africa's growing population: a labour market perspective

Janneke Pieters

1 Introduction

African farmers face many constraints to increasing their productivity and output, including factors related to agricultural technology, the environment, and missing markets for inputs, credit, insurance, and outputs. Lifting these constraints will be important for ensuring food security for Africa's growing population. In this chapter, however, I argue that this cannot succeed without a transition of employment from agriculture to the non-agricultural sector. This requires policies beyond farming and rural development that address labour market frictions, as well as trade and industrial policy.

Productive employment in the non-agricultural sectors is essential to address what Giller et al. (2021a) refer to as the *crisis in the economic and social sustainability of (family) farming*. In most low- and middle-income countries (LMICs) outside Africa, including Bangladesh and India, both the share and the absolute number of people working in agriculture (i.e. workers whose primary sector of work is agriculture) are on the decline. But in several highly populated African countries, while the agricultural share of total employment has declined, the number of people employed in agriculture is still growing (Figure 14.1).[1] At the same time, output per

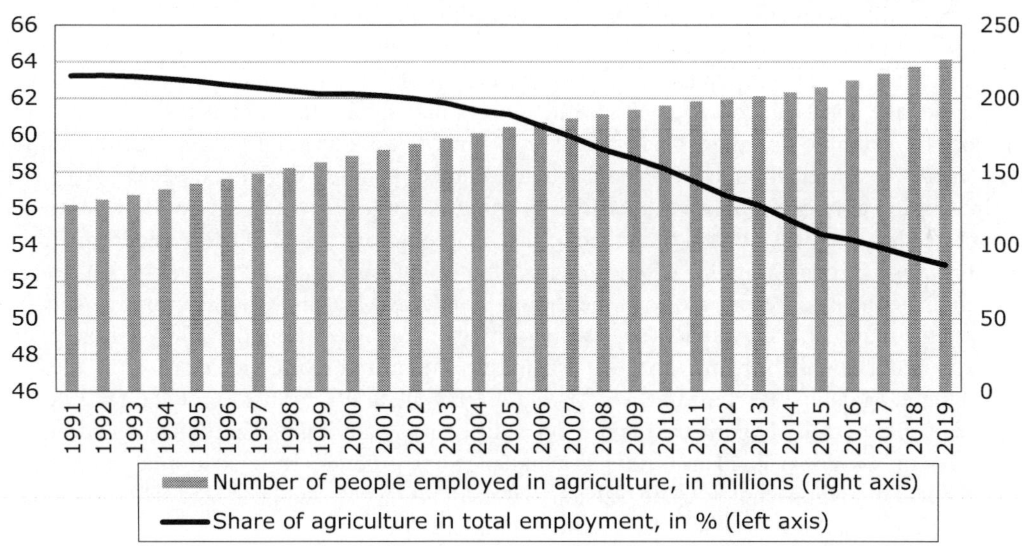

Figure 14.1 Employment in agriculture, in low- and middle-income countries in sub-Saharan Africa.

Source: World Development Indicators and Our World in Data.

DOI: 10.4324/9781032649696-18

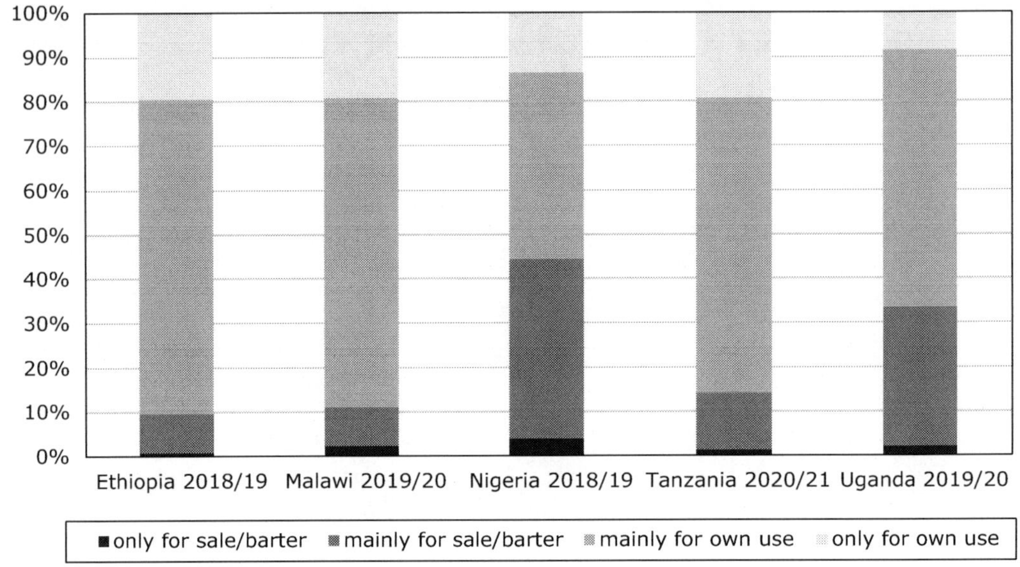

Figure 14.2 Self-reported intended destination of output from family farming.

Source: Author's calculations using data from the Living Standards Measurement Study.

worker is low in agriculture compared to other sectors (Gollin et al., 2014; McCullough, 2017). Using individual-level data from the Living Standards Measurement Survey-Integrated Surveys in Agriculture (LSMS-ISA), McCullough (2017) finds that on average, agricultural workers in Ethiopia, Malawi, Tanzania, and to a lesser extent Uganda work substantially fewer hours per year than non-agricultural workers.[2] The agricultural sector in these countries is described as a "large reservoir of underemployed workers" (McCullough, 2017, p. 134).

Seasonality in agricultural production is one reason for the underemployment of agricultural workers. One may also argue that underemployment masks the fact that family farm workers also undertake non-farm activities. Yet, while it is not uncommon for farmers to work a secondary job, for example running a small non-farm business, these account for very few hours per person per year (McCullough, 2017) and provide limited income (Nagler and Naudé, 2018).[3]

In addition to underemployment, Africa's farming sector is characterized by a low degree of commercialization. It is beyond the scope of this chapter to discuss the factors underlying this, but it is worth documenting the extent to which family farmers produce for the market. Based on data from the most recent waves of the Living Standards Measurement Study in five African countries, Figure 14.2 shows that the vast majority of persons working on family farms report that they produce only or mainly for own use, in line with rates reported in Gaddis et al. (2023). In Ethiopia, Malawi, and Tanzania, only around 10% of family farmers produce for sale. In Uganda and especially Nigeria the picture looks different, but still most family farmers produce mainly or only for own use. This is not just a matter of low productivity. Due to small farm sizes, many smallholder households in sub-Saharan Africa (SSA) would not be able to achieve a living income even if yield gaps were closed (Giller et al., 2021b).

Given high underemployment and low levels of income, Africa's small-scale farming sector will not be able to absorb the expected increase in the number of workers over the coming decades without significant deteriorations in rural living standards. Furthermore, the large number of people involved in family farming may hold back agricultural productivity and output growth through its effect on land fragmentation. Farm sizes are often below 1 ha, making investments in new technologies unprofitable (Giller et al., 2021a, 2021b, Giller & Andersson, this volume). Even if they were profitable, and small farmers had sufficient access to credit, the technological change required for productivity growth can be labour-saving. In India, for example, mechanization of tilling contributed to a reduction in women's agricultural labour (Afridi et al., 2023) – although some types of technology are labour-complementary. The key point is that farm land consolidation and technology investments are unlikely to succeed without a decline in the number of people relying on family farming for their livelihoods.

This implies that enabling the outflow of workers from family farming is an important prerequisite for agricultural productivity growth (and broad-based economic development) in many parts of Africa. While approaches to improving food security often take investments in agricultural productivity and output as the starting point, these are unlikely to succeed without a simultaneous acceleration of the movement of labour out of agriculture. Governments need to play an active role to speed up this labour transition.

2 Employment outside agriculture

Cross-country data show a clear relationship between economic development and the nature of work. As countries grow richer, work shifts from unpaid to paid work due to the emergence of markets, and then from self-employment to wage employment due to the emergence of firms (Bandiera et al., 2022a. These shifts coincide with the outflow of agricultural workers to non-agricultural occupations (Bandiera et al., 2022a). While this pattern is clearly visible when comparing countries across different income levels, enabling the outflow of workers from farming remains a major challenge for many African countries nowadays. Compared to LMICs in other regions, the transition has been much slower in Africa, where non-agricultural labour demand continues to be weak. This appears from firm-level employment and productivity data as well as from household survey data.

Data from the GGDC/UNU-WIDER Economic Transformation Database (ETD), which includes 20 African countries (Kruse et al., 2022), show a decline in the share of agricultural employment in all African countries covered since the start of the 21st century, with wide variation in the degree of reduction (Figure 14.3). Manufacturing employment remains relatively limited, however, accounting for 8.3% of total employment in 2015, across the 20 countries covered.

Between 2000 and 2015, the manufacturing share of total employment grew fastest in Burkina Faso, Ethiopia, Senegal, and Cameroon.[4] In most countries, employment growth was concentrated in services. In 2015, the services sector accounted for 20% (in Ethiopia, Rwanda, Mozambique) to around 60% (in South Africa, Namibia, Mauritius) of total employment.

The relative shift of employment out of agriculture has contributed to rising aggregate labour productivity, as workers are increasingly employed in sectors with higher labour productivity (Diao et al., 2019; McMillan & Zeufack, 2023). However, labour productivity growth within the manufacturing sector has been close to zero (McMillan & Zeufack, 2023). If employment growth in manufacturing is accompanied by rising labour productivity within manufacturing, as was the case in many Asian countries (see Fox et al., 2017), this is a signal that productivity growth in manufacturing is attracting resources from other parts of the economy. However, this is not the pattern we observe in Africa.

In Ethiopia and Tanzania, for which detailed data at the firm level are available, manufacturing employment growth in the early 2000s was largely concentrated in small, informal firms

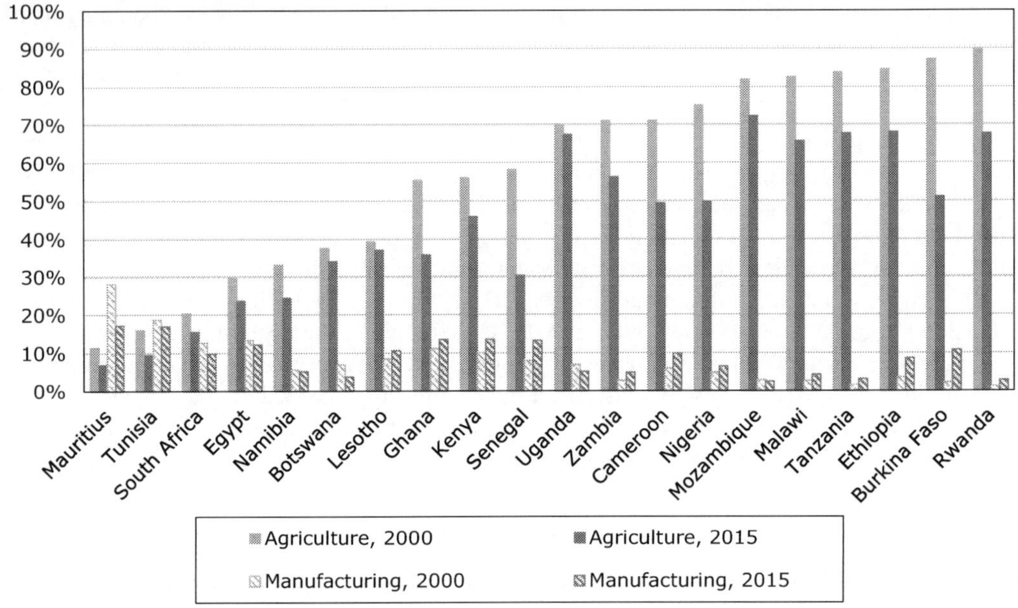

Figure 14.3 Share of total employment in agriculture and manufacturing, 2000 and 2015.

Source: Author's calculations based on GGDC/UNU-WIDER Economic Transformation Database.

Note: Countries are ordered by the share of employment in agriculture in 2000.

(Diao et al., 2021).[5] McMillan and Zeufack (2023) show a similar picture for Ghana, Kenya, Nigeria, and Senegal – and how this contrasts with the experience in Bangladesh, Laos, Sri Lanka, and Vietnam, where a large share of manufacturing employment growth was concentrated in formal firms. While manufacturing *output* growth in many African countries was driven by large firms, these are highly capital-intensive and not generating much employment. In the services sector, too, employment growth has been concentrated in low-productivity jobs, mostly in trade and transportation (Baccini et al., 2023).

We thus see a particular type of structural change in many LMICs in Africa, in which employment opportunities outside agriculture are growing, but they are concentrated in low-productivity activities. Self-employment and unpaid family work continue to be the dominant types of work, among men and especially among women (Lo Bue et al., 2022). For rural households, off-farm work is often used to provide additional income outside the agricultural peak season and is therefore limited to flexible activities that are easy to enter and exit (Christiaensen and Maertens, 2022; Nagler and Naudé, 2018).

Data on labour market dynamics, or the changes in people's employment status over time, indicate that self-employment in low-income countries is mostly used to generate income while searching for wage employment, rather than a means to climb the job ladder (Donovan et al., 2023). While there are relatively high rates of job entry and exit in poorer countries compared to richer countries, most people switch between self-employment and unemployment or low-wage work. Self-employment is not a stepping stone to better paid wage employment (Donovan et al., 2023). If small, informal businesses are run as a form of unemployment insurance, we should not be surprised to see the lack of productivity growth documented by Diao et al. (2021) and McMillan and Zeufack (2023).

The concentration of work in low-productivity self-employment has been persistent. A particular pattern seen in Africa's LMICs is that young and older cohorts of workers are equally (un)likely to hold a salaried job, while outside of Africa there has been a shift from self-employment and micro-enterprise work among older cohorts to salaried employment in larger firms among younger cohorts (Bandiera et al., 2022b). Compared to youth in other LMICs, African youth are more likely to be in unpaid work and self-employment rather than paid work and salaried employment, and more likely to work in agriculture (Bandiera et al., 2022b). Hence, while the concentration of work in self-employment partly reflects the fact that African countries are poorer than developing countries in other regions, the growth and structural change that have taken place do not translate into productive wage employment opportunities for young workers.

Figure 14.4 plots the share of wage and salaried workers in total employment in 2000 and 2015 for all LMICs in SSA and Asia with available data. In 2000, wage and salaried work accounted for 18.9% of total employment in Vietnam, similar to the rates in Zambia, Cameroon, Ghana, Senegal, and Cote d'Ivoire in that year. By 2015, the rate had increased to almost 40% in Vietnam, much more than in the African countries. Only Senegal came close with 33%, and stands out as an exception across all African countries in this plot (see also Hathie, this volume). Bangladesh and Pakistan were similar to Kenya and Zimbabwe in 2000, but also saw a stronger increase in the rate of wage and salaried employment (which even declined in Zimbabwe; Nyandoro, this volume).

It is evident that the lack of wage employment growth poses a real constraint to the non-agricultural employment opportunities of young generations of Africans. The labour market challenges facing Africa's youth have long been recognized (see, for example, Filmer and Fox, 2014 and Sumberg et al., 2021). Given the lack of (formal) wage employment opportunities, Fox et al. (2016) and others emphasized the importance of investing in the productivity

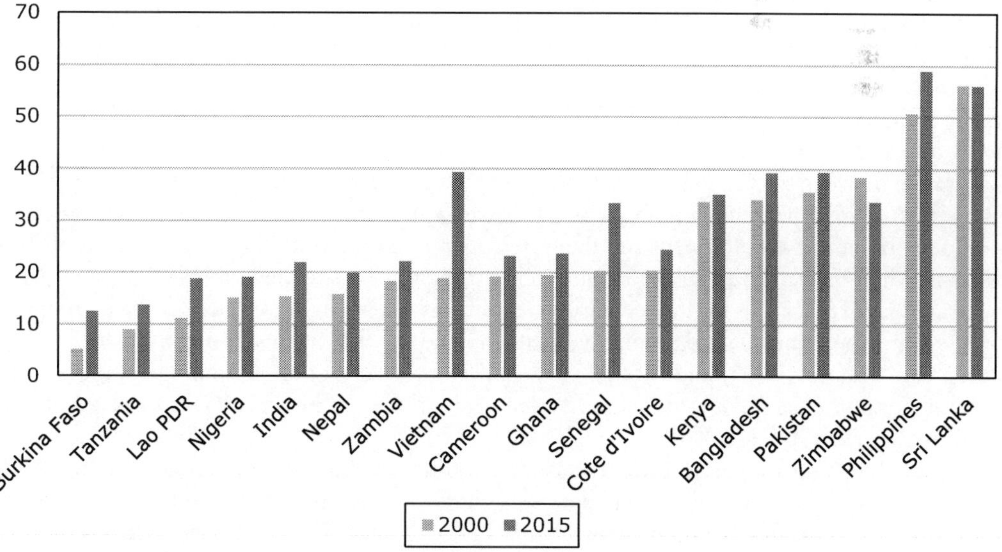

Figure 14.4 Share of wage and salaried workers in total employment, 2000 and 2015.

Source: ILO Modelled Estimates database, accessed through the World Development Indicators database.

Note: Countries are ordered by the share of wage and salaried workers in 2000.

of self-employment. Yet from evaluations of programmes and policies targeting entrepreneurs in LMICs it appears they are typically ineffective in achieving sustained increases in income, let alone job creation (McKenzie and Woodruff, 2014; Jayachandran, 2021). Furthermore, policies that are effective at increasing productivity and income among the self-employed may actually hinder the entrance of firms and creation of wage employment (Rud and Trapeznikova, 2021).

Given the low rates of wage employment, poverty reduction will necessarily involve policies that aim to raise the incomes of farmers and non-farm self-employed. But these should be considered anti-poverty policies as part of social protection programmes. They are unlikely to contribute to the creation of productive wage employment.

3 A vicious cycle? Employment and fertility

Productive non-agricultural employment is not only essential to facilitate agricultural productivity growth and inclusive structural change. It is also related to long-term future food security through its effects on fertility.

Employment, earnings, and the labour market returns to education affect people's income, the opportunity cost of their time, and their incentives to invest in the education of their children. All of these factors contribute to shaping decisions around childbearing. According to economic theory, rising incomes stimulate higher fertility, as people derive well-being from having children and can afford to have more children as their income increases. At the same time, higher wages constitute an increase in the opportunity cost of time, which makes child care more expensive because of the foregone earnings associated with time spent on care. The overall effect of employment opportunities on fertility depends in particular on women's wage rates and on the importance of women's earnings in total household income. In addition, how employment affects fertility is shaped by the constraints it puts on parents' (and especially mothers') time.

In now-rich countries, industrialization contributed to declining fertility through the sharpened separation between work and home spheres, which made combining market work and child care more difficult (Doepke et al., 2023). While women's labour force participation is relatively high in sub-Saharan African countries, their participation in wage employment is much lower compared to LMICs in other regions (Büttner et al., 2023). The prevalence of farm work and non-farm self-employment means that most workers have flexible work hours and the separation between work and home is limited. In a recent experimental study, Berge et al. (2022) show that entrepreneurship training for adolescent girls in Tanzania increased not only their participation in self-employment and their income, but also fertility through the effect of rising income. They further find that the labour market flexibility associated with self-employment is an important explanation for the increase in fertility, as the positive association between income and fertility was concentrated among women who had income from self-employment.[6] Indeed, according to Zipfel (2023) and Büttner et al. (2023), the lack of non-agricultural wage employment opportunities in most low-income African countries is an important factor contributing to high fertility rates in SSA.

Zipfel (2023) shows that in LMICs in Africa, desired fertility is much higher than in LMICs in other regions. Moreover, the gap in desired fertility between poor and rich households within African countries is exceptionally large. Using Demographic and Health Survey data, she shows that in provinces with lower participation of women in wage employment, desired fertility is higher. In addition to the flexibility of farm work and self-employment, the associated economic insecurity can also contribute to high fertility, especially when women depend on their children for support in old age or widowhood (Donald et al., 2023; Rossi and Godard, 2022).

Another link between wage employment and fertility is related to the labour market returns to education. If returns to education increase, parents may start to desire fewer children in order to be able to invest more in the education of each child (the so-called quantity-quality trade off; Becker and Lewis, 1973; Schultz, 1997). Evidence from South Asia shows that more skilled wage employment opportunities can spur changes in educational investments in girls – with associated impacts on marriage and child bearing. The boom in export-oriented garment production in Bangladesh provided new factory employment opportunities to women. It offered young women an opportunity to contribute to their families and increase their bargaining power, stimulating more investments in girls' education and changing ideas about the ideal age of marriage and childbearing (Kabeer & Mahmud, 2004; Heath and Mobarak, 2015). In India, job opportunities for young women in the business outsourcing industry led to increased participation of girls in schooling, training, and employment, while it reduced the likelihood of young women getting married and starting childbearing (Jensen, 2012).

Despite the links between labour markets and fertility, labour market and industrial policy are not typically considered as policy tools for managing population growth. The focus has rather been on increasing access to contraceptives, training, and education to enhance (women's) knowledge on sexual and reproductive health, and programmes aimed at improving women's bargaining position. Yet the available evidence suggests that policies to stimulate productive wage employment outside agriculture, especially for women, could contribute to a fertility transition in Africa. With the additional impacts on agricultural development, it is clear that such policies should be part of an approach to future food security in the region.

4 What can be done?

As I have argued, improving agricultural productivity and output, and rural livelihoods more broadly, will require off-farm employment growth. Productive non-agricultural wage employment can also contribute to fertility declines. Currently, however, non-agricultural employment growth is limited and concentrated in low-productivity activities and self-employment. What can be done to stimulate non-agricultural wage employment growth in Africa's LMICs?

Part of the answer will come from understanding why African countries have developed differently from LMICs in Asia. Why has the share of employment in agriculture remained so high, and why is employment growth in formal manufacturing so limited? These are major questions in the economic development literature that I will not discuss in much depth here, but I will highlight two factors.

A first key factor is the lack of competitiveness in African manufacturing. As Fox et al. (2017) describe, the share of manufacturing goods in the exports of LMICs in Africa is very low compared to East and South Asian countries, indicating that Africa's industrial sector is less competitive and more dominated by nontraded activities. In LMICs in SSA, per capita GDP growth during the past decades was largely driven by agricultural commodity and mineral rents, and less by manufacturing, resulting in "jobless growth" (also see Asmal et al., 2023, on SSA's de-industrialization). In the period 2000–2010, Bangladesh, Cambodia, and Vietnam experienced much more labour-intensive industrial growth than LMICS in Africa (Asmal et al., 2023). Constraints to SSA's manufacturing competitiveness include high-wage costs (Gelb et al., 2020; Frankema, this volume).

A second factor constraining structural change in employment is high population growth itself. Africa's fast-growing and young population forms a rapidly growing labour force. To reduce relative (and certainly absolute) employment in agriculture, and increase the share of wage employment in total employment, requires very high growth of labour demand in (formal) manufacturing and services. Putting the required number of productive jobs in a global perspective, Paice (2021) calculates African countries would need to add the equivalent of 80% of Europe's current total labour force during the 2020s: "a challenge of mammoth proportions" (Paice, 2021, Chapter 17). Anything less means that many new labour market entrants will continue to depend on agriculture or non-agricultural self-employment.

If we focus on the labour demand side, trade and industrial policy will be needed to actively stimulate non-agricultural labour demand. By enhancing firms' access to technology and intermediate inputs, international trade can stimulate manufacturing labour demand. Foreign direct investment in Ethiopia's manufacturing industry is associated with increased productivity and employment in domestic plants, which upgrade their technology due to competitive pressure and through imitation (Abebe et al., 2022). Similarly, increased exposure to Chinese imports contributed to employment growth in Ethiopian manufacturing during the period 2002–2017, due to productivity enhancing effects of imported intermediate inputs (Ngoma, 2023). These positive employment effects were concentrated in large (>20 workers) and relatively labour-intensive firms. Across 20 African countries, Owusu et al. (2022) find that intermediate imports from developed countries are associated with employment creation, though it disproportionately benefits skilled workers. They also find, however, that intermediate imports from China were negatively related to manufacturing employment, conflicting with the evidence for Ethiopia.

Getting international trade to work for employment growth is difficult. Given that large manufacturing firms in Tanzania and Ethiopia are very capital-intensive (more so than in low-income countries in the past), Rodrik and others warn that today's global value chains stimulate the spread of labour-saving technology. Technological development in the manufacturing sector is contributing to reduced labour intensity as it primarily caters to the needs of rich countries, where workers are relatively expensive. As labour-saving technology spreads more easily with the increased international fragmentation of value chains, it is more difficult now than in the past for low-income countries to benefit from their comparative advantage in low-wage labour (Rodrik, 2018; Diao et al., 2021).

More may be expected from intra-Africa trade, after the implementation of the African Continental Free Trade Area. The World Bank estimates substantial positive impacts of the agreement – and the foreign direct investment it is expected to generate – on employment and wages by 2035, and also emphasizes that governments must prioritize distributional and social impacts during the transition (Echandi et al., 2022). We know from other contexts that labour market adjustments to trade policy reform can involve major costs in terms of income and employment loss, especially for poorer segments of the population (Topalova, 2010; Dix-Carneiro and Kovak, 2017).

This means labour market policies that ease job search and matching have to be part of the policy package. Recent research on labour market dynamics points at an important role for market distortions that constrain labour reallocation (Donovan and Schoellman, 2023). Analyses based on panel data, tracking people's work activities over time, suggest that labour market frictions are especially affecting high-wage employment in LMICs. Although people easily move into and out of low-wage jobs and self-employment, there is little movement into higher-wage employment while the probability of quitting early remains high (Donovan et al., 2023).

Labour market frictions exist when it is difficult for job seekers and employers to find one another (for example, due to poor digital infrastructure) or to learn about the characteristics of the worker or the job. As a result, finding a good job or a good employee is costly and takes time. Experimental studies in Ethiopia, South Africa, and Uganda have shown that increasing workers' skills as well as their ability to signal their skills through reference letters or certificates can contribute to a better functioning wage labour market (Abel et al., 2020; Alfonsi et al., 2020; Abebe et al., 2021; Carranza et al., 2022). Rud and Trapeznikova (2021) provide evidence for six African countries that reducing labour market frictions can spur private sector wage employment, especially in the poorest economies, and more so when it is accompanied by policies that make it easier for firms to enter the market.

Labour reallocation will have an important spatial dimension, too. Studies focusing on rural-urban income gaps suggest that rural-urban migration is held back by information frictions (rural families do not have accurate information on urban wages), limited access to formal insurance and social safety nets (due to which families rely on informal risk-sharing arrangements), and land market failures, among others (De Brauw et al., 2014; Lagakos, 2020). Policies that address these frictions could speed up the transition of workers out of agriculture by stimulating rural-urban migration.

5 Concluding remarks

In this chapter, I provide an overview of the role of the labour market in Africa's future food security challenge. As is clear, this relates to many big questions of economic development, such as why Africa's pattern of structural change has been different from Asia's and why Africa's labour-intensive manufacturing sector is relatively uncompetitive. While this chapter only scratches the surface of these issues, I hope it conveys the importance of understanding the links between agricultural productivity and the growth in output, employment, and productivity in other sectors. Given the huge numbers of youth entering the labour force in the near future, policies encouraging productive wage employment outside agriculture are vital. Food security therefore requires policies in the areas of labour market frictions, international trade, and industrial development.

Notes

1 Due to a change in statistical standards adopted by the International Conference of Labour Statisticians in 2013, the official definition of employment used by the International Labour Organization (ILO) has been adjusted (ILO, 2013). A key change is that employment no longer includes production of goods primarily intended for own consumption. The recent decline in the share of agriculture in total employment could partly reflect this change in statistical standards, because subsistence farmers are not counted as part of total employment any longer. Because changes in survey design have been implemented at different points in time across countries, cross-country comparisons in recent years should be interpreted with caution. See ILO (2022) and Gaddis et al. (2023).

2 The ratio of mean hours per year by non-agricultural workers in non-agriculture to hours by agricultural workers in agriculture is 2.3–2.5 in Malawi, 2.4–2.6 in Ethiopia, 2.1–2.2 in Tanzania, and 1.0–1.6 in Uganda.

3 In fact, individuals whose primary sector of work is manufacturing or services work more hours in agriculture than vice versa (McCullough, 2017, Figure 3).

4 The data for Burkina Faso should be interpreted cautiously, because of a change in the timing of the Labour Force Surveys from the harvest season to the dry season since 2014 (see De Vries et al., 2021).

5 Sida et al. (this volume) further note that Ethiopia's manufacturing sector's contribution to GDP has been disappointing. They argue that more investment in education and training is necessary to enhance the transition of workers into productive enterprises.

6 A positive effect of income on fertility may seem at odds with the fact that fertility rates are lower in urban than in rural areas, but the urban-rural difference reflects many factors, including higher rates of wage employment in urban areas. The positive income effect in Berge et al. (2022) is conditional on being in self-employment.

References

Abebe, G., Caria, S., Fafchamps, M., Falco, P., Franklin, S., & Quinn, S. (2021). Anonymity or distance? Job search and labour market exclusion in a growing African City. *The Review of Economic Studies*, *88*(3), 1279–1310.

Abebe, G., McMillan, M., & Serafinelli, M. (2022). Foreign direct investment and knowledge diffusion in poor locations. *Journal of Development Economics*, *158*, 102926.

Abel, M., Burger, R., & Piraino, P. (2020). The value of reference letters: Experimental evidence from South Africa. *American Economic Journal: Applied Economics*, *12*(3): 40–71.

Afridi, F., Bishnu, M., & Mahajan, K. (2023). Gender and mechanization: Evidence from Indian agriculture, *American Journal of Agricultural Economics*, *105*(1), 52–75.

Alfonsi, L., Bandiera, O., Bassi, V., Burgess, R., Rasul, I., Sulaiman, M., & Vitali, A. (2020). Tackling youth unemployment: Evidence from a labor market experiment in Uganda. *Econometrica*, *88*, 2369–2414.

Asmal, Z., Bhorat, H., Rooney, C., & Steenkamp, F. (2023). Manufacturing in sub-Saharan Africa: Deindustrialisation or a renaissance? DPRU working paper 202303. University of Cape Town, Development Policy Research Unit.

Baccini, L., Fiorini, M., Hoekman, B., & Sanfilippo, M. (2023). Services, jobs, and economic development in Africa. *The World Bank Research Observer*, *38*(1), 147–178.

Bandiera, O., Elsayed, A., Heil, A., & Smurra, A. (2022a). Presidential address 2022: Economic development and the organisation of labour: Evidence from the jobs of the world project. *Journal of the European Economic Association*, *20*(6), 2226–2270.

Bandiera, O., Elsayed, A., Smurra, A., & Zipfel, C. (2022b). Young adults and labor markets in Africa. *Journal of Economic Perspectives*, *36*(1), 81–100.

Becker, G.S., & Lewis, H.G. (1973). On the interaction between the quantity and quality of children. *Journal of Political Economy*, *81*(2), 279–288.

Berge, L., Bjorvatn, K., Makene, F., Helgesson Sekei, L., Somville, V., & Tungodden, B. (2022). On the doorstep of adulthood: Empowering economic and fertility choices of young women. Working papers 2022-035, Human Capital and Economic Opportunity Working Group.

Büttner, N., Grimm, M., Günther, I., Harttgen, K., & Klasen, S. (2023). The fertility transition in sub-Saharan Africa: The role of structural change. IZA discussion paper, No. 15966.

Carranza, E., Garlick, R., Orkin, K., & Rankin, N. (2022). Job search and hiring with limited information about workseekers' skills. *American Economic Review*, *112*(11), 3547–3583.

Christiaensen, L., & Maertens, M. (2022). Rural employment in Africa: Trends and challenges. *Annual Review of Resource Economics*, *14*(1), 267–289.

De Brauw, A., Mueller, V., & Lee, H.L. (2014). The role of rural–urban migration in the structural transformation of sub-Saharan Africa. *World Development*, *63*, 33–42.

De Vries, G., Arfelt, L., Drees, D., Godemann, M., Hamilton, C., Jessen-Thiesen, B., … & Woltjer, P. (2021). The economic transformation database (ETD): Content, sources, and methods. WIDER technical note, 2/2021. UNU-WIDER, Helsinki.

Diao, X., Ellis, M., McMillan, M.S., & Rodrik, D. (2021). Africa's manufacturing puzzle: Evidence from Tanzanian and Ethiopian firms. NBER working paper No. 28344.

Diao, X., McMillan, M.S., & Rodrik, D. (2019). The recent growth boom in developing economies: A structural-change perspective. In M. Nissanke & J.A. Ocampo (Eds.), *The Palgrave handbook of development economics* (pp. 281–334). Palgrave Macmillan.

Dix-Carneiro, R., & Kovak, B.K. (2017). Trade liberalization and regional dynamics. *American Economic Review*, *107*(10), 2908–2946.

Doepke, M., Hannusch, A., Kindermann, F., & Tertilt, M. (2023). Chapter 4: The economics of fertility: A new era. In S. Lundberg & A. Voena (Eds.), *Handbook of the economics of the family* (vol. 1, pp. 151–254). Elsevier.

Donald, A., Koroknay-Palicz, T., Goldstein, M., & Sage, M. (2023). The fertility impacts of development programs. Unpublished paper.

Donovan, K., Lu, W.J., & Schoellman, T. (2023). Labor market dynamics and development. *Quarterly Journal of Economics*, *138*(4), 2287–2325.

Donovan, K., & Schoellman, T. (2023). The role of labour market frictions in structural transformation. *Oxford Development Studies*, forthcoming.

Echandi, R., Maliszewska, M., & Steenbergen, V. (2022). *Making the most of the African continental free trade area: Leveraging trade and foreign direct investment to boost growth and reduce poverty*. World Bank.

Filmer, D., & Fox, L. (2014). *Youth employment in sub-Saharan Africa*. Africa Development Forum. World Bank and Agence Française de Développement.

Fox, L., Thomas, A. & Haines, C. (2017). *Structural transformation in employment and productivity. What can Africa hope for? IMF, African Department*. International Monetary Fund.

Gaddis, I., Oseni, G., Palacios-Lopez, A., & Pieters, J. (2023). Who is employed? Evidence from sub-Saharan Africa on redefining employment. *Journal of African Economies*, *32*(2), 151–174.

Gelb, A., Ramachandran, V., Meyer, C., Wadhwa, D., & Navis, K. (2020). Can sub-Saharan Africa be a manufacturing destination? Labor costs, price levels, and the role of industrial policy. *Journal of Industry, Competition and Trade*, *20*, 335–357.

Giller, K.E., Delaune, T., Silva, J.V., Descheemaeker, K., van de Ven, G., Schut, A.G.T., … & van Ittersum, M.K. (2021a) The future of farming: Who will produce our food? *Food Security*, *13*, 1073–1099. https://doi.org/10.1007/s12571-021-01184-6

Giller, K.E., Delaune, T., Silva, J.V., van Wijk, M., Hammond, J., Descheemaeker, K., … & Andersson, J.A. (2021b) Small farms and development in sub-Saharan Africa: Farming for food, for income or for lack of better options? *Food Security*, *13*, 1431–1454. https://doi.org/10.1007/s12571-021-01209-0

Gollin, D., Lagakos, D., & Waugh, M.E. (2014). The agricultural productivity gap. *The Quarterly Journal of Economics*, *129*(2), 939–993.

Heath, R., & Mushfiq Mobarak, A. (2015). Manufacturing growth and the lives of Bangladeshi women. *Journal of Development Economics*, *115*, 1–15.

ILO. (2013). Resolution I: Resolution concerning statistics of work, employment and labour underutilization. Adopted resolution at the 19th international conference of labour statisticians. International Labour Organization, Geneva. https://www.ilo.org/wcmsp5/groups/public/---dgreports/---stat/documents/normativeinstrument/wcms_230304.pdf

ILO. (2022). Quick guide to understanding the impact of the new statistical standards on ILOSTAT databases. International Labour Office, Geneva. https://www.ilo.org/sites/default/files/wcmsp5/groups/public/@dgreports/@stat/documents/publication/wcms_854830.pdf

Jayachandran, S. (2021). Micro-entrepreneurship in developing countries. In K.F. Zimmermann (Ed.), *Handbook of labor, human resources and population economics*. Springer.

Jensen, R. (2012). Do labor market opportunities affect young women's work and family decisions? Experimental evidence from India. *The Quarterly Journal of Economics*, *127*, 753–792.

Kabeer, N., & Mahmud, S. (2004). Globalization, gender and poverty: Bangladeshi women workers in export and local markets. *Journal of International Development*, *16*, 93–109. https://doi.org/10.1002/jid.1065

Kruse, H., Mensah, E., Sen, K., & de Vries, G.J. (2022). A manufacturing renaissance? Industrialization trends in the developing world. *IMF Economic Review*. https://doi.org/10.1057/s41308-022-00183-7

Lagakos, D. (2020). Urban-rural gaps in the developing world: Does internal migration offer opportunities? *Journal of Economic Perspectives*, *34*(3), 174–192.

Lo Bue, M.C., Thi Ngoc Le, T., Santos Silva, M., & Sen, K. (2022). Gender and vulnerable employment in the developing world: Evidence from global microdata. *World Development*, *159*. https://doi.org/10.1016/j.worlddev.2022.106010

McCullough, E.B. (2017). Labor productivity and employment gaps in sub-Saharan Africa. *Food Policy*, *67*, 133–152.

McKenzie, D., & Woodruff, C. (2014). What are we learning from business training and entrepreneurship evaluations around the developing world? *The World Bank Research Observer*, *29*(1), 48–82.

McMillan, M., & Zeufack, A. (2023). Labour productivity growth and industrialization in Africa. Policy research working paper, No. 10294.

Nagler, P., & Naudé, W. (2018). Chapter 9: Nonfarm enterprises in rural Africa. In L. Christiaensen & L. Demery (Eds.), *Agriculture in Africa – Telling myths from facts*. World Bank.

Ngoma, M.M. (2023). Chinese imports and industrialization in Africa: Evidence from Ethiopia. Unpublished paper.

Owusu, S., Ndubuisi, G., & Mensah, E.B. (2022). Import penetration and manufacturing employment: Evidence from Africa. UNU-MERIT working paper series, No. 2022-007.

Paice, E. (2021). *Youthquake: Why African demography should matter to the world.* Head of Zeus.

Rodrik, D. (2018). New technologies, global value chains, and developing economies. NBER working papers, No. 25164, National Bureau of Economic Research

Rossi, P., & Godard, M. (2022). The old-age security motive for fertility: Evidence from the extension of social pensions in Namibia.

Rud, J.P., & Trapeznikova, I. (2021). Job creation and wages in least developed countries: Evidence from sub-Saharan Africa. *The Economic Journal*, *131*(635), 1331–1364.

Schultz, T.P. (1997). Chapter 8: Demand for children in low income countries. In M. Rosenzweig & O. Stark (Eds.), *Handbook of population and family economics* (vol. 1, pp. 349–430). Elsevier.

Sumberg, J., Fox, L., Flynn, J., Mader, P. & Oosterom, M. (2021). Africa's "youth employment" crisis is actually a "missing jobs" crisis. *Development Policy Review*, *39*, 621–643.

Topalova, P. (2010). Factor immobility and regional impacts of trade liberalization: Evidence on poverty from India. *American Economic Journal: Applied Economics*, *2*(4), 1–41.

Zipfel, C. (2023). The demand side of Africa's demographic transition: Desired fertility, wealth, and jobs. Unpublished paper, Stockholm School of Economics.

15 How small is beautiful? Farm size and economic development in Africa

Ken E. Giller and Jens A. Andersson

1 Introduction

Small is beautiful (Schumacher, 1973), but how small? Schumacher railed against the trend towards economic domination by big business monopolies and argued for smaller, locally owned and locally adapted intermediate technologies and approaches. An abundant literature builds on Schumacher's arguments for appropriate scale of economic organization. Smallholder farming is held to exemplify how small businesses can be highly productive and provide employment, while being more benign to the environment. Yet the reality is far from romantic, as exemplified by Stephen Carr's (2017) essay entitled "Surviving on Half a Hectare of Land", which explores the plight of rural households in southern Malawi. Farms can become too small to provide for a family's basic needs. This is a far cry from the tendency to romanticize and idealize life on a small farm as self-sufficient, food-sovereign, independent of market forces and in harmony with nature.[1] So, at what point do farms become too small? That is the focus of this chapter.

We draw inspiration from Schumacher's ideas in our research on smallholder farming in sub-Saharan Africa. The huge diversity of African farming systems led Kofi Annan (2008) to call for a uniquely African Green Revolution that recognizes the rich diversity of culture, history, climates and soils in the urgent endeavour to improve agricultural productivity sustainably. Our own research has focused on unravelling and exploring this diversity to tailor technologies to the needs of farmers (Giller et al., 2011). An abundance of agricultural technologies exists to diversify and enhance agricultural productivity – comprising a 'basket of options' that could match the diversity of farmers (Ronner et al., 2021). The sluggish uptake of such technologies has led to the realization that smallholder farmers' investment in agriculture is not only a question of lack of cash or competing demands on labour but also a lack of sufficient land. Small farm size precludes self-sufficiency in food (Hengsdijk et al., 2014) and income (Giller et al., 2021b). Productivity-enhancing technologies, when adopted on small farms, can only generate small absolute gains that are insufficient to lift rural families out of poverty (Harris & Orr, 2013).

This leads us to question at what point small farm size becomes a binding constraint for agricultural development, and whether this is already the case in much of Africa. This question is particularly pertinent given that, despite rapid urbanization, the rural population will continue to grow until 2050 and beyond (van Wissen, this volume). Population pressure on land, although unevenly distributed due to differences in agro-ecological potential and historical patterns of settlement, is thus unlikely to ease. We start by reviewing the literature on farm size in Africa, arguing that farm sizes are often already too small to sustain farming families, let alone generate the investment capital needed to scale up operations and increase farm size, land and labour productivity. We then highlight how smallholder farming has for many become a resource sink,

DOI: 10.4324/9781032649696-19

an activity subsidized and sustained by household income earned in other economic sectors, locally, nationally and internationally.

Our analysis questions the value of the structural transformation model of economic development for understanding current developments in Africa. Based on economic two-sector models (i.e. Lewis, 1954) and theories of modern economic growth (i.e. Kuznets, 1966), structural change or transformation refers to a wide set of changes in economies associated with changes in the distribution of economic activity across sectors of the national economy – agriculture, industry and services. Largely developed through extensive analyses of a limited number of early 'developed economies', the structural change model's applicability to the large and highly diverse economies of the global south is, however, far from straightforward (Kuznets, 1973).

Fundamentally based on the notion that labour productivity differences between economic sectors push or pull labour out of low-productivity sectors – e.g. agriculture – to higher-productivity sectors, the model postulates that this movement also results in (labour) productivity growth in agriculture. Such labour productivity growth in agriculture is associated with technological advancement (i.e. high yield crop varieties, irrigation) and, particularly, labour-saving technologies (i.e. mechanization), which imply increasing farm sizes or an expansion of agricultural lands (per worker). Yet, the structural transformation model provides few clues on the mechanisms that drive this change process: does agricultural productivity growth lead the process? Is the labour pulled out of agriculture, enabling productivity and farm size growth in agriculture? And what role do land- and capital-constrained small farms play?

2 Farm size and its implications

2.1 How large are farms in Africa?

Unsurprisingly, different approaches to estimating farm size yield different results, yet the divergence of estimates is alarming. Based on the World Programme for the Census of Agriculture from 2010, Lowder et al. (2021) conclude that the average farm size for sub-Saharan Africa (excluding South Africa) is 1.6 ha. To provide more current estimates of the number of farms, Erenstein et al. (2021) harmonized rural population and farm data sources to produce estimates of the number of farms for 2020 and projections for 2030. However, they estimated farm size by dividing the area of "agricultural land" (from FAO, 2021) by the number of farms – suggesting an average farm area of 8.2 ha! This is misleading, as the category "agricultural land" includes communal grazing land, which is often unsuitable for crop production. Samberg et al. (2016), using a similar approach based on a dataset harmonized for 2010, did not provide an average value but estimated a similar number of farms in sub-Saharan Africa fell into the 2–5 ha category as those smaller than 2 ha.

Analyses based on household surveys support the conclusions of Lowder et al. (2021). A large proportion of rural households have farm sizes far smaller than 1 ha (Giller et al., 2021b; Julien et al., 2019). A disadvantage of using household surveys to assess farm size is that they do not include the large estates captured by the agricultural census. For instance, in Tanzania, c. 1,000 large estates account for 7% of the agricultural land. They produce 80% of the country's wheat and, respectively, 63%, 34% and 15% of the country's tea, tobacco and coffee (Lowder et al., 2021). Household surveys also tend to miss 'middle-sized' farms, which are increasing in number, mainly due to wealthier urban households investing in land (Jayne et al., 2016). Nonetheless, most rural households in Africa have much less than 2 ha of land for crop production. Indeed, for eight out of nine countries for which we could access large datasets, we found median farm sizes well below 1 ha (Giller et al., 2021b) (Figure 15.1).

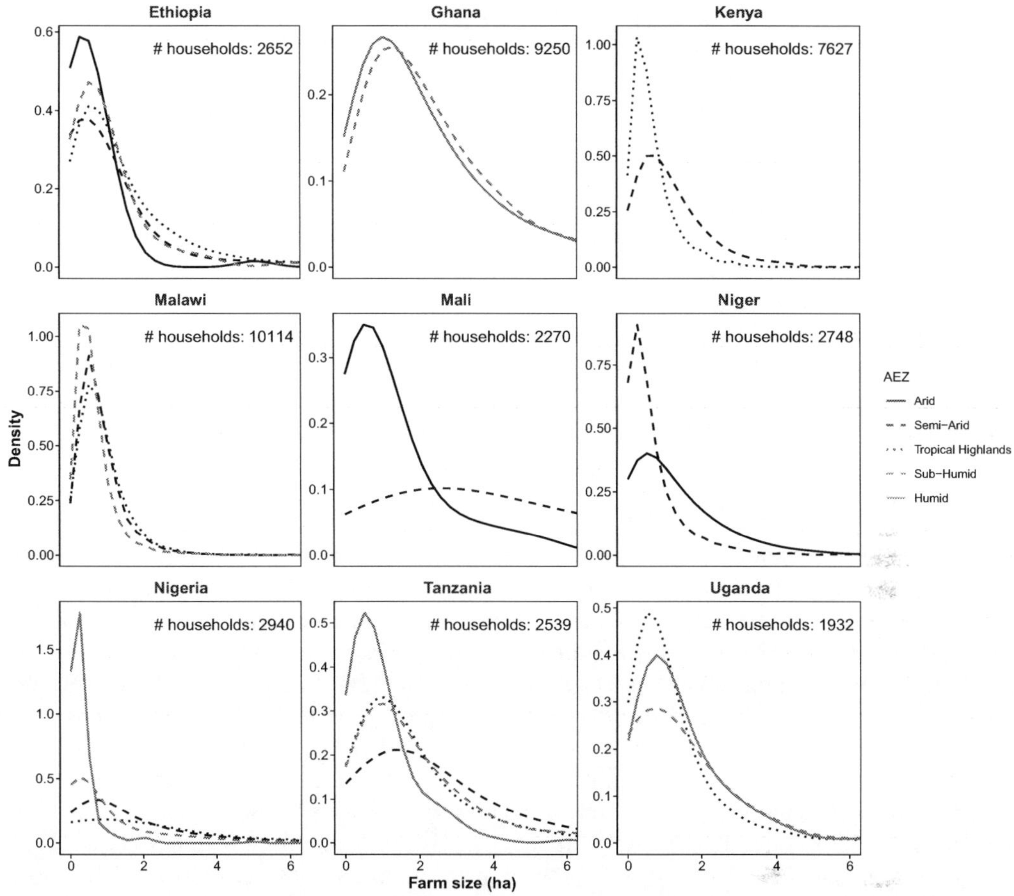

Figure 15.1 Density distributions of farm size across different agro-ecological zones for selected countries in sub-Saharan Africa based on LSMS-ISA (World Bank, 2020) and the FAO RuLIS data (FAO, 2018) – from Giller et al. (2021b).

In many Asian countries, where farm sizes may be even smaller than in Africa, smallholder production is often highly subsidized through complex price and input support arrangements (i.e. India) or is more productive due to the widespread use of irrigation, allowing for two or more cropping cycles per year. In Africa, less than 2% of the cultivated land is said to be irrigated (Sheahan & Barrett, 2017), although this is certainly an underestimate (see Veldwisch et al., this volume). In other words, small farms in Africa are far less productive than similar-sized small farms in Asia. In addition, there is an important historical difference: unlike in the past, Africa's small farms now have to compete in global markets, with farms that can produce at lower costs due to lower labour and input costs (i.e. in Asia), agricultural subsidies and import restrictions (i.e. the European Union), and/or through considerable economies of scale (Giller et al., 2021a)

2.2 *Small farm size and low productivity as poverty traps*

As farm sizes shrink, the opportunity disappears to allow land to lie fallow to recover its soil fertility. The link between shortening of the length of the fallow period and declining soil fertility as a result of increasing human population has long been recognized (Nye & Greenland, 1960). In particular, poorer households with small land areas do not have the luxury of leaving land fallow to regenerate their soils. If the nutrients removed are not replenished, crop productivity inevitably declines. Smallholder farmers have four principal sources of nutrients: (1) the recycling of crop residues; (2) animal manures; (3) symbiotic nitrogen fixation by legumes and (4) mineral fertilizers. Animal manures are often used but are limited in supply. Further, the poorest households generally lack livestock to generate manure that can be recycled to maintain soil fertility (Zingore et al., 2007). Coupled with their lack of cash for investment in fertilizers, this leads to their land being managed with few inputs (Tittonell et al., 2005; Zingore et al., 2007). Continuous cultivation without the return of nutrients in organic manures or fertilizer leads to decline in soil organic matter and nutrient depletion (Zingore et al., 2007). Thus small farm size and poor soil fertility among the poorest households represent a double poverty trap[2]: they can neither produce enough for household needs nor restore the productive capacity of the soil (Hengsdijk et al., 2014; Tittonell & Giller, 2013).

2.3 For whom does farm size become a binding constraint?

Recognition of the diversity of rural households, and their diverse needs, has stimulated a substantial body of research. Considerable focus was placed on characterizing the diversity of farms and creating typologies that could be used for targeting or tailoring of agricultural technologies (Giller et al., 2011). Often four or five different farm types were identified to inform further development interventions. With the benefit of hindsight, it became clear that farm types that could benefit the most from productivity-enhancing agricultural technologies were those with sufficient land, labour and cash to invest (Franke et al., 2014). Yet, such farms typically represent only a small fraction of the overall rural population. Moving away from farm typologies, research has focused more on understanding the frequency distributions of households in terms of farm size (Giller et al., 2021b; Hengsdijk et al., 2014) or food self-sufficiency (Frelat et al., 2016). Such studies reveal surprisingly similar distribution patterns across diverse locations, confirming that the 'better-off' households who can make a 'living income' from farming are a small fraction of the rural population (Giller et al., 2021b). Examples that support this conclusion include not only farming systems focused on annual crops (Franke et al., 2014; Marinus et al., 2022) but also perennial cash crops such as cocoa (van Vliet et al., 2021).

3 Smallholder farming as teleological pathway to proletarianization?

In agrarian economies, farm sizes tend to decrease over time as population grows. However, the movement of labour from agriculture into more productive economic sectors has not followed a 'classic' structural change process during Africa's colonial past. As population densities on the continent were low, the colonial expansion of capitalist-organized economic sectors did not spontaneously draw large numbers of Africans away from agriculture (Amin, 1972). On the contrary, in many areas, the expansion of non-agricultural sectors and commodity trade posed new opportunities for smallholder farmers, who expanded their operations into new lands, producing food and cash crops for upcoming (export) markets, sometimes assisted by the introduction of new technologies such as the ox-drawn plough (Baudron et al., 2012). Contrary to early structural change thinking (Lewis, 1954), underemployment in agriculture and the emergent colonial economy thus did not create a situation of 'unlimited supply of labour' for the expanding modern economic sectors. African labour had to be mobilized, through taxation, forced labour services and the undermining of smallholder farmers' productive capacity.

How the productive capacity of smallholder farmers was undermined during the colonial era became a major theme in political economy-inspired historical analyses in the 1970s and 1980s. Viewing smallholder agriculture as a transitional phase in the development of capitalist agriculture, Marxist-inspired scholars studied how discriminatory policies of colonial states and capitalist market forces 'secured' the unlimited supply of labour for the modern capitalist sector (Arrighi, 1970; Palmer & Parsons, 1977). In the settler economies of East and Southern Africa, land alienation policies and the concentration of African smallholder farmers in designated rural areas called 'reserves' were, arguably, the most visible attempts to undercut smallholder farmers' productive capacity, and to increase their dependency on other sources of income; the 'reserves' were often dubbed 'labour reserves'. Although these land policies further reduced smallholders' holdings, actual farm sizes received scant attention in these political economy analyses – possibly because farm sizes were not regarded a binding constraint (cf. Phimister, 1993). Rather, scholarly attention was focused on what was seen as the inevitable process of proletarianization, the moment when smallholder farmers became unable to reproduce themselves through farming – e.g. when they became proletarians. Yet, smallholder farms did not disappear, making way for large-scale capitalist farms, neither in the colonial era, nor thereafter. Rather, they increased in number and became increasingly smaller.

More recently, development economists have started to pay explicit attention to farm size distributions, land (mis)allocation and the productivity of large versus small farms (e.g. Adamopoulos & Restuccia, 2014). Perhaps this interest is informed by the experiences of some structurally transforming Asian economies which do not witness clear farm size growth (Giller et al., 2021a; see below). Different perspectives can be discerned in this economic literature. Many scholars seem to view productivity growth on small farms as the (necessary) prime driver of poverty reduction and the process of structural transformation, building on the inverse productivity-farm size relation (McMillan & Headey, 2014). Farm size growth is seen as a function of labour productivity growth in agriculture. Yet, some development economists contest this perspective for Africa, contending that the evidence for smallholder-led growth and transformation is weak (Barrett et al., 2017; Collier & Dercon, 2014). Observing that 'African smallholders have not chosen to be entrepreneurs' but are farming 'by default', they argue for a shift in policy focus towards 'the desirable size and composition of the large farm sector, (…), and on facilitating migration' to achieve 'a shrinking of the labour force dependent on agriculture' (Collier & Dercon, 2014). But while these perspectives differ on whether small farms can drive farm size growth, they both view labour productivity growth and farm size increase as linked and desirable outcomes of development policies that aim to achieve rural poverty reduction and structural transformation.

Although ideologically opposed, development economists and the political economists of the 1970/1980s view a movement of labour out of agriculture as necessary for structural transformation to occur. Both assume that a movement of labour out of agriculture will either enable farm consolidation or result from farm size and labour productivity growth in agriculture. Yet, neither scholarly tradition provides clues on how small farms persist despite productivity and wage differences between sectors. Nor do they consider how persistently small farm size impedes (labour) productivity growth in agriculture and structural transformation.

4 Households rely on diverse income streams

Although it has long been recognized that rural households rely on highly diverse sources of income (Ellis, 1998), our persistence in referring to them as 'smallholder farmers' perpetuates the idea that they *depend solely* on farming. Farming is undoubtedly a central part of household food and income security, but income diversification is the norm rather than an exception. In a cross-sectional study (>13,000 households, 93 locations in 17 countries), the 'better off' households had more income from all sources – sale of food and cash crops, livestock and off-farm (Frelat et al., 2016). This raises the 'chicken and egg' question as to whether investment in farming leads to increased earnings on-farm, which in turn leads to better opportunities to find employment outside agriculture. Or conversely, whether increased earnings off-farm allow for greater investment in agriculture to boost production and income on-farm. No doubt, both are true in different situations, and there is a clear generational dimension to this; rural households generally prioritize children's education to enhance their ability to find secure employment, and older farming households, rather than young ones, are often the technologically more advanced. They can invest in farming (technologies) using their off-farm income or pension (i.e. Mutungi et al., 2023). In migrant labour societies, rural households' livelihood security is often located in (family) networks spanning both rural and urban areas, with income earned in different economic sectors (Andersson, 2002).

In Africa, the question 'What farm size sustains a living?' (Marinus et al., 2022) – for food self-sufficiency, food security or a living income – is thus merely a hypothetical one, rather than a reflection of a common reality. This pattern is widely visible throughout the continent: the most successful and technologically advanced smallholder farmers are those who earned their investment capital outside agriculture, working in service and industrial sectors of the economy. Investment in agriculture may thus be regarded as a reflection of the common adage of 'eating one's money'.

5 Food system change, structural transformation and African agriculture

In the opening chapter of this book, de Haas et al. (this volume) discuss four possible paths within which African food systems might develop in the coming decades. The defining dimensions – the axes of Figures 1.5 and 1.6 – of these different pathways are sustained changes in agricultural productivity and the transition of labour out of agriculture, both major features of the classic economic structural transformation model. As the discussion above has shown, increases in agricultural productivity, the movement of labour out of agriculture *and* increases in farm size are generally seen as interrelated developments which follow productivity growth in non-agricultural (urban) sectors. Yet, despite sustained, albeit slow economic growth of non-agricultural sectors in many African economies (de Haas et al., this volume), we find ongoing growth of urban *and* rural populations, and remarkably little evidence of a movement of people out of agriculture (Bandiera et al., 2022).

5.1 Rural population growth, agricultural intensification and marginalization

Although urban sectors – both formal and informal – absorb increasing numbers of labourers, so does the agricultural sector in many places. For instance, Sida et al. (this volume) report significant increases in agricultural productivity in Ethiopia over the past decade, but no evidence for a decrease in the number of people involved in agriculture. This represents the upper left quadrant of "intensive agricultural growth" of Figures 1.5 and 1.6. In terms of future trends, the rural population of Ethiopia will continue to grow, despite even faster urban growth, suggesting a further fragmentation of farms. The potential repercussions of such fragmentation are highlighted by studies on land pressure in the densely populated southern highlands of Ethiopia. In response to the scarcity of grazing land and fodder for their animals, households scaled-down their livestock holdings, shifting from cattle to small ruminants (Mellisse et al., 2017). In turn, less manure was available to maintain soil fertility, which led to declining yields of the perennial crop enset (*Enset ventricosum*). Households could maintain their food security by switching to cultivation of khat (*Catha edulis*) in areas with good market linkages, and to cereal crops in remoter areas (Mellisse et al., 2018). Given the already dense population of >500 people/km^2, to what extent such farming systems will be able to absorb and feed an even larger population is unclear. A similar pattern is repeated across the East African highlands where farms are smallest in areas that have long been settled – parts of western Kenya had population densities exceeding 300 people/km^2 already a century ago (Crowley & Carter, 2000). But unlike in Ethiopia, agricultural productivity is hardly rising in western Kenya (Giller et al., 2021b).

5.2 Rural population growth, and expansion into fragile lands

In relatively sparsely populated areas, such as Bougouni in Mali (Ollenburger et al. (2019), ongoing rural population growth generally results in an expansion of the agricultural area, rather than agricultural productivity growth, suggesting that Africa's Sleeping Giant (Morris et al., 2009) will remain in hibernation (Ollenburger et al., 2016). Baudron et al. (2012) concluded that African farmers are predisposed to extensification rather than intensification, based on a historical analysis of technology adoption.

Given that much of Africa's fertile land is already used for agriculture, this raises the alarm that further agricultural expansion will extend into increasingly fragile and marginal lands, making farming-dependent livelihoods increasingly risky and vulnerable to droughts and climate change (Andersson, 2007). The push for expansion into marginal land is strongly driven by lack of employment opportunities elsewhere (Giller et al., 2013), suggesting that future expansion of cropland and livestock grazing will also be at the expense of biodiversity (see Sassen, this volume).

6 Rural-urban linkages: the internationalization of remittance flows

While during the colonial era, labour for the expanding wage labour economy (mines, farms, urban centres) gradually increased and was sometimes actively mobilized through taxation, the outflow of labour from agriculture was limited as urban workers returned to their rural homes in case of unemployment, or upon retirement. Although urbanization has progressed and become more permanent in nature in many areas, the links of farming households with urban areas and the wage labour economy have persisted. For instance, international migration within Africa and beyond its borders has resulted in massive remittance flows, by which a diverse category of migrants sustain both rural and urban livelihoods of family members *and* agricultural production on smallholder farms. While figures on the importance of remittances for sustaining smallholder agricultural production are hard to come by and often hidden from view by a burgeoning industry of money transfers (Mpesa, Moneygram, Western Union, etc.), numerous African economies and, notably, rural and urban households are increasingly dependent on international money transfers (Figure 15.2). Such transfers, formal and informal, international and local, are likely to increase in significance for sustaining rural livelihoods (Crush, 2013), food and nutrition security (Zezza et al., 2011). Small farms and their production will continue to, and increasingly, be subsidized by such off-farm incomes.

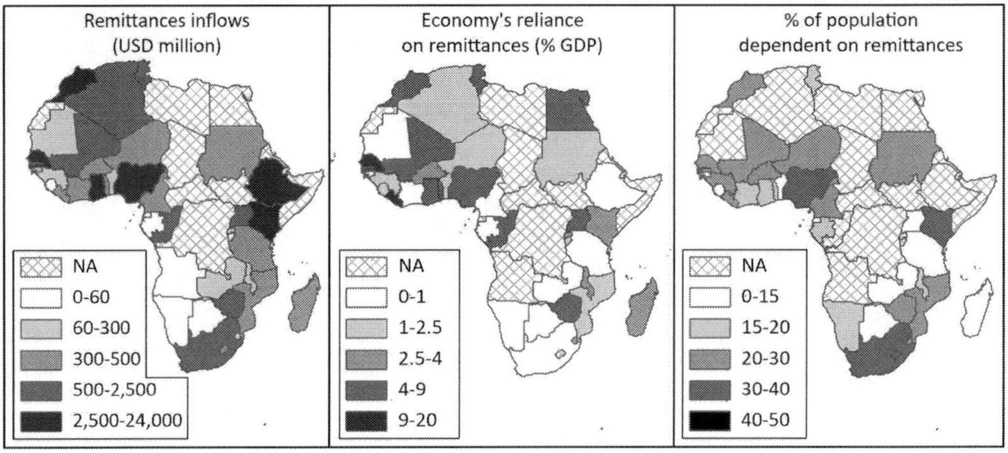

Figure 15.2 Remittance inflows, percentage of GDP and percentage of the population dependent on (inter)national remittances.

Data source: https://remitscope.org/africa/themes/ (Accessed: 23 November 2022).

7 What will lead to consolidation of land?

So the question remains, what are the likely implications of the shift of livelihood focus towards non-agricultural sources of income on farm size? Whereas in Europe's process of structural transformation migration to expanding urban centres, sometimes assisted by deliberate government policies, led to consolidation of small, unviable farms into larger units, it seems unlikely that the same trends will be observed in Africa. An important reason for this is that the farm often functions as a fall-back option – albeit insufficient – given the uncertainty in (international) migration and of urban jobs, as well as a safe-haven for retirement given the lack of pensions or other forms of social security. It is noteworthy to mention that agricultural mechanization, scale enlargement and land consolidation in post-World War II Europe took place in a situation of both rapid growth of non-agricultural sectors *and* the emergence of welfare states, providing livelihood security to pensioners and the unemployed. Second, land and the rural home is much more than a productive resource – a place of belonging and tradition, where ancestors are buried (Andersson, 2002). Given the time frame of our analysis, the prospects for substantial consolidation of land before 2050 thus appear small.

8 Conclusion – revisiting the model of structural transformation

It seems that the concept of structural transformation is so deeply ingrained in our thinking that we interpret developments in its categories of thought and are unable to question the assumptions on which it is based. Africa is unique in many ways, and its late demographic transition, coming at a time when other economies are already industrialized and the world has globalized, blocks many avenues for export-oriented economic development. Structural transformation assumes labour productivity growth in agriculture and movement of people out of agriculture, with consolidation of land into larger, more economic units, a natural accompanying process. However, in Asia there is limited evidence that substantial land consolidation has taken place

(Giller et al., 2021a). No doubt, the Green Revolution led to increases in agricultural (labour) productivity, which was an essential part of ensuring the food security of countries in South and South-east Asia. But despite rapid industrialization and urbanization it did so without a concomitant change in farm size (Liu et al., 2020). Concerning the future of farm size, the speed of transition of the population into employment outside agriculture will be key. A recent estimate suggests that 420 million of the 428 million prime-age workers entering the job market between 2020 and 2040 will be in Africa, 98% of the global labour force growth (Goldstone & May, 2023). The potential "demographic dividend" this presents can only be realized if labour markets thrive, can absorb these workers and provide for salaries that enable workers to sustain themselves and their families (Paice, 2021). Many African countries are experiencing, or have experienced, growth in agricultural productivity. But this has not led to the expected transition out of agriculture, not least due to the lack of jobs to pull people away (Sumberg et al., 2021). Rapid urbanization no doubt creates substantial informal employment in all manner of service roles (hospitality, transport, domestic help, etc.) – but not necessarily the means to grow the economy. Pieters (this volume) shows that work outside agriculture in Africa is concentrated in low-productivity, self-employment. She further emphasizes that, given the technological advances in manufacturing leading to labour saving, it is difficult for low-income countries to benefit from low-wage labour.

Given that Africa is facing rapid demographic growth in both urban and rural areas, coupled with sluggish non-agricultural employment growth (see Pieters, this volume), let alone the lack of welfare-state-like securities for pensioners and the unemployed, the prospects for consolidation of land into larger farms are bleak. The arguments raised in this chapter lead us to question the explanatory power of the structural change model and its value as a roadmap for rural development in Africa.

In terms of food security in Africa, the past and current gains in agricultural productivity are unable to keep pace with the rate of population growth (Schut & van Ittersum, this volume). This is due, at least in part, to the slow uptake of agricultural technology in Africa (Suri & Udry, 2022). But small farm size is also a cause of such slow uptake; productivity gains often translate into only small gains in household food security. Thus, small farm size becomes a major barrier to achieving greater food self-sufficiency at a *national* scale. As there are few incentives for households to invest in agriculture, this leads to stagnation in agricultural productivity – i.e. the African 'Food Security Conundrum' (Giller, 2020). Rural households reluctant to invest in boosting agricultural productivity occupy the majority of the land. As Hazell et al. (2010) bemoan, governments tend to focus on short-term horizons with support such as subsidies without sufficient attention to long-term investment in agricultural infrastructure. Meanwhile, we keep searching for the magic (suite of) technology improvements to improve the yields of smallholders, increasing our knowledge and the benefits of tailoring to local circumstances while forgetting the big picture. At least equal, but perhaps more attention needs to be given to creating opportunities for employment and livelihood security outside agriculture to create the economic incentives for longer-term investments in agriculture.

The time frame we have chosen for the analyses presented in this book – to 2050 – is fairly predictable in terms of future trends in farm size. What will happen after that is impossible to foresee. No doubt, farming will continue to be an important source of food and income for rural households, alongside other income streams. But trends in farm size suggest that rural poverty will become even more entrenched, both in absolute terms and in the proportion of households. Re-orientation of farming towards ensuring more healthy diets through diversification is perhaps more important than a singular focus on raising productivity.

Notes

1 Alliance for Food Sovereignty in Africa – https://afsafrica.org/; FAO Family Farming Network – https://www.fao.org/family-farming/en/.
2 A poverty trap is "any self-reinforcing mechanism which causes poverty to persist".

References

Adamopoulos, T., & Restuccia, D. (2014). The size distribution of farms and international productivity differences. *The American Economic Review*, *104*(6), 1667–1697. https://www.jstor.org/stable/42920862

Amin, S. (1972). Underdevelopment and dependence in Black Africa: Historical origin. *Journal of Peace Research*, *9*(2), 105–119. https://doi.org/10.1177/002234337200900201

Andersson, J.A. (2002). Going places, staying home: Rural-urban connections and the significance of land in Buhera District, Zimbabwe (PhD thesis). Wageningen University.

Andersson, J.A. (2007). How much did property rights matter? Understanding food insecurity in Zimbabwe: A critique of Richardson. *African Affairs*, *106(425)*, 681–690.

Annan, K.A. (2008). Forging a uniquely African green revolution. Address by Mr. Kofi A. Annan, Chairman of AGRA, 30 April 2008.

Arrighi, G. (1970). Labour supplies in historical perspective: A study of the proletarianization of the African peasantry in Rhodesia. *Journal of Development Studies*, *6*, 197–234.

Bandiera, O., Elsayed, A., Smurra, A., & Zipfel, C. (2022). Young adults and labor markets in Africa. *Journal of Economic Perspectives*, *36*(1), 81–100. https://doi.org/10.1257/jep.36.1.81

Barrett, C.B., Christiaensen, L., Sheahan, M., & Shimeles, A. (2017). On the structural transformation of rural Africa. *Journal of African Economies*, *26*(suppl_1), i11–i35. https://doi.org/10.1093/jae/ejx009

Baudron, F., Andersson, J.A., Corbeels, M., & Giller, K.E. (2012). Failing to yield? Ploughs, conservation agriculture and the problem of agricultural intensification: An example from the Zambezi Valley, Zimbabwe. *Journal of Development Studies*, *48*, 393–412.

Carr, S.J. (2017). *Surviving on half a hectare of land: An introduction to the issues surrounding smallholder farming in Malawi*. Plant Production Systems, Wageningen University and Research https://doi.org/10.18174/417831.

Collier, P., & Dercon, S. (2014). African agriculture in 50 years: Smallholders in a rapidly changing world? *World Development*, *63*, 92–101. https://doi.org/10.1016/j.worlddev.2013.10.001

Crowley, E.L., & Carter, S.E. (2000). Agrarian change and the changing relationships between toil and soil in Maragoli, western Kenya (1900–1994). *Human Ecology*, *28*, 383–414.

Crush, J. (2013). Linking food security, migration and development. *International Migration*, *51*. https://doi.org/10.1111/imig.12097.

Ellis, F. (1998). Household strategies and rural livelihood diversification. *Journal of Development Studies*, *35*(1), 1–38. https://doi.org/10.1080/00220389808422553

Erenstein, O., Chamberlin, J., & Sonder, K. (2021). Farms worldwide: 2020 and 2030 outlook. *Outlook on Agriculture*, *50*(3), 221–229. https://doi.org/10.1177/00307270211025539

FAO. (2018). *Rural livelihoods information system (RuLIS)*. Statistics Division, September. Available from https://www.fao.org/in-action/rural-livelihoods-dataset-rulis/data-application/bulk-download/indicators/en

FAO. (2021). *FAOSTAT crop production [Online]*. Available from https://www.fao.org/faostat/en/#home.

Franke, A.C., van den Brand, G.J., & Giller, K.E. (2014). Which farmers benefit most from sustainable intensification? An *ex-ante* impact assessment of expanding grain legume production in Malawi. *European Journal of Agronomy*, *58*, 28–38.

Frelat, R., Lopez-Ridaura, S., Giller, K.E., Herrero, M., Douxchamps, S., Djurfeldt, A.A., … & van Wijk, M.T. (2016). Drivers of household food availability in sub-Saharan Africa based on big data from small farms. *Proceedings of the National Academy of Sciences, USA*, *113*, 458–463. https://doi.org/10.1073/pnas.1518384112

Giller, K.E. (2020). The food security conundrum of sub-Saharan Africa. *Global Food Security*, *26*, 100431. https://doi.org/10.1016/j.gfs.2020.100431

Giller, K.E., Baudron, F., Matema, S., Milgroom, J., Murungweni, C., Guerbois, C., & Twine, W. (2013). Population and livelihoods on the edge. In J.A. Andersson, M. de Garine-Wichatitsky, D.H.M. Cumming, V. Dzingirai, & K.E. Giller (Eds.), *Transfrontier conservation areas: People living on the edge* (pp. 62–88). Routledge.

Giller, K.E., Delaune, T., Silva, J.V., Descheemaeker, K., van de Ven, G., Schut, A.G.T., … & Van Ittersum, M.K. (2021a). The future of farming: Who will produce our food? *Food Security*, *13*, 1073–1099.

Giller, K.E., Delaune, T., Silva, J.V., Descheemaeker, K., van de Ven, G., Schut, A.G.T., … & Andersson, J.A. (2021b). Small farms and development in sub-Saharan Africa: Farming for food, for income or for lack of better options? *Food Security*, *13*, 1431–1454. https://doi.org/10.1007/s12571-021-01209-0

Giller, K.E., Tittonell, P., Rufino, M.C., van Wijk, M.T., Zingore, S., Mapfumo, P., … & Vanlauwe, B. (2011). Communicating complexity: Integrated assessment of trade-offs concerning soil fertility management within African farming systems to support innovation and development. *Agricultural Systems*, *104*, 191–203. https://doi.org/10.1016/j.agsy.2010.07.002

Goldstone, J.A., & May, J.F. (2023). The global economy's future depends on Africa: As others slow, a youthful continent can drive growth. *Foreign Affairs*. https://www.foreignaffairs.com/africa/global-economys-future-depends-africa

Harris, D., & Orr, A. (2013). Is rainfed agriculture really a pathway from poverty? *Agricultural Systems*, *123*, 84–96.

Hazell, P., Poulton, C., Wiggins, S., & Dorward, A. (2010). The future of small farms: Trajectories and policy priorities. *World Development*, *38*, 1349–1361. https://doi.org/10.1016/j.worlddev.2009.06.012

Hengsdijk, H., Franke, A.C., van Wijk, M.T., & Giller, K.E. (2014). How small is beautiful? Food self-sufficiency and land gap analysis of smallholders in humid and semi-arid sub Saharan Africa. Plant Research International, Wageningen UR, 68.

Jayne, T.S., Chamberlin, J., Traub, L., Sitko, N., Muyanga, M., Yeboah, F.K., … & Kachule, R. (2016). Africa's changing farm size distribution patterns: The rise of medium-scale farms. *Agricultural Economics*, *47*(S1), 197–214. https://doi.org/10.1111/agec.12308

Julien, J.C., Bravo-Ureta, B.E., & Rada, N.E. (2019). Assessing farm performance by size in Malawi, Tanzania, and Uganda. *Food Policy*, *84*, 153–164. https://doi.org/10.1016/j.foodpol.2018.03.016

Kuznets, S. (1966). *Modern economic growth: Rate, structure, and spread*. Yale University Press, 1966.

Kuznets, S. (1973). Modern economic growth: Findings and reflections. *The American Economic Review*, *63*(3), 247–258.

Lewis, W.A. (1954). Economic development with unlimited supplies of labor. *The Manchester School*, *22*, 139–191.

Liu, Y., Barrett, C.B., Pham, T., & Violette, W. (2020). The intertemporal evolution of agriculture and labor over a rapid structural transformation: Lessons from Vietnam. *Food Policy*, *94*, 101913–101914. https://doi.org/10.1016/j.foodpol.2020.101913

Lowder, S.K., Sánchez, M.V., & Bertini, R. (2021). Which farms feed the world and has farmland become more concentrated? *World Development*, *142*, 105455. https://doi.org/10.1016/j.worlddev.2021.105455

Marinus, W., Thuijsman, E.S., van Wijk, M.T., Descheemaeker, K., van de Ven, G.W.J., Vanlauwe, B., & Giller, K.E. (2022). What farm size sustains a living? Exploring future options to attain a living income from smallholder farming in the East African Highlands. *Frontiers in Sustainable Food Systems*, *5*, 759105. https://doi.org/10.3389/fsufs.2021.759105

McMillan, M., & Headey, D. (2014). Introduction - Understanding structural transformation in Africa. *World Development*, *63*(C), 1–10. https://doi.org/10.1016/j.worlddev.2014.02.007

Mellisse, B.T., Descheemaeker, K., Giller, K.E., Abebe, T., & van de Ven, G.W.J. (2018). Are traditional home gardens in southern Ethiopia heading for extinction? Implications for productivity, plant species richness and food security. *Agriculture, Ecosystems and Environment*, *252*, 1–13. https://doi.org/10.1016/j.agee.2017.09.026

Mellisse, B.T., van de Ven, G.W.J., Giller, K.E., & Descheemaeker, K. (2017). Home garden system dynamics in southern Ethiopia. *Agroforestry Systems*. https://doi.org/10.1007/s10457-017-0106-5

Morris, M., Binswanger-Mkhize, H.P., & Byerlee, D. (2009). *Awakening Africa's sleeping giant: Prospects for commercial agriculture in the Guinea Savannah zone and beyond*. The World Bank. https://doi.org/10.1596/978-0-8213-7941-7

Mutungi, C., Manda, J., Feleke, S., Abass, A., Bekunda, M., Hoeschle-Zeledon, I., & Fischer, G. (2023). Adoption and impacts of improved post-harvest technologies on food security and welfare of maize-farming households in Tanzania: A comparative assessment. *Food Security*, *15*(4), 1007–1023.

Nye, P.H., & Greenland, D.J. (1960). *The soil under shifting cultivation*. Commonwealth Agricultural Bureaux.

Ollenburger, M., Crane, T., Descheemaeker, K., & Giller, K.E. (2019). Are farmers searching for an African Green Revolution? Exploring the solution space for agricultural intensification in southern Mali. *Experimental Agriculture*, *55*, 288–310. https://doi.org/10.1017/S0014479718000169

Ollenburger, M.H., Descheemaeker, K., Crane, T.A., Sanogo, O.M., & Giller, K.E. (2016). Waking the sleeping giant: Agricultural intensification, extensification or stagnation in Mali's Guinea Savannah. *Agricultural Systems*, *148*, 58–70. https://doi.org/10.1016/j.agsy.2016.07.003

Paice, E. (2021). *Youthquake: Why African Demography Should Matter to the World*. Head of Zeus.

Palmer, R.H., & Parsons, N. (1977). *The roots of rural poverty in Central and Southern Africa* (Vol. 11977). Univ of California Press.

Phimister, I. (1993). Rethinking the reserves: Southern Rhodesia's land husbandry act reviewed. *Journal of Southern African Studies*, *19*, 225–239.

Ronner, E., Sumberg, J., Glover, D., Descheemaeker, K.K.E., Almekinders, C.J.M., Haussmann, B.I.G., … & Giller, K.E. (2021). Basket of options: Unpacking the concept. *Outlook on Agriculture*, *50*(2), 116–124. https://doi.org/10.1177/00307270211019427

Samberg, L.H., Gerber, J.S., Ramankutty, N., Herrero, M., & West, P.C. (2016). Subnational distribution of average farm size and smallholder contributions to global food production. *Environmental Research Letters*, *11*(12), 1–11. https://doi.org/10.1088/1748-9326/11/12/124010

Schumacher, E.F. (1973). *Small is beautiful: A study of economics as if people mattered*. Blond & Briggs Ltd.

Sheahan, M., & Barrett, C.B. (2017). Ten striking facts about agricultural input use in sub-Saharan Africa. *Journal of Food Policy*, *67*, 12–25. https://doi.org/10.1016/j.foodpol.2016.09.010

Sumberg, J., Fox, L., Flynn, J., Mader, P., & Oosterom, M. (2021). Africa's "youth employment" crisis is actually a "missing jobs" crisis. *Development Policy Review*, *39*(4), 621–643. https://doi.org/10.1111/dpr.12528

Suri, T., & Udry, C. (2022). Agricultural technology in Africa. *Journal of Economic Perspectives*, *36*(1), 33–56. https://doi.org/10.1257/jep.36.1.33

Tittonell, P., & Giller, K.E. (2013). When yield gaps are poverty traps: The paradigm of ecological intensification in African smallholder agriculture. *Field Crops Research*, *143*, 76–90. https://doi.org/10.1016/j.fcr.2012.10.007

Tittonell, P., Vanlauwe, B., Leffelaar, P.A., Rowe, E.C., & Giller, K.E. (2005). Exploring diversity in soil fertility management of smallholder farms in western Kenya - I. Heterogeneity at region and farm scale. *Agriculture, Ecosystems and Environment*, *110*, 149–165. https://doi.org/10.1016/j.agee.2005.04.001

van Vliet, J.A., Slingerland, M.A., Waarts, Y.R., & Giller, K.E. (2021). A living income for cocoa producers in Côte d'Ivoire and Ghana? *Frontiers in Sustainable Food Systems*, *5*, 350. https://doi.org/10.3389/fsufs.2021.732831

World Bank. (2020). *Microdata library, living standards measurements study*. Available from https://microdata.worldbank.org/index.php/catalog/lsms

Zezza, A., Carletto, C., Davis, B., & Winters, P. (2011). Assessing the impact of migration on food and nutrition security. *Food Policy*, *36*(1), 1–6. https://doi.org/10.1016/j.foodpol.2010.11.005

Zingore, S., Murwira, H.K., Delve, R.J., & Giller, K.E. (2007). Influence of nutrient management strategies on variability of soil fertility, crop yields and nutrient balances on smallholder farms in Zimbabwe. *Agriculture, Ecosystems and Environment*, *119*, 112–126. https://doi.org/10.1016/j.agee.2006.06.019

Part 4

Opportunities

16 Transformative trees: forests for water and food security

Douglas Sheil

1 Introduction

Water is essential for food security, life and wellbeing. Access to reliable water shapes food security, child nutrition and development (Heino et al., 2018). Reliable access to adequate water, in turn, depends on trees and forests. Recent research indicates that this dependence is stronger and more powerful than generally recognised.

Trees provide shade, cool the local climate, draw carbon dioxide from the air and store it and can repair and replicate themselves while running on little more than sunlight and rainwater. They also contribute numerous goods and services—including shade, fruit, fodder, wood and soil improvement—with a wide variety of suitable species and varieties for different needs and conditions. While many benefits derived from trees are clear, the implications for water availability have been controversial. Recent science provides new insights. While including trees and forests in landscapes offers many benefits—here, I focus on water security. Water security is essential to food security.

The challenges associated with water are considerable. Much of the world is experiencing seasonal water scarcity, while global per capita freshwater reserves halved between 1960 and 2016, and consumption has continued to double approximately every two decades (Ripple et al., 2017). Between 1900 and 2013, more than 11 million people died due to drought, more than 7 million died due to floods and more than 5,000 million required emergency assistance due to one or other (Sheil, 2014). Climate change and further disruption of the global water cycle exacerbate these challenges.

Much of Africa already suffers water insecurity (Veldwisch et al., this volume). Agriculture, employment and exports are all dependent on unreliable rains. Moreover, many regions experience increasing aridity, temperatures and extreme weather events (Descheemaeker et al., this volume). Observations indicate alarming shifts in rainfall patterns, including reductions in key river basins and expansions of arid regions. The spectra of mass displacement loom.

Across Africa, the area used for food production has generally increased at the expense of natural areas, including forests and woodlands (Sassen et al., this volume). There are major uncertainties in the estimates of forest cover change in Africa. One recent overview of closed forests suggested that considerable areas had already been lost by 1,900 and that an additional ~21.7% of tropical African forests were lost by 2,000 (3,527,307–2,760,003 km^2). Patterns varied across the continent: the forests in West and East Africa were the most impacted (over 80% reductions), while those in Central Africa, the world's second-largest tropical forest, actually expanded (~1.4%) (Aleman, Jarzyna, & Staver, 2018)—see Figure 16.1. In recent decades, losses of closed forests have continued. Small-scale agriculture is considered the main cause of

DOI: 10.4324/9781032649696-21

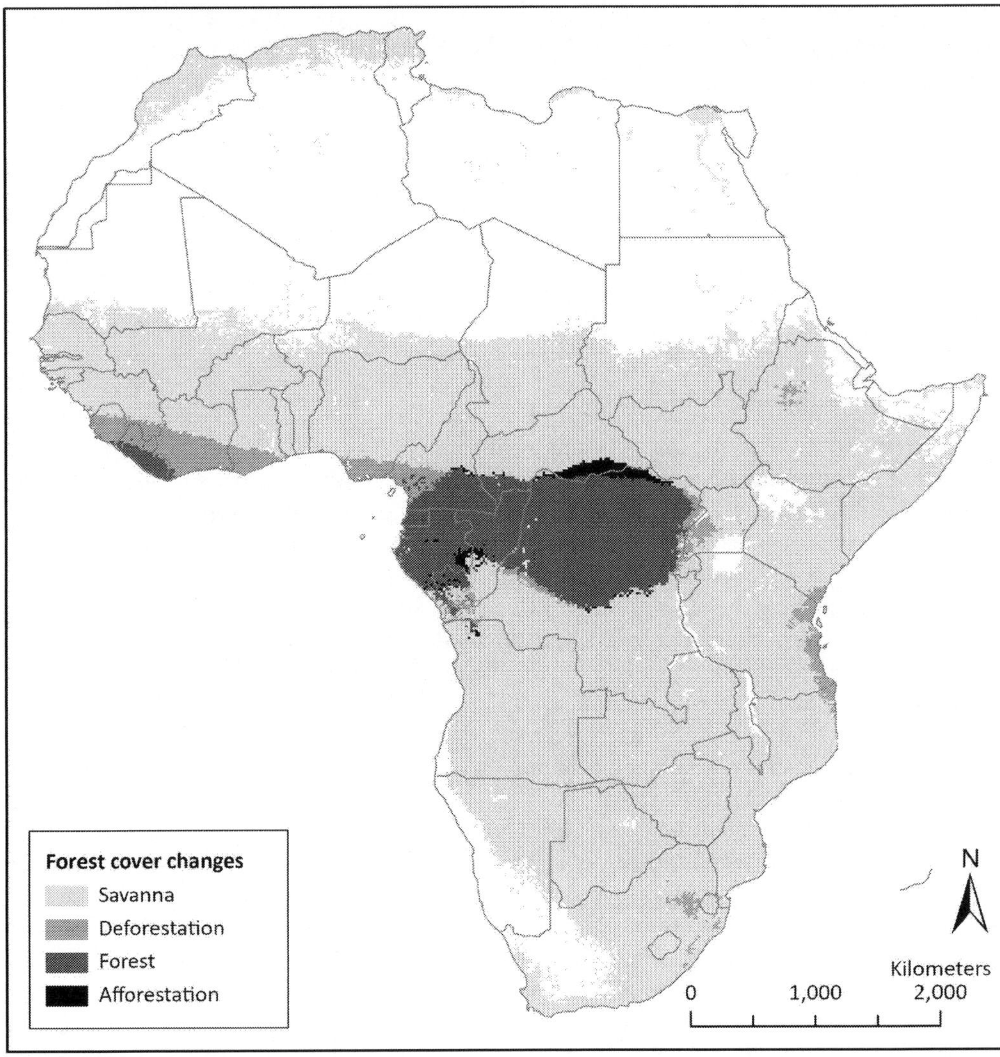

Figure 16.1 Modelled distribution of closed forest (1900–2000) based on a modelling approach using various reference data—following methods and sources detailed in Aleman, Jarzyna and Staver (2018). Redrawn by A Verhoek.

these losses. We know less about the long-term trends in open tree cover, although there appear to be mixed patterns with increasing tree cover in many savanna regions.

Forest loss is often linked to deterioration in access to water (e.g., Mapulanga & Naito, 2019). There are also positive stories. Africa possesses considerable potential for increased tree cover and is the focus of various ambitious tree-based restoration initiatives (Ellison & Speranza, 2020). Many local regreening initiatives have arisen across Niger, Mali, Senegal, Burkina Faso, Ethiopia and Malawi, in which farmers have permitted more trees to grow on their land (Magrath, 2020). Increased tree cover appears to be associated with improved crop yields in dry climates; for example, Reij, Tappan and Smale (2009) noted that the 5 million ha of increased tree cover in agricultural lands that they documented in southern Niger involved increased grain yields of 100 kg or more per hectare. In wetter climates, such as the East African highlands, agroforestry is ubiquitous on small farms, providing firewood, poles, fruits and fodder, although there are clear trade-offs with crop yields close to the trees (Ndoli et al., 2017).

In this essay, I examine the role of trees and forests in securing reliable water, a fundamental cornerstone of food production. Such a brief overview is necessarily limited in scope. I draw from some accessible reviews and encourage reading of these sources for further details (Bruijnzeel et al., 2023; Ellison et al., 2017; Sheil, 2018, 2019; Sheil & Bargués Tobella, 2020). I focus on dispelling the common misconception that there is necessarily a trade-off between trees and water. I then summarise the current understanding of how tree cover can enhance water availability, both locally (primarily infiltration, while the interception of fog and cloud droplets is also important in some locations) and at larger scales (via atmospheric processes). I conclude with a brief discussion of the implications.

2 Misconceptions

The topics I present appear neglected. I suspect that part of the reason lies in their technical nature, the fact that they lie between disciplines and the prevalence of misconceptions. I will address these misconceptions directly.

Many believe that trees necessarily reduce local water availability. This belief remains common even in expert-led publications. Where does it come from? The many decades of debate are well covered elsewhere (see, e.g., Bonan, 2023). One key event was when, in 2005, two top journals endorsed a misleading message. A review article in *Science* reported that more than 500 areas where new forests were planted had become drier, with local stream flow declining by more than 50% (Jackson et al., 2005). An editorial in *Nature* quoted the first author, directly, "It does not matter where you are in the world, when you grow trees on croplands, you use more water" (Hopkin, 2005) (see trade-off in Figure 16.2, line A). The simple message couldn't have been clearer and was also easy to repeat and share: "more trees means less water". Unfortunately, this message was wrong.

While some experts have raised doubts immediately, few people have heard them. Few were aware of the comparisons of streamflow data derived from a few long-term catchment studies (not the "more than 500 places" claimed in the *Nature* summary; further reading suggests that this was a mistaken reference to the 26 sites, with 506 contrasts used in the article), of which just two were in the tropics and none were in drylands. Other limitations were also ignored, such as the fact that the study only compared treeless versus densely planted forest stands, did not assess locations with heavy grazing or degraded soils, included sites where the native vegetation was not closed forest and neglected intermediate tree densities.

Counter examples were already well known such as when trees capture moisture from clouds or fog. In tropical mountains, such cloud capture can contribute as much as 1,990 mm yr^{-1} and three quarters of the total runoff (Bruijnzeel et al., 2011). Many drylands receive streams from upland regions through such processes (Viviroli et al., 2020). Nonetheless, in Africa, these linkages remain poorly recognised.

The often positive role of trees in soil moisture and local water availability has also long been indicated through local observations. Loss of tree cover coincided with springs drying, while increased cover improved water availability; indeed, many such observations are known from Africa and are often recognised by local people (Carey, 2020).

In recent decades, we have gained many insights into the ways in which trees can influence soil moisture, such as hydraulic lift, interception and infiltration (see Table 1 in Sheil & Bargués Tobella, 2020). At larger scales, the processes that determine rain are also important (see later sections).

There is now good evidence that landscapes with trees can sometimes capture several times more water to replenish soil and groundwater than they would without trees. Three years of careful measurements in a landscape involving scattered trees, crops and grazing in Burkina Faso have shown how trees increase infiltration, reduce runoff and soil surface evaporation and thus substantially bolster groundwater recharge (Ilstedt et al., 2016). According to observation-based estimates, only some 10 mm of rain a year reaches groundwater in treeless areas while around trees this amount increases dramatically due to soil macropores such as the channels created by roots and soil fauna (Bargués Tobella et al., 2014). A few trees per hectare can substantially improve recharge, which is maximised, with an estimated 5–6 times more water being captured than when treeless (Ilstedt et al., 2016) (see Figure 16.2, line B).

The right density of trees, with a value above zero, on a given piece of land can sometimes improve groundwater recharge severalfold (Ilstedt et al., 2016). Both the optimal tree cover and magnitude of the resulting benefits depend on the context: soil conditions, terrain and rainfall patterns. Observations in Morogoro, Tanzania, have clarified how the benefits can depend on grazing intensities: the positive relationship between tree cover and infiltration capacity was clearest when grazing intensity was low but appeared less apparent at higher grazing pressure (Lulandala et al., 2022).

Studies of farmer-managed tree regeneration in the Sahel indicate substantial increases in soil organic matter content and infiltration rates. Indeed, new data from 3,573 plots distributed across 18 countries in sub-Saharan Africa show a consistent positive relationship between infiltration rates and woody cover (as well as between soil sand content and soil organic carbon), especially when sites with soils compacted by heavy grazing are excluded (Bargués-Tobella et al., 2024). Agronomists have various other methods to improve infiltration, such as terracing, organic inputs and contour planting; but tree cover provides a cheap and simple option over vast regions, especially in grazing lands.

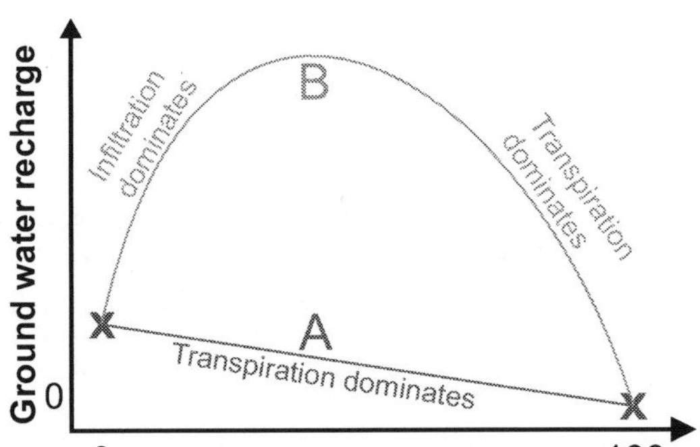

Figure 16.2 Schematic relationship between tree cover and groundwater recharge. In the past, many assumed, as in line A, that transpiration is proportional to tree cover and dominates over the entire range of tree densities; hence, more trees means less water. More recently, we recognise that in some contexts, a small number of trees can have a major positive influence on groundwater recharge (mainly through infiltration and preferential flow), which can dominate at low tree densities, leading to a strongly nonlinear peak distribution, as in line B. The mean annual recharge values for the Burkina Faso study are 10 mm for a landscape with no trees and more than 50 mm for the peak in curve B.

Source: Drawn by the author.

A closely related technical claim that 'more forest implies less total streamflow' is true and widely repeated too but is again misleading. There are two components of streamflow: *stormflow*, which can involve large stormflow volumes in short periods and can thus be destructive, causing erosion, flooding and siltation; and *baseflow*, which is the water-sustaining dry season that flows from groundwater and slow subsurface flows. Most people want and rely on the baseflow. Recent studies have shown that increasing tree cover can often bolster baseflow and thus provide more useful water. The largest absolute gains are expected in degraded areas with poor infiltration and deep soils capable of holding water, where most of the rain falls over short periods (Bruijnzeel et al., 2023).

The misconception that removing trees would make Africa wetter is dangerous. A treeless continent would be considerably drier, and life for most would be extremely challenging. Crucial to this understanding are the local processes that retain water in the landscape through infiltration, soil storage and redistribution. We will now explore the atmospheric water cycle and how trees present opportunities.

3 Atmospheric water recycling

Most people in Africa depend on reliable rainfall to meet their water needs. This rainfall depends on the behaviour of atmospheric moisture. Understanding this behaviour is vital. In recent years, it has become clear that trees play a greater role in determining this behaviour than was previously recognised (Sheil, 2018).

Trees make various contributions to atmospheric water and multiple mechanisms are involved (Sheil, 2018; Sheil & Bargués Tobella, 2020). Trees can access and store water unavailable to other vegetation types and return it to the atmosphere (much more per unit area than most crops and other low vegetation types, although irrigated agriculture can also be of local significance) (Sheil, 2014; Sheil & Murdiyarso, 2009). Another set of mechanisms relates to the aerosol particles that determine when and where water vapour condenses. I have summarised such topics for nonspecialist readers elsewhere (Sheil, 2018). Despite their significance, we know little about these particles and their dependencies and implications in Africa. Trees and forests provide both much of the water vapour and likely most of the aerosol particles that contribute to and determine rainfall downwind—and trees and forests also play a role in drawing in moisture from one region from another and in stabilising the processes that maintain rain.

Forest loss at smaller scales has been shown to influence circulation patterns, sometimes leading to local increases in the frequency of rainfall, as observed in West Africa (Taylor et al., 2022). At larger scales, net declines in rainfall follow forest loss (Lawrence & Vandecar, 2015; Smith, Baker, & Spracklen, 2023). The simulated impacts of reduced forest cover in Central Africa with climate models generally indicate increased local temperatures and declines in rainfall, yet detailed predictions are often very inconsistent among models (Smith, Baker, & Spracklen, 2023). While most publications around these topics rely on such simulations, I doubt their value. The models neglect or simplify key mechanisms (see also Sheil, 2018; Smith, Baker, & Spracklen, 2023), and model calibration remains poor in Africa, where data remain sparse (Creese, Washington, & Jones, 2019; Washington et al., 2013).

Atmospheric scientists can trace moisture and identify where rainfall originates. Globally, of the approximately 117,600 km^3 of rain that falls on land each year, approximately 71,800 km^3 (just over 60%) evaporates or transpires from land (Schneider et al., 2017). At least half of this land-derived moisture is transpired by trees and other perennial vegetation (Wei et al., 2017). The local dependence of precipitation on land-derived moisture varies with location and season (van der Ent et al., 2010). In Africa, the Congo Basin and the Great Lakes Region are

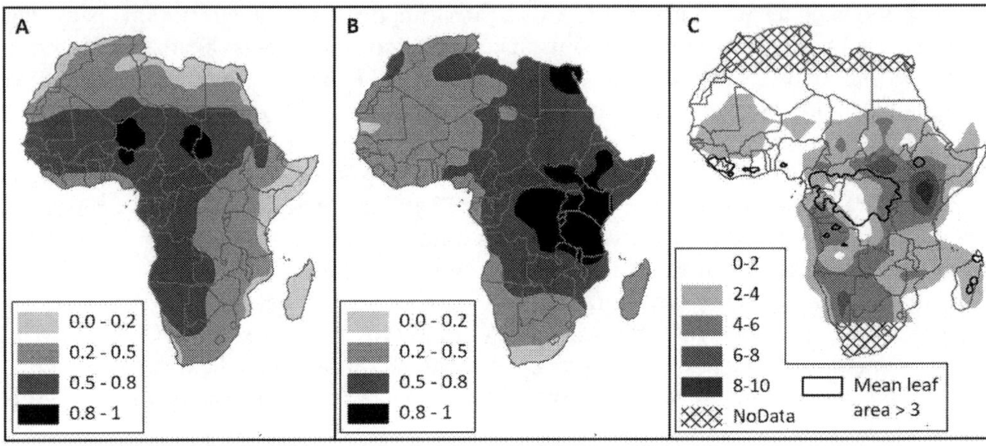

Figure 16.3 Proportion of rainfall recycled from land by where it ends up (a) and where it comes from (b) and the inferred role of forest (c). (a, b) Annual average continental precipitation recycling ratio in which darker colours denote that more than 50% of this rainfall moisture is derived from land—redrawn with modifications from van der Ent (2014). (c) Relationships between daily rainfall and cumulative exposure of 10-day back trajectories to canopy density (measured by the cumulative leaf area index) for 2001–2007. The number of calendar months (shade scale) with significant (P<0.01) positive relationships between precipitation and canopy density. Stippling denotes regions where rain is a factor of at least two greater where the wind arrives from dense forest versus open areas. The contour delimits areas with >3 m^2 m^2 mean leaf area. The materials were redrawn with modifications from A Verhoek (Spracklen, Arnold, & Taylor, 2012).

major sources of rain within the continent (van der Ent et al., 2010). There are large regional moisture flows from source to receiver areas differing (Figure 16.3a, b). One tracing study estimated that 40% of South Sudan's rain came from Central African evaporation (Salih, Zhang, & Tjernström, 2015), while another estimated that 32% of the rain in the northern Ethiopian highland arrived via humid West and Central Africa (Viste & Sorteberg, 2013). Comparable values are found elsewhere in the Sahel and have been judged as "vulnerable to deforestation in the Congo" (Keys et al., 2012). The circulation patterns are seasonal, making the resulting relationships and dependencies seasonal. For example, most January rainfall in West Africa depends on recycled moisture, making it particularly susceptible to upwind deforestation (Keys et al., 2014).

Studies that track air flows and rainfall show that air that passes over tropical tree cover captures more water and produces more rain than does air that passes over sparse vegetation (Spracklen et al., 2012); this pattern is also observed in sub-Saharan Africa (Figure 16.3c). When combining techniques and a few simplifying assumptions, it is also possible to estimate how much rain depends directly on vegetation. One recent study summarised this for major catchments across Africa and concluded that approximately half of Africa's total rain originates from transpiration, although this varies among sites (Figure 16.4) (Te Wierik et al., 2022). Any changes in continental transpiration thus impact rainfall.

A recent pantropical study examining changes in tropical forests and rainfall between 2003 and 2017 noted that the magnitude of the measured impact increased as the scale of observation increased; at the largest scale considered, two degrees or approximately 200 km, a 1% decrease in forest cover led to an average decrease in rainfall of more than 2 cm per year in Central

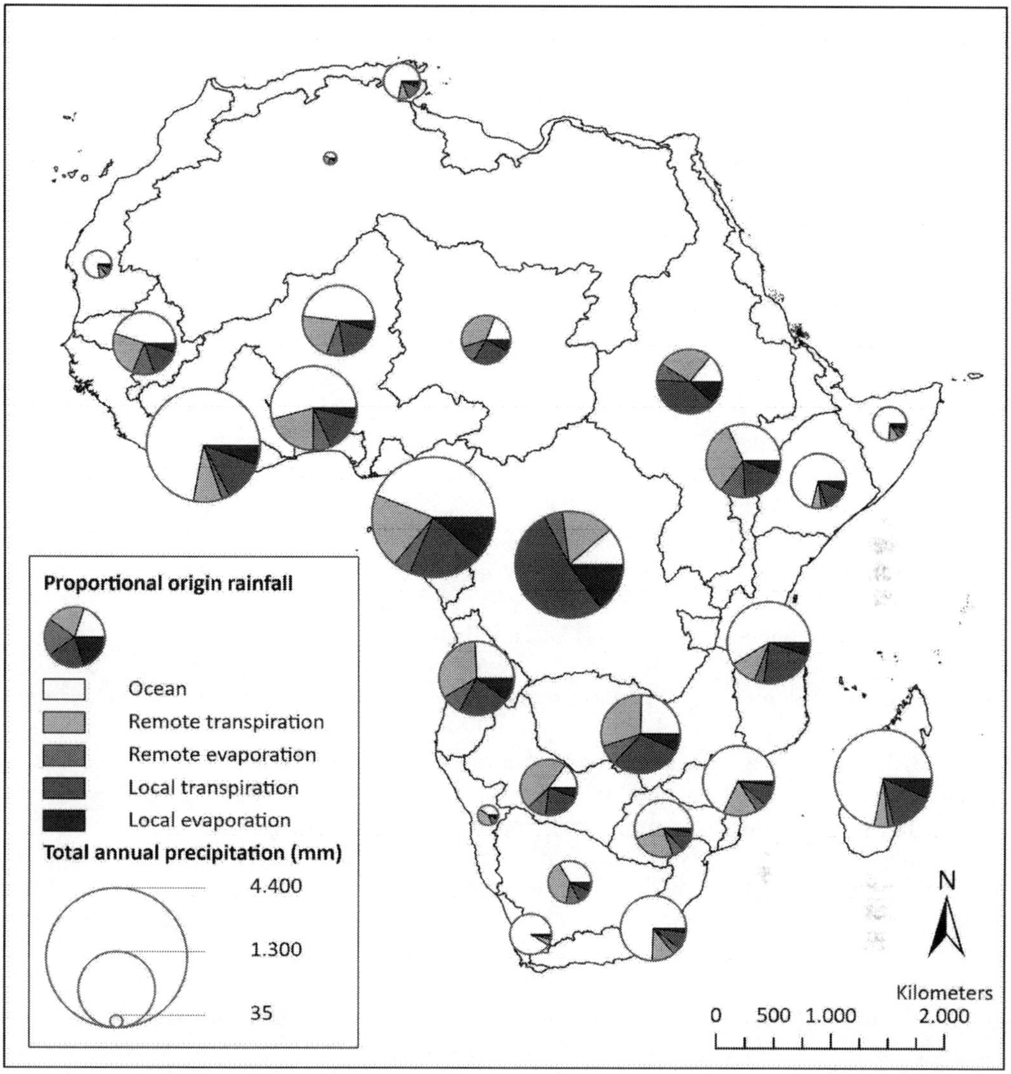

Figure 16.4 The estimated sources by volume contributed to the mean annual precipitation across major African watersheds (1981–2016). The size of each pie chart represents the magnitude of the mean annual precipitation. Redrawn by A Verhoek from the data and analysis from Te Wierik et al. (2022).

Africa. This is a conservative estimate, as it neglects feedback and outcomes across wider regions (Smith, Baker, & Spracklen, 2023).

The water cycle also influences local temperatures. Like evaporation, transpiration not only contributes to rainfall but also dissipates heat, reducing local surface temperatures. This leads to cloud formation, affecting planetary albedo and influencing how heat energy moves in the atmosphere. Drier places can typically become much hotter than wetter places. Such moderation of temperatures is important when we are confronted with dangerous heat events that threaten crops, livestock and people.

4 Importing water, feedbacks and tipping points

A recent conceptual shift has been to distinguish rainfall recycling from the processes that draw water vapour to specific locations. Both aspects can be influenced by changes in tree cover. This second process, often neglected in the past, is often referred to as "convergence" and can be thought of as the reason why some areas experience more rain than others in similar settings. Often, the role of terrain and other factors may obscure these local effects. One recent study combined various observations to distinguish the role of moisture recycling and the import of water vapour and clarified how changes in tree cover may impact rain (Makarieva et al., 2023). One key implication is that forest loss not only impacts rainfall recycling but also reduces the import of water (moist air) from elsewhere. Another key implication underlines how local conditions influence such outcomes. When the atmosphere is sufficiently moist, small increases in transpiration can amplify imports, precipitation and total water yield. In contrast, when the atmosphere is sufficiently dry, the benefit of additional vegetation is confined to precipitation recycling, while the import of atmospheric moisture may be reduced (even favouring moisture export). These implications explain why increased tree cover can bolster or diminish rain in different contexts (see Makarieva et al., 2023).

Such observations indicate powerful feedbacks that, as observed in monsoon climates, can result in alternative wet or dry climate states. When a system is close to the tipping point, any change in the contribution from trees and forests (or reservoirs, irrigation or drainage) may tip the system from one state to another. These influences are nonlinear, as both the likelihood and volume of rain are highly sensitive to humidity; for example, under some conditions, a 10% decrease in relative humidity may reduce precipitation by 50% or more. In such contexts small changes in tree cover may have unexpectedly large impacts. A satellite observation study (1988–2013) of the Congo Basin indicated that the first (March–May) dry season had already lengthened, starting sooner and ending later, by 6.4–10.4 days per decade—a result ascribed to low soil moisture delaying the return of the wet season (Jiang et al., 2019).

By drawing on water accessible only to deep roots and stored in their stems, trees can maintain transpiration when most other vegetation cannot. Access to such moisture permits trees to develop new leaves and transpire even after a protracted dry season, a phenomenon well established in African woodlands and the continent more generally (Adole, Dash, & Atkinson, 2018). Such a "green up" contributes to the atmospheric moisture needed to trigger rains in monsoon climates. Weakening monsoons have been linked to deforestation in various parts of the world. Vegetation thus does not simply respond to climatic events but is active in the processes that maintain them.

Among the ideas advanced to explain such feedbacks, the "biotic pump" has gained particular attention (Makarieva & Gorshkov, 2007; Sheil & Murdiyarso, 2009). This theory explains how forests create low-pressure regions that draw in moist air from surrounding higher-pressure areas, sustaining high rainfall far within continents (Makarieva et al., 2013; Makarieva & Gorshkov, 2007). This theory can explain the rapid back and forth switching that determines the start and end of tropical monsoons. Importantly, this theory indicates that changes in tree cover impact not only moisture entering the atmosphere but also the patterns of atmospheric circulation; thus, land-users should not only be concerned about forests as a source of upstream moisture but also as downstream drivers of the winds that carry this moisture. While sometimes labelled "controversial" (Pearce, 2020), this idea is substantiated by observations (Makarieva et al., 2023).

Tipping points arise from more than the inherent day-to-day or seasonal aspects of the climate system but have lasting impacts through their links to, and consequences for, land cover.

Droughts can promote forest loss through tree death, fires and increased forest clearing (see, e.g., Leblois, 2021). Thus, once a system reaches a threshold, it can switch from wet (water importing) to drought prone (water exporting). Suitable reforestation can reduce and reverse these impacts and increase moisture availability in regions that are now dry and unproductive.

Our understanding of the atmospheric water cycle and how it is affected by land cover conditions is incomplete but improving. One recent study based on observations across African drylands concluded that more extensive and later dry season fires lead to reductions in rainfall of up to 30 mm or approximately 10% of the total—likely a result of multiple mechanisms (Saha, Scanlon, & D'Odorico, 2016).

5 Implications for land use

Tree cover offers major opportunities for safeguarding and improving water availability and thus contributing to food security. While what is desirable and acceptable varies with context, we know enough to provide constructive guidance to policymakers, planners and practitioners. In the context of ensuring food security, trees and forests are not minor afterthoughts but rather life-support systems. Forest loss jeopardises water resources and food production. Conversely, maintaining and expanding tree cover can stabilise and enhance water supply, supporting food security. Policymakers, planners and practitioners must recognise trees as essential to these systems, not afterthoughts.

Various studies have shown that restoring tree cover in open landscapes improves the hydrological functioning of rivers and streams. The effectiveness of restoration efforts depends on various factors, including climate, soil, terrain and vegetation type. It is important to note that both the goals and the scale of observations can affect what is desirable and judged successful. Some may want to increase groundwater recharge, while others require at least some runoff for filling reservoirs or wish to lower groundwater to avoid salt flows or reduce mosquitoes. Nonetheless, some generalisations seem possible for those who wish to maintain groundwater and sustain or bolster rainfall. As shown in Burkina Faso, the optimal level of tree cover can sometimes boost soil water and groundwater by several hundred percent when compared to treeless areas. Tree and land management—such as tree species selection, land-use practices, grazing and pruning—will also influence soil moisture and water availability. For example, tree pruning reduces transpiration but has little effect on infiltration.

Not all species have similar properties in terms of accessing, using and influencing water—as well as the other roles and services that people need. Here, we lack space to review these options, and much remains unknown but underlines their importance. Native species are typically adapted to local conditions, while some exotic species may be maladapted—for example, by increasing drought and the risk of fire. We should look at natural systems to provide both template and inspiration (Sheil & Bargués Tobella, 2020). Where natural systems are not an option, local practices that include tree cover offer a useful guide to what works in specific locations (Carey, 2020; Reij & Winterbottom, 2015).

The regreening seen in parts of the Sahel shows that improvement is not only possible but also happening. While agencies have helped in some locations, the scale of impact results from individual farmers choosing to favour and manage the regrowth of trees on their fields. There is much to learn from this process. "In a world grappling with the challenges of food insecurity, climate change, landscape degradation and rural poverty, regreening offers a path forward, especially in dryland areas" (Reij & Winterbottom, 2015). These authors proposed six steps to help scale up this regreening while noting that not all may be applicable in every situation—they start by identifying and analysing existing successes and emphasising support for grassroots movements and peer-to-peer learning along with attention given to research and communication (Reij & Winterbottom, 2015).

Trees have transformative powers. At larger scales, maintaining tree cover upwind, especially near oceans, lakes, mountains and forests (especially wet forests), should bolster rainfall across regions and continents, while downwind tree cover protects the atmospheric flows on which moisture import occurs. Protecting and restoring tree cover in and around all such regions is fundamental. Isolated, or extended, tree cover projects such as the Great Green Wall (Ellison & Speranza, 2020) are unlikely to be useful without more general regional tree cover. Protecting and restoring tree cover and other natural vegetation is crucial, as is promoting agroforestry and scattered tree cover within landscapes dedicated to food production. Trees and forests sustain hydrological processes that these systems depend on, and trade-offs may need to be identified and managed at local and regional levels. A balanced landscape that preserves water is important for safeguarding the future for everyone.

Despite our improved knowledge, the critical role of tree cover in safeguarding Africa's water security remains undervalued by policymakers. Action is necessary, requiring supportive policies. Communication is paramount. We must actively dispel misconceptions and emphasise the economic, social and environmental values that tree cover offer. To secure and enhance reliable water access for future generations, prioritising the protection and, wherever possible, the expansion of natural tree cover is non-negotiable. Where agriculture and livestock are necessary, strategies to integrate trees within and between fields, and across the wider landscape, must be actively pursued. The challenges are undoubtedly considerable, but so are the opportunities to ensure water security for future generations.

References

Adole, T., Dash, J., & Atkinson, P.M. (2018). Large-scale prerain vegetation green-up across Africa. *Global Change Biology*, *24*(9), 4054–4068. https://doi.org/10.1111/gcb.14310

Aleman, J.C., Jarzyna, M.A., & Staver, A.C. (2018). Forest extent and deforestation in tropical Africa since 1900. *Nature Ecology & Evolution*, *2*(1), 26–33. https://doi.org/10.1038/s41559-017-0406-1

Bargués-Tobella, A., Winowiecki, L.A., Sheil, D., & Vågen, T.G. (2024). Determinants of soil field-saturated hydraulic conductivity across sub-Saharan Africa: Texture and beyond. *Water Resources Research*, *60*(1), e2023WR035510.

Bargués Tobella, A., Reese, H., Almaw, A., Bayala, J., Malmer, A., Laudon, H., & Ilstedt, U. (2014). The effect of trees on preferential flow and soil infiltrability in an agroforestry parkland in semiarid Burkina Faso. *Water Resources Research*, *50*(4), 3342–3354. https://doi.org/10.1002/2013WR015197

Bonan, G. (2023). *Seeing the forest for the trees.* Cambridge University Press.

Bruijnzeel, L.A., Mulligan, M., & Scatena, F.N. (2011). Hydrometeorology of tropical montane cloud forests: Emerging patterns. *Hydrological Processes*, *25*, 465–498. https://doi.org/10.1002/hyp.7974

Bruijnzeel, L.A., Peña-Arancibia, J.L., Ziegler, A.D., Birkel, C., Sun, G., Wang, Y., … & Sheil, D. (2023). Potential for improved hydrological functioning and stream baseflows through forest landscape restoration in the tropics. In *Preprints*: Preprints.

Carey, J. (2020). News feature: The best strategy for using trees to improve climate and ecosystems? Go natural. *PNAS*, *117*(9), 4434–4438. https://doi.org/10.1073/pnas.2000425117%

Creese, A., Washington, R., & Jones, R. (2019). Climate change in the Congo Basin: Processes related to wetting in the December–February dry season. *Climate Dynamics*. https://doi.org/10.1007/s00382-019-04728-x

Ellison, D., Morris, C.E., Locatelli, B., Sheil, D., Cohen, J., Murdiyarso, D., … & Sullivan, C.A. (2017). Trees, forests and water: Cool insights for a hot world. *Global Environmental Change*, *43*, 51–61. https://doi.org/10.1016/j.gloenvcha.2017.01.002

Ellison, D., & Speranza, C.I. (2020). From blue to green water and back again: Promoting tree, shrub and forest-based landscape resilience in the Sahel. *Science of The Total Environment*, *739*, 140002.

Heino, M., Puma, M.J., Ward, P.J., Gerten, D., Heck, V., Siebert, S., & Kummu, M. (2018). Two-thirds of global cropland area impacted by climate oscillations. *Nature Communications*, *9*(1), 1257. https://doi.org/10.1038/s41467-017-02071-5

Hopkin, M. (2005). Tree planting not always green. *Nature*. https://doi.org/10.1038/news051219-14

Ilstedt, U., Bargués-Tobella, A., Bazié, H.R., Bayala, J., Verbeeten, E., Nyberg, G., … & Malmer, A. (2016). Intermediate tree cover can maximize groundwater recharge in the seasonally dry tropics. *Scientific Reports*, *6*(1), 21930. https://doi.org/10.1038/srep21930

Jackson, R.B., Jobbágy, E.G., Avissar, R., Roy, S.B., Barrett, D.J., Cook, C.W., … & Murray, B.C. (2005). Trading water for carbon with biological carbon sequestration. *Science*, *310*(5756), 1944–1947.

Jiang, Y., Zhou, L., Tucker, C.J., Raghavendra, A., Hua, W., Liu, Y.Y., & Joiner, J. (2019). Widespread increase of boreal summer dry season length over the Congo rainforest. *Nature Climate Change*, *9*(8), 617–622. https://doi.org/10.1038/s41558-019-0512-y

Keys, P.W., Barnes, E., Van Der Ent, R., & Gordon, L.J. (2014). Variability of moisture recycling using a precipitationshed framework. *Hydrology and Earth System Sciences*, *18*(10), 3937–3950.

Keys, P.W., Van der Ent, R., Gordon, L.J., Hoff, H., Nikoli, R., & Savenije, H. (2012). Analyzing precipitationsheds to understand the vulnerability of rainfall dependent regions. *Biogeosciences*, *9*(2), 733–746.

Lawrence, D., & Vandecar, K. (2015). Effects of tropical deforestation on climate and agriculture. *Nature Climate Change*, *5*(1), 27–36.

Leblois, A. (2021). Mitigating the impact of bad rainy seasons in poor agricultural regions to tackle deforestation. *Environmental Research Letters*, *16*(5), 054003. https://doi.org/10.1088/1748-9326/abdbf0

Lulandala, L., Bargués-Tobella, A., Masao, C.A., Nyberg, G., & Ilstedt, U. (2022). Excessive livestock grazing overrides the positive effects of trees on infiltration capacity and modifies preferential flow in dry miombo woodlands. *Land Degradation & Development*, *33*(4), 581–595. https://doi.org/10.1002/ldr.4149

Magrath, J. (2020). *Regreening the Sahel: A quiet agroecological evolution*. Oxfam, GB.

Makarieva, A., Gorshkov, V., Sheil, D., Nobre, A., & Li, B.-L. (2013). Where do winds come from? A new theory on how water vapor condensation influences atmospheric pressure and dynamics. *Atmospheric Chemistry and Physics*, *13*(2), 1039–1056.

Makarieva, A.M., & Gorshkov, V.G. (2007). Biotic pump of atmospheric moisture as driver of the hydrological cycle on land. *Hydrology and Earth System Sciences*, *11*(2), 1013–1033. Retrieved from <Go to ISI>://000246812000025.

Makarieva, A.M., Nefiodov, A.V., Nobre, A.D., Baudena, M., Bardi, U., Sheil, D., … & Rammig, A. (2023). The role of ecosystem transpiration in creating alternate moisture regimes by influencing atmospheric moisture convergence. *Global Change Biology*, *29*, 2536–2556. https://doi.org/10.1111/gcb.16644

Mapulanga, A.M., & Naito, H. (2019). Effect of deforestation on access to clean drinking water. *Proceedings of the National Academy of Sciences*, *116*(17), 8249–8254. https://doi.org/10.1073/pnas.1814970116

Ndoli, A., Baudron, F., Schut, A.G.T., Mukuralinda, A., & Giller, K.E. (2017). Disentangling the positive and negative effects of trees on maize performance in smallholdings of Northern Rwanda. *Field Crops Research*, *213*, 1–11.

Pearce, F. (2020). Weather makers. *Science*, *368*(6497), 1302–1305. https://doi.org/10.1126/science.368.6497.1302

Reij, C., Tappan, G., & Smale, M. (2009). Agroenvironmental transformation in the Sahel: Another kind of 'green revolution'. IFPRI Discussion Paper 00914. International Food Policy Research Institute Washington, DC.

Reij, C., & Winterbottom, R. (2015). Scaling up regreening: Six steps to success. World Resources Institute, Washington DC, United States, p. 66. https://www.wri.org/research/scaling-regreening-six-steps-success

Ripple, W.J., Wolf, C., Newsome, T.M., Galetti, M., Alamgir, M., Crist, E., … & 15,364 scientist signatories from 184 countries (2017). World scientists' warning to humanity: A second notice. *BioScience, 67*(12), 1026–1028.

Saha, M.V., Scanlon, T.M., & D'Odorico, P. (2016). Suppression of rainfall by fires in African drylands. *Geophysical Research Letters, 43*(16), 8527–8533. https://doi.org/10.1002/2016GL069855

Salih, A.A., Zhang, Q., & Tjernström, M. (2015). Lagrangian tracing of Sahelian Sudan moisture sources. *Journal of Geophysical Research: Atmospheres, 120*(14), 6793–6808.

Schneider, U., Finger, P., Meyer-Christoffer, A., Rustemeier, E., Ziese, M., & Becker, A. (2017). Evaluating the hydrological cycle over land using the newly-corrected precipitation climatology from the Global Precipitation Climatology Centre (GPCC). *Atmosphere, 8*(3), 52.

Sheil, D. (2014). How plants water our planet: Advances and imperatives. *Trends in Plant Science, 19*(4), 209–211.

Sheil, D. (2018). Forests, atmospheric water and an uncertain future: The new biology of the global water cycle. *Forest Ecosystems, 5*(19), 1–22.

Sheil, D. (2019). Importance of Central Africa's forests for regional climateand rainfall. Briefing Paper. Tropenbos International, Wageningen, the Netherlands. https://www.tropenbos.org/file.php/2314/18092019-briefingafricancentralrainforests.pdf

Sheil, D., & Bargués Tobella, A. (2020). More trees for more water in drylands: Myths and opportunities. *ETFRN News*.

Sheil, D., & Murdiyarso, D. (2009). How forests attract rain: An examination of a new hypothesis. *BioScience, 59*(4), 341–347.

Smith, C., Baker, J.C.A., & Spracklen, D.V. (2023). Tropical deforestation causes large reductions in observed precipitation. *Nature*. https://doi.org/10.1038/s41586-022-05690-1

Spracklen, D.V., Arnold, S.R., & Taylor, C.M. (2012). Observations of increased tropical rainfall preceded by air passage over forests. *Nature, 489*(7415), 282–286. https://doi.org/10.1038/nature11390

Taylor, C.M., Klein, C., Parker, D.J., Gerard, F., Semeena, V.S., Barton, E.J., & Harris, B.L. (2022). "Late-stage" deforestation enhances storm trends in coastal West Africa. *Proceedings of the National Academy of Sciences, 119*(2), e2109285119. https://doi.org/10.1073/pnas.2109285119

Te Wierik, S.A., Keune, J., Miralles, D.G., Gupta, J., Artzy-Randrup, Y.A., Gimeno, L., … & Cammeraat, L.H. (2022). The contribution of transpiration to precipitation over African watersheds. *Water Resources Research, 58*(11), e2021WR031721. https://doi.org/10.1029/2021WR031721

van der Ent, R.J. (2014). A new view on the hydrological cycle over continents (Doctorate). Technische Universiteit Delft, Netherlands.

van der Ent, R.J., Savenije, H.H.G., Schaefli, B., & Steele-Dunne, S.C. (2010). Origin and fate of atmospheric moisture over continents. *Water Resources Research, 46*. https://doi.org/10.1029/2010wr009127

Viste, E., & Sorteberg, A. (2013). Moisture transport into the Ethiopian highlands. *International Journal of Climatology, 33*(1), 249–263.

Viviroli, D., Kummu, M., Meybeck, M., Kallio, M., & Wada, Y. (2020). Increasing dependence of lowland populations on mountain water resources. *Nature Sustainability*. https://doi.org/10.1038/s41893-020-0559-9

Washington, R., James, R., Pearce, H., Pokam, W.M., & Moufouma-Okia, W. (2013). Congo Basin rainfall climatology: Can we believe the climate models? *Philosophical Transactions of the Royal Society B: Biological Sciences, 368*(1625), 20120296. https://doi.org/10.1098/rstb.2012.0296

Wei, Z., Yoshimura, K., Wang, L., Miralles, D.G., Jasechko, S., & Lee, X. (2017). Revisiting the contribution of transpiration to global terrestrial evapotranspiration. *Geophysical Research Letters, 44*(6), 2792–2801.

17 A water angle on Africa's 2050 food security challenge

Gert Jan Veldwisch, Jonathan Denison, Gabriella Izzi, Jean Kamwamba-Mtethiwa, Hans Komakech, Bancy Mati, and Pieter Waalewijn

1 Introduction

The projected significant population growth in Africa poses substantial challenges for achieving food security by 2050. Ongoing and anticipated climate changes are seriously impeding the potential for increasing agricultural production, particularly due to water insecurities. Among all the water utilized by humans, a staggering 92% is estimated to be used in agriculture, largely as the 'water we eat' (Hoekstra and Mekonnen, 2012; Hoekstra, 2020). For Africa too, by far most water is used for agriculture, though only a fraction of its water resources are utilized in comparison to the potential capacity. While data on agricultural water use across Africa is known to underestimate reality due to unrecorded farmer-led irrigation development, expansion is needed and possible (African Union, 2020). This observation might direct decision-maker's attention towards expansion and production increase through large infrastructure investments and technological advancements. However, a significant portion of undernourishment in Africa (and other regions) stems from uneven wealth distribution, inadequate social protection systems, and inequitable market access. An irrigation expansion drive, or "expanding the size of the pie", will on its own fail to reduce the prevailing exclusion of many farmers from the food availability and financial benefits of irrigation access. A more nuanced approach is therefore needed alongside, and as part of, an expansion agenda. Such an approach would strive to enable irrigation water and technology access for smallholder farmers who are currently excluded yet have the capability to drive an irrigation farming business. It must also address their land-tenure insecurities, particularly for women farmers who face higher risks, and include a focus on increased productivity and profitability. Irrigated agriculture contributes to food security in many ways; for farmers it increases food self-sufficiency and increases earnings allowing purchase of a broader food basket while the urban population can benefit from greater food availability and a more diverse food basket. Still, it is important to realize that an increase in food availability does not by itself lead to sustainable healthy diets (Melse-Boonstra et al., this volume). Beyond improving food availability for consumers there are also multipliers in the wider economy, such as in jobs created along the value chains and a decrease in food prices.

The pivotal question is then how do we get the most benefits for most people from irrigation investments that combine an increase of irrigated areas with a refocus on inclusion and optimization. Such an approach would mean that land and water productivity improvements and profitability-increasing measures would need to focus on complementary tracks: one track focusing on existing irrigation aiming to achieve the quickest and most significant gains in terms of food availability and income generation; the second track seeking irrigation expansion through inclusive interventions that address socio-economic mechanisms that constrain the access to irrigation technology and water. In Africa many opportunities exist for such a shift.

DOI: 10.4324/9781032649696-22

Farming optimization and profitability interventions on the millions of existing irrigated smallholder farms can reduce the large yield gap, increase water productivity, cut labour and pumping costs, and strengthen value chains. Inclusive expansion is justified by the substantially available water for irrigation. While the expansion potential in most sub-Saharan African countries is significant, where and how it exists also depends on increasing cross-sectoral competition, catchment degradation, and high regional variability in water availability.

In Section 2 we first explain and define some key concepts regarding water's crucial relations to food security systems, as well as the transformative processes in which water occupies a central role. This is followed (in Section 3) by an analysis of existing development models related to agricultural water management, framed within broader concepts of agricultural development, and their intricate connection to water use and the distribution of benefits. Section 4 concludes this chapter by presenting a perspective on how a water angle may contribute to a strategy for Africa to achieve food security by 2050.

Throughout this chapter, rather than confining ourselves to enumerating pragmatic "opportunities," we argue from the vantage point of a moral imperative that necessitates a paradigm shift in thinking. We underscore that substantial investment in water infrastructure and farm-system optimization is necessary and unavoidable in the forthcoming decades to ensure resilient farming systems.

2 Water and agricultural production: some key concepts

The risks of crop failure in many African contexts are notably high due to erratic rainfall patterns, which is one of the largest production risk for farmers and is likely to be exacerbated by climate change (Descheemaeker et al., this volume). This unpredictability implies that investing substantial capital in agriculture is often unreasonable without some form of assured water security. Paying for high-quality seeds, artificial fertilizer, and mechanization does not make sense if the looming threat of crop failure resulting from drought remains substantial. Consequently, agricultural water management becomes a vital prerequisite for farmers to embark on agricultural intensification and without it Africa is unlikely to reach self-sufficiency (Van Ittersum et al., 2016).

In practice agricultural water management transcends the dichotomy of irrigated versus rain-fed agriculture, although policy frameworks often maintain this distinction. A more adequate perspective involves considering these practices as a diverse array of combinations encompassing both green and blue water use. Green water corresponds to soil moisture derived from precipitation, while blue water refers to freshwater in flows and storage. Water management practices include flood recession, water harvesting, drainage management, and supplementary and full irrigation from surface and groundwater. Food and Agriculture Organization (FAO, 2024) provides a useful categorization of the agricultural water management array under two main groupings, namely areas equipped for irrigation (blue water practices) and areas with other forms of management (green water enhancement practices, comprising mainly on-field water harvesting, bunding, mulching).

In reality, farmers often combine green and blue water use on the same plot calling for a wider perspective on plant water availability and best solutions. Large numbers of smallholder farmers across the blue-green spectrum actively manage their agricultural water to facilitate intensive farming, often targeted at urban markets (Woodhouse et al., 2017; Veldwisch et al., 2019). These endeavours usually involve a blend of "rainfall-based water management" and resourceful utilization of other water sources such as mountain streams, lakes, and groundwater. We concur with the widely held notion that dependence on rain-fed agriculture renders

Africa vulnerable to escalating climate-induced risks, yet we note that rain-fed production still contributes a larger share of the continent's food production compared to irrigated agriculture (Rockström et al., 2010); with 3–6% of the agricultural land irrigated it likely contributes about 25% of agricultural production value. However, the label "rain-fed agriculture" frequently encompasses agricultural water management practices involving green water enhancement at the field level. Increasing the impact of rainfall where it lands, primarily through concentrating runoff, retaining soil moisture, and managing soil nutrients, stands as a pivotal approach to bolstering food security in Africa (Warner et al., 2012). The reality of diverse practices across the green and blue water spectrum implies that we need an approach that enables the promotion and adoption of mixed practices to support viable economic farming activities rather than technical solutions that simply artificially wet the soil.

While FAO's AQUASTAT (2024) puts the existing 'total agricultural water managed area' in Africa at 21.8 million ha (reference year 2020), there are important indications that the actual numbers are likely (much) larger, depending on how you define and count irrigated area and other forms of agricultural water management. A large share of the area identified in AQUASTAT (43%) is located in Northern Africa, where irrigation practices often more clearly fall with the narrow definition of irrigation. AQUASTAT does include categories for water harvesting, flood recession agriculture, and wetland cultivation, but it depends on countries to report on them. Many countries do not report on them, even though we know these areas exist. Countries that do report on farmer-managed areas seem to structurally under-report on the extent; for Central Mozambique Beekman et al. (2014) demonstrated that the area under irrigation could be about 17 times bigger than in the official statistics, mostly due to irrigation from mountain stream diversion by farmers not being counted. Venot et al. (2021) show that in the region of Shinyanga the area of irrigated rice could be about around 100 times bigger than in the official FAO AQUASTAT statistics and between 7 and 10 times larger than official Tanzania census data suggest, both due to rice paddies in managed wetland cultivation not being counted as irrigation. Drechsel and Keraita (2014) document an area of about 40,000 ha of small-scale urban irrigation in and around Accra, Ghana, which is largely undocumented in official statistics. These are only a few examples that illustrate the extent to which the official statistics fail to capture and represent actual irrigation and other agricultural water management practices across the continent.

The existence of this gap is now well established, and also the reasons underlying such under-reporting have been explored extensively (see for instance Woodhouse et al., 2017; De Bont and Veldwisch, 2020; De Bont et al., 2019; Venot et al., 2021). This realization has formed the basis for many new investment programmes taking actually existing water use practices as the starting point (African Union, 2020; Izzi et al., 2021).

While we thus know that the extent of irrigation in Africa is much larger than official statistics indicate, still it is vulnerable and too little. The irrigation potential of Africa is a contentious concept, often expressed as a single static number of hectares that depends on an assessment of the availability of water and suitable land (Siebert et al., 2010; Altchenko and Villholth, 2015; Xie et al., 2014). From an economic perspective it is more realistic to think of the potential area as dynamically growing and shrinking in relation to economic opportunity related to fluctuating demand of agricultural produce and cost of irrigation, both the cost for equipment and the unit costs of pumping (see for instance Beekman et al., 2014; Izzi et al., 2021). Also, in thinking beyond the dichotomy of rain-fed versus irrigated agriculture, the meaning of 'irrigation potential' stretches even further; where irrigation is supplementary or as a protection in case of dry spells, the overall water use is not very high and the area that can potentially be irrigated is much higher than in case of counting with full irrigation in extremely dry conditions. In thinking about where and for what irrigation can be developed, it should not be about the total area that can be irrigated, but rather about the share of cropped area that can be irrigated, and what value can be generated with it.

Meanwhile, calls for avoiding 'traditional' and 'inefficient' water use practices in African agriculture to 'conserve water' abound. There are two reasons why we want to push back against these arguments. The first reason is strategic. If you are not using about 90–95% of your renewable water resources, focusing on the improvement of the performance of the 5–10% is erroneously importing a foreign paradigm (useful in South Asia, California, or the Middle East) that distracts from focusing on the real issues in most of Africa (for some parts of Southern and for North Africa this does apply). The second reason is technical, a matter of defining losses and efficiencies. Water lost at the field level often gets recaptured downstream, and at the level of the river basin may not amount to losses at all (Lankford et al., 2020). 'Modern irrigation technologies' like drip and sprinkler irrigation might not necessarily 'conserve water'; instead, they often redirect water usage and paradoxically result in less water available downstream (Perry and Steduto, 2017; Lankford and Scott, 2023). For instance, of the irrigation water that is inefficiently applied to a field through surface irrigation methods more than half may leave the field again as a loss from that field. This tail water flow may actually be used by a downstream user. This is also why irrigation schemes in Egypt generally have very low irrigation efficiencies at scheme level, but at the level of the Nile River Basin, Egypt has a very high irrigation efficiency through multiple cycles of water re-use.

For increasing production through real water savings, rather than focusing on improving field-level efficiencies, the focus must shift to a three-pronged agro-social-technical approach that includes: (i) addressing water losses that flow into the ocean by reducing the non-recoverable fraction of return flows; (ii) improving the transmission systems and the physical control of water distribution on schemes, through both technical and management improvements; and (iii) improving irrigation agronomy such as cultivar selection, weeding, and planting times. Thus to achieve 'water conservation' objectives a response is needed that goes well beyond irrigation technologies at field level (Van Opstal et al., 2021; Giordano et al., 2017).

In thinking about the possibilities for agricultural water use in Africa it is important to look at the fluctuations in rainfall and hydrological flows across both time and space, beyond the total annual rainfall or flow. The water flows that can be utilized do not only depend on natural factors but also on: (1) the extent of water resource development and (2) watershed-level degradation and its subsequent impact on hydrology. Water availability varies between seasons and both within and between watersheds, it has upstream-downstream dynamics and varies with distance from the river. Natural variation in space and time is amplified by catchment degradation; peak runoff is increasing, while recharge, baseflow, and moisture feedback loops from trees are reducing (Sheil, this volume).

The prevailing uncertainty surrounding water availability in Africa will, in most locations, exacerbate due to ongoing climate change. This amplification does not primarily stem from absolute shifts in annual precipitation, but rather from an accelerated hydrological cycle powered by increased evapotranspiration due to temperatures increasing and from heightened rainfall variability, particularly intra-annual variability. Unpredictability and delayed onset of the rains along with longer dry spells within the rainy season trigger crop losses with a high impact on vulnerable farmers. Climate change introduces additional stress, both directly by altering physical water availability dynamics and more significantly by exposing the institutional and economic limitations to deal with these dynamics. The latter is referred to as economic water scarcity, which in most of Africa is a greater challenge than hydrologic constraints (Rosa et al., 2020).

Concluding this section, which introduces key concepts for incorporating a water-centric perspective into food security discussions, we turn to the notions of water footprints and virtual water. These concepts pertain to the amount of water inherently tied to agricultural products. They can aid in contemplating the water aspect of food imports and exports, facilitating our thinking of what should be produced where. Simultaneously, these concepts raise ethical concerns about high-value export-oriented production in water-scarce regions, such as whether it is ethical to virtually transfer water to high-end customers across the globe through production and export of roses by international companies from semi-dry areas in Kenya, Ethiopia, and Tanzania, where there is frequently insufficient water for local food production (Vos et al., 2019). The water footprint concepts also shed light on the water implications of food loss, food waste, and diverse diets. For instance, the concept of "resource-inefficient diets" (as referred by de Haas et al. in the introductory chapter of this volume) prompts a water-focused examination. Globally, animal production utilizes 70% of agricultural land, of which roughly half for grazing and half for feed production. Hoekstra (2020) shows that in many countries dairy and meat consumption account for 30–45% of all indirect water use. How exactly this relates to questions of water distribution varies with feed source, type of animal, and food type, with some heavily depending on blue/irrigation water and others on rainfall that would have been difficult to be put to use for anything else but pastures. Within the context of food security, it is imperative to consider the opportunity cost of diverting water to producing animal protein. From a resource standpoint, it is not leaky irrigation canals but rather dripping steaks that should be of greater concern, especially where they come from industrial/specialized farming, as these make use of high-quality feed inputs with high indirect water consumption. Theoretically, fully embracing plant-based diets could instantaneously address the majority of food security concerns, including water scarcity issues. This thought experiment does not only hold for the African continent, but equally so for our global society. It exposes how the rich, through meat production and consumption, divert calories, high-quality proteins, and water, creating scarcity for less endowed people (Hoekstra, 2020). In many places of the world, and certainly in Africa, animal keeping of course has important functions beyond milk and meat production, such as providing a buffer against drought, provision of manure, status and wealth accumulation, and especially in agro-pastoral system the competition with crop production is minimal (Oosting et al., this volume).

3 Development and agricultural water

In this section, we examine three ongoing processes and policies guiding the further development of agricultural water in Africa: (1) farmer-led irrigation development, (2) expansion of corporate/industrial agriculture, and (3) the African Union's investment framework for agricultural water (African Union, 2020).

First, numerous smallholder farmers across Africa display entrepreneurial creativity by engaging in intensive agriculture in response to factors like market demand and technology availability. This is evident in the widespread development of practices such as irrigation, rainwater harvesting, and regenerative agriculture by African smallholder farmers themselves, often with minimal or no external support. Such endeavours not only enhance water security but also improve the efficiency with which other resources are used – such as fertilizers, pesticides, seeds, and labour. Consequently, when African farmers pursue agricultural intensification, it inherently encompasses a water development component, regardless of where their practices fall on the blue-green spectrum. However, while some farmers embrace this intensification, a significant portion continues to view agriculture as a survival strategy ("hanging in"), preferring to exit agriculture ("stepping out") rather than entering the risky domain of agricultural intensification ("stepping up") (Dorward et al., 2009). There are strong indications that farmer-led irrigation development is dominated by farmers who aim to step up or intensify agriculture, whereby they also step up with regard to the technology they use, the amount they produce as well the income generated (Kamwamba-Mtethiwa et al., 2021; Scoones et al., 2019; Veldwisch and Woodhouse, 2022). But also poor farmers focused on subsistence and those aiming to move out of agriculture do take part in farmer-led irrigated development when opportunities arise.

While non-farm income is on the rise in Africa, it remains considerably lower than in other global regions. Instead of viewing non-farm income as a radical alternative to agriculture, it often serves as a source of partial income (see Giller and Andersson, this volume; Pieters, this volume), potentially aiding in balancing the risks posed by unreliable rainfall. It is worth noting that while agricultural income is diminishing for an increasing number of African households, it is being supplemented by non-farm income sources. This highlights that processes of structural change in Africa follow diverse trajectories, as outlined by de Haas et al. in the introductory chapter of this volume.

Second, corporate or industrial agriculture, often organized in estates, serves as a distinct model of agricultural development characterized by intensive water usage. A nuanced understanding of water use by industrial versus smallholder agriculture extends beyond biophysical water productivity to looking at what crops are being produced, for what purpose, and who controls decisions. In terms of economic development, a critical aspect involves assessing the value generated per unit of water, ownership of this value, and its integration into or extraction from the local and national economy in which the agricultural production takes place. While industrial agriculture might exhibit higher water use efficiencies and biophysical water productivity, when water is used on estates it often provides less jobs, produce is exported, there is little local capacity building, and the value created ends up in the pockets of local elites and/or foreign investors (De Bont et al., 2016). The 'social water productivity' (Bee, 2000; Solbes, 2003), seen as the broader societal value being created per unit of water used, tends to be much lower for industrial agriculture than for more locally embedded forms of agricultural intensification.

Finally, it is relevant to analyse the African Union (AU)'s (2020) framework for irrigation development and agricultural water management, which serves as a reference for many national irrigation policies. The framework outlines four pathways for smallholder farmers' agricultural water use development:

1 Improved water management in rain-fed environments.
2 Farmer-led irrigation development.
3 Irrigation scheme development and modernization.
4 Unconventional water use for irrigation (e.g., treated wastewater in urban settings).

To some extent, this aligns with the four paths delineating food system change, as presented in Figure 1.1 of de Haas et al. (this volume). Classic irrigation scheme development (AU's pathway 3) predominantly contributes to the "exclusive structural change" path, fostering a dual system featuring pockets of prosperity (irrigation schemes) within a stagnant broader agricultural sector. AU's pathways 1 and 2, which focus on enhanced rain-fed practices and farmer-led irrigation development, respectively, promote inclusivity and complementarity. The characteristics of irrigation development by farmers themselves make it amenable to private sector investments for smallholders who are self-driven (Mati, 2023), yet it happens quietly outside the realm of policy support. Capital-intensive forms of smallholder intensification appear to expedite the transition of labour out of agriculture, channelling profits into the education of younger generations. While none of AU's pathways explicitly address labour exit from agriculture, pathway 4 could be viewed as a component of an urban migration pattern, where individuals who have left rural areas find new (part-time) opportunities in urban irrigated agriculture.

4 Building blocks for a strategy to attaining food security in 2050

Irrigated agriculture enhances food security, both for farmers that directly benefit from increased production and income, and for the urban population that benefits from increased food availability and a more diverse food basket. There are also wider economic effects such as through job creation along the value chain and through reduced food imports (see also Frankema, this volume). Irrigated agriculture contributes to diversified diets by providing fresh fruits and vegetables, but equally plays an important role in the production of staple crops and a protection against crop failure due to droughts. We thus argue that irrigated agriculture importantly contributes towards addressing Africa's food security challenges. For maximum impact both an improvement of existing irrigated agriculture and expansion of the irrigated area is needed in ways that makes the benefits accessible to a large group of people. Rather than solely identifying practical opportunities for water-driven food security development, we underscore a moral imperative that urges a fundamental shift in the mindset about what is attainable and should be pursued. A business-as-usual approach will not suffice to support a much faster development of water-based agricultural intensification. Certainly there is room to improve on the basis of existing technical approaches and implementation processes, but there is also a need for radically different approaches. We sketch five areas where work can be done differently.

First, irrigation investments should be part of comprehensive, multifaceted government programmes rather than isolated infrastructure investments. Such programmes should encompass all aspects of agricultural water development and incorporate various financing streams. Programmes should strategically combine public investment, community development, and facilitation of farmer-led irrigation development, in relation to available water resources and agricultural markets with the aim to enable farmers to initiate or scale up their irrigation farming. This involves a blend of micro-financing innovations, incentives to enhance irrigation suppliers' reach and service quality, equipment price reductions, expansion of technical knowledge, and adherence to equipment quality standards. Encouraging the adoption and continued use of irrigation technologies relies on providing incentives to farmers and ensuring sustained external support. Additionally, safeguarding the availability and maintenance of irrigation infrastructure is essential for the longer term (Kamwamba-Mtethiwa et al., 2021).

Second, Operation, Management and Maintenance (OMM) performance in public irrigation schemes should be improved. Public schemes across Africa widely suffer from low service delivery performance that is compounded by poor maintenance, low fee collections, and an inability to sustain themselves technically and financially. A performance turnaround demands a fresh look at co-management arrangements between the Government and Irrigation Organizations, potentially involving private sector actors as maintenance and/or operational service providers. A critical service threshold has to be achieved to ensure reliable water to irrigation farmers so that they can farm profitably, and then pay for the services. Recent strides in tackling the service conundrum on public schemes, building on performance approaches of the last 20 years, have been made with comprehensive approaches such as in the Irrigation Operator of the Future (iOF) Toolkit (Waalewijn et al., 2022). These and similar combined governance and technical responses are crucial to move to operational and financial sustainability on public schemes. The approaches include scheme-level diagnostics with organizational capability development to improve operational systems and skills, infrastructure rehabilitation with measurement instrumentation, and with possible operational roles for private companies.

Third, the role of irrigation representatives in local catchment management organizations should be strengthened, placing them at the forefront. Irrigation farmers, often the largest water user group in the watershed, rely heavily on a secure water supply. Trends of catchment degradation and shifting rainfall patterns negatively impact irrigation water availability and quality. To address this better, sub-catchment planning is needed, along with the implementation of reforestation, soil conservation, and sustainable farming practices (also see Sheil, this volume). Involving irrigation farmers and scheme operators in catchment management is therefore essential. Irrigation water users' associations should be a distinct component of catchment planning and management.

Fourth, a paradigm shift is needed to achieve optimal farm-level water management practices, particularly in the plant root zone. Soil-water management, water productivity, and farm profitability can be significantly enhanced through hi-tech innovations and basic farmer training. Emphasizing irrigation application methods alone risks missing the primary objective when high-cost irrigation technologies are used in sub-optimal knowledge or technical contexts. Accessible, intuitive soil-water management technology, including internet-connected soil-water sensors, can yield massive financial returns. This technology reduces pumping and fertilization costs and increases water availability in the root zone, leading to more robust crops, resilience to pests and diseases, increased land productivity, enhanced water productivity, reduced groundwater pollution, and ultimately increased income for farmers.

Fifth and finally, private capital should be mobilized through public-private partnerships for financing irrigation infrastructure, focusing on a cost-sharing approach between farmers and the government. This is different from the traditional public-private partnership model of a major commercial or agri-business co-investing or operating large irrigation schemes. Direct cost sharing between farmers and the government has demonstrated the potential for significant private capital mobilization, with farmers contributing 50% of infrastructure costs on smaller community schemes owned by the farmers themselves. Additionally, individual small irrigation farmers already contribute the majority of capital for their equipment. A mix of proven interventions, including accessible micro-finance, targeted demand-side and supply-side subsidies, digital farmer-information platforms, and private sector incentives that help to cut equipment costs, can facilitate private capital mobilization at scale.

It is important to note that these considerations extend beyond the realms of irrigation and agricultural water management policies. Instead, they form a broader framework that acknowledges the multifaceted nature of achieving sustainable food security in Africa.

References

African Union. (2020). *Framework for irrigation development and agricultural water management in Africa*. African Union.

Altchenko, Y., & Villholth, K.G. (2015). Mapping irrigation potential from renewable groundwater in Africa–A quantitative hydrological approach. *Hydrology and Earth System Sciences*, *19*(2), 1055–1067.

Bee, A. (2000). Globalization, grapes and gender: Women's work in traditional and agro-export production in northern Chile. *Geographical Journal*, *166*(3), 255–265.

Beekman, W., Veldwisch, G.J., & Bolding, A. (2014). Identifying the potential for irrigation development in Mozambique: Capitalizing on the drivers behind farmer-led irrigation expansion. *Physics and Chemistry of the Earth, Parts A/B/C*, *76*, 54–63.

De Bont, C., Veldwisch, G.J., Komakech, H.C., & Vos, J. (2016). The fluid nature of water grabbing: The on-going contestation of water distribution between peasants and agribusinesses in Nduruma, Tanzania. *Agriculture and Human Values*, *33*, 641–654.

De Bont, C., Liebrand, J., Veldwisch, G.J., & Woodhouse, P. (2019). Modernisation and African farmer-led irrigation development: Ideology, policies and practices. *Water Alternatives*, *12*(1), 107–128.

De Bont, C., & Veldwisch, G.J. (2020). State engagement with farmer-led irrigation development: Symbolic irrigation modernisation and disturbed development trajectories in Tanzania. *The Journal of Development Studies*, *56*(12), 2154–2168.

Dorward, A., Anderson, S., Bernal, Y.N., Vera, E.S., Rushton, J., Pattison, J., & Paz, R. (2009). Hanging in, stepping up and stepping out: Livelihood aspirations and strategies of the poor. *Development in practice*, *19*(2), 240–247.

Giordano, M, Turral, H., Scheierling, S. M., Treguer, D. O., McCornick, P. G. (2017). *Beyond 'More Crop Per Drop': Evolving thinking on agricultural water productivity*. International Water Management Institute (IWMI).

Drechsel, P., & Keraita, B. (2014). *Irrigated urban vegetable production in Ghana characteristics, benefits and risk mitigation* (2nd ed). International Water Management Institute.

FAO. (2024). AQUASTAT Database. AQUASTAT Website. Food and Agriculture Organization of the United Nations. https://data.apps.fao.org/aquastat/ Accessed on 20/11/2024.

Hoekstra, A.Y., & Mekonnen, M.M. (2012). The water footprint of humanity. *Proceedings of the National Academy of Sciences*, *109*(9), 3232–3237.

Hoekstra, A.Y. (2020). *The water footprint of modern consumer society* (2nd ed.). Earthscan, from Routledge.

Izzi, G., Denison, J., & Veldwisch, G.J. (Eds.) (2021). The farmer-led irrigation development guide: A what, why and how-to for intervention design. World Bank.

Kamwamba-Mtethiwa, J., Wiyo, K., Knox, J., & Weatherhead, K. (2021). Diffusion of small-scale pumped irrigation technologies and their association with farmer-led irrigation development in Malawi. *Water International*, *46*(3), 397–416.

Lankford, B., Closas, A., Dalton, J., Gunn, E.L., Hess, T., Knox, J.W., … & Zwarteveen, M. (2020). A scale-based framework to understand the promises, pitfalls and paradoxes of irrigation efficiency to meet major water challenges. *Global Environmental Change*, *65*, 102182.

Lankford, B.A., & Scott, C.A. (2023). The paracommons of competition for resource savings: Irrigation water conservation redistributes water between irrigation, nature, and society. *Resources, Conservation and Recycling*, *198*, 107195.

Mati, B. (2023). Farmer-led irrigation development in Kenya: Characteristics and opportunities. *Agricultural Water Management*, *277*, 108105. https://doi.org/10.1016/j.agwat.2022.108105

Perry, C., & Steduto, P. (2017). *Does improved irrigation technology save water? Discussion paper on irrigation and sustainable water resources management in the Near East and North Africa*. Food and Agriculture Organization of the United Nations. Cairo.

Rockström, J., Karlberg, L., Wani, S.P., Barron, J., Hatibu, N., Oweis, T., … & Qiang, Z. (2010). Managing water in rainfed agriculture—The need for a paradigm shift. *Agricultural Water Management*, *97*(4), 543–550.

Rosa, L., Chiarelli, D.D., Rulli, M.C., Dell'Angelo, J., & D'Odorico, P. (2020). Global agricultural economic water scarcity. *Science Advances*, *6*(18), eaaz6031.

Scoones, I., Murimbarimba, F., & Mahenehene, J. (2019). Irrigating Zimbabwe after land reform: The potential of farmer-led systems. *Water Alternatives*, *12*(1), 88–106.

Siebert, S., Burke, J., Faures, J.M., Frenken, K., Hoogeveen, J., Döll, P., & Portmann, F.T. (2010). Groundwater use for irrigation–a global inventory. *Hydrology and Earth System Sciences*, *14*(10), 1863–1880.

Solbes, R.V. (2003). Economic and social profitability of water use for irrigation in Andalucia. *Water International*, *28*(3), 326–333.

Van Ittersum, M.K., Van Bussel, L.G., Wolf, J., Grassini, P., Van Wart, J., Guilpart, N., … & Cassman, K.G. (2016). Can sub-Saharan Africa feed itself? *Proceedings of the National Academy of Sciences*, *113*(52), 14964–14969.

Van Opstal, J., Droogers, P., Kaune, A., Steduto, P., & Perry, C. (2021). Guidance on realizing real water savings with crop water productivity interventions. Wageningen, FAO and FutureWater. https://doi.org/10.4060/cb3844en

Veldwisch, G.J., Venot, J.P., Woodhouse, P., Komakech, H.C., & Brockington, D. (2019). Re-introducing politics in African farmer-led irrigation development: Introduction to a special issue. *Water Alternatives*, *12*(1), 1–12.

Veldwisch, G.J., & Woodhouse, P. (2022). Formal and informal contract farming in Mozambique: Socially embedded relations of agricultural intensification. *Journal of Agrarian Change*, *22*(1), 162–178.

Venot, J.-P., Bowers, S., Brockington, D., Komakech, H., Ryan, C., Veldwisch, G.-J., & Woodhouse, P. (2021). Below the radar: Data, narratives and the politics of irrigation in sub-Saharan Africa. *Water Alternatives*, *14*(2): 546–572.

Vos, J., van Oel, P., Hellegers, P., Veldwisch, G.J., & Hoogesteger, J. (2019). Four perspectives on water for global food production and international trade: Incommensurable objectives and implications. *Current Opinion in Environmental Sustainability*, *40*, 30–36.

Waalewijn, P., Onimus, F., de Jong, I., Trier, R., Denison, J., Malerbe, F.D., … & Valieva, S. (2022). *The irrigation operator of the future: A toolkit; Information pack for irrigation service delivery performance assessment and planning*. World Bank.

Warner, K., Afifi, T., Henry, K., Rawe, T., Smith, C., & De Sherbinin, A. (2012). *Where the rain falls: Climate change, food and livelihood security, and migration*. Global policy report of the where the rain falls project. CARE France and UNU-EHS.

Woodhouse, P., Veldwisch, G.J., Venot, J.P., Brockington, D., Komakech, H., & Manjichi, Â. (2017). African farmer-led irrigation development: Re-framing agricultural policy and investment? *The Journal of Peasant Studies*, *44*(1), 213–233.

Xie, H., You, L., Wielgosz, B., & Ringler, C. (2014). Estimating the potential for expanding smallholder irrigation in Sub-Saharan Africa. *Agricultural Water Management*, *131*, 183–193.

18 Harnessing biotechnology to release the potential of the bioeconomy for Africa

Enoch Kikulwe and Justus Wesseler

1 Introduction

Approximately 123 million people (12% of the population) in sub-Saharan Africa (SSA) are acutely food insecure, suffering from chronic malnutrition and unable to meet their minimum food consumption needs (Diogo et al., 2022). Climate change is increasingly threatening to reverse the achievements made in the fight against hunger and malnutrition across SSA with lasting adverse macroeconomic effects on economic growth and poverty (FAO, 2015; Diogo et al., 2022). Moreover, about 70% of the very poor live in rural areas, depending almost entirely on agriculture, a sector severely affected by climate change for their livelihoods, producing approximately 80% of the food consumed in Asia and SSA (IFAD, 2011; Antwi, 2013). Climate change, therefore, negatively impacts Africa's natural ecological systems retarding the capacity for human development and food security, and poses a huge risk to millions of people who are reliant on climate-sensitive resources for their livelihoods (Descheemaeker et al., 2016). For instance, expected changes in temperature, rainfall, and wind patterns attributed to climate change are predicted to have a dramatic effect on the prevalence of one of the most dangerous migratory pests (desert locusts) in Africa and will potentially have devastating impacts on crop yields (Tang et al., 2023). The area suitable to banana bunchy top disease, an invasive and transboundary disease is expected to expand from the low and humid zones of west and central Africa to the currently cooler altitude in the eastern Africa with global warming (Bouwmeester et al., 2023). Similarly, impacts of climate change on animal health especially for vector-borne diseases have also been well documented (Lacetera, 2019). Diseases including West Nile virus, bluetongue or Lyme's disease and schistosomiasis are expected to expand into new areas with outbreaks of Rift Valley fever in East Africa being associated with increased rainfall and flooding due to El Niño-Southern Oscillation events (Caminade et al., 2019; FAO, 2015; Porter et al., 2014).

Climate change is not the only threat to food security. There are number of other issues including regional conflicts, depletion of natural resources, institutional factors, and more (von Braun et al., 2023). The issues mentioned are interlinked with each other and are characterized by a high degree of complexity complicating identifying appropriate strategies for improving food security. Nevertheless, there is general agreement that technological change is important for the strategies of improving food security and that food security is more than just increasing the quantity and availability of food produced (Directorate-General for Research and Innovation of the European Commission et al., 2022; German Science and Humanities Council, 2023; Trigo et al., 2023). Further, economic research has shown food security at household level depends on the purchasing power of households which can include many sources in addition to the production of food as already pointed out by among other Nobel Laureates Theodor W. Schultz (1980) and Amartya Sen (1983) several decades ago. One of the solutions for increasing the

DOI: 10.4324/9781032649696-23

purchasing power of households via technological change is increasing the efficiency of using biological resources or the bioeconomy. Bioeconomy, here, is understood as the conversion of biological resources into food, feed, fibre, fuel, and further products (4+1Fs). The bioeconomy (Figure 18.1) has been considered to have a high potential to address a number of the challenges and contributing to sustainable development (Trigo et al., 2023). However, the contribution of the bioeconomy is complex and depends on several factors as discussed below.

2 Bioeconomy potential for Africa

Africa is rich in biological resources. Investing in the bioeconomy has a high potential for generating sustainable development in Africa. The performance of the bioeconomy is complex as illustrated in Figure 18.1. The performance of the bioeconomy is influenced by policies, national strategies, and legislation at national and international level. There are a number of driving forces for change that have an impact on the supply and use of biomass. The supply and use of biomass depends on the availability in quantitative terms including the amount of land and water available and their qualities, the quality and quantity of labour supply, and the kind of biomass production systems including the by-products and waste.[1] Policies, public and private sector strategies, and the legal environment further have an impact on the use and supply of biomass. The results that the supply and use of biomass generates can be measured against a number of societal objectives. These societal objectives are implemented via policies, strategies,

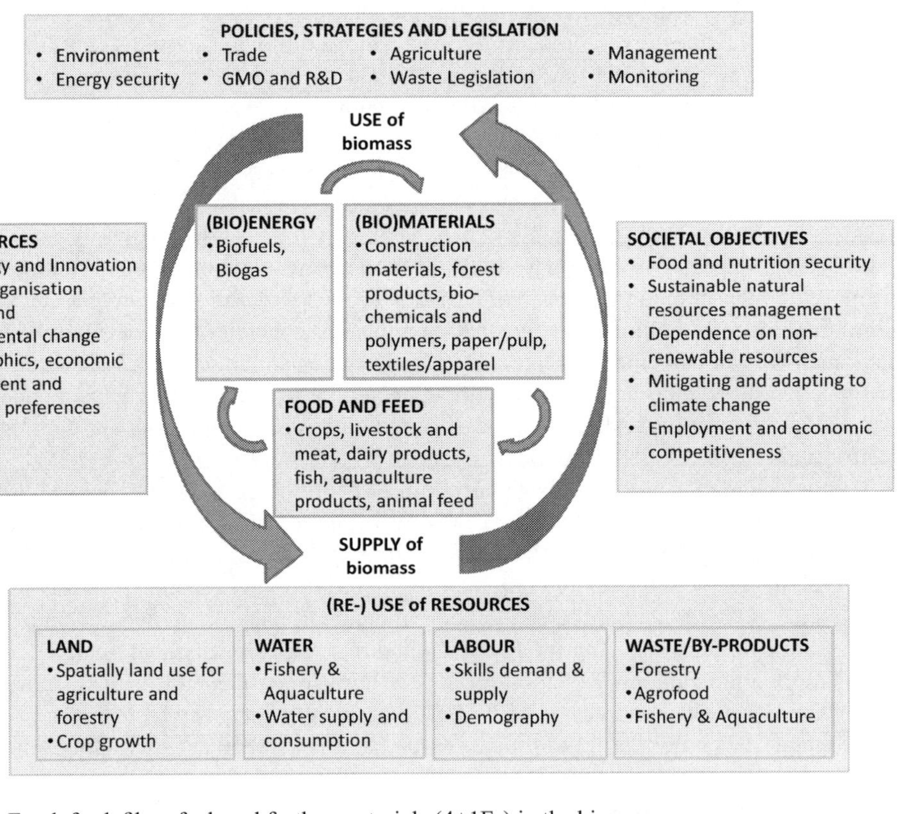

Figure 18.1 Food, feed, fibre, fuel, and further materials (4+1Fs) in the bioeconomy.

Source: Adapted from Kardung et al. (2021).

and legislation. Further, the whole system has to be seen as an evolutionary system developing over time. The system depicted in Figure 18.1 allows bioeconomy strategies to be assessed.

Several member states of the European Union and Latin American countries have developed national bioeconomy strategies with the expectation to improve the efficiency of the development of the bioeconomy towards achieving societal objectives. Only a few African countries such as South Africa and the East African Community under BioInnovate initiative have a dedicated bioeconomy strategy (M'Barek & Wesseler, 2023; EAC, 2022). Such a strategy is not essential, but having a national strategy indicates to what extent the bioeconomy is considered to be of importance (Poku et al., 2018).

Producing food remains the most important role of the African bioeconomy and fuel (energy) from biological resources is of economic importance in many parts of Africa in particular for food preparation. In these sectors there is an endogenous growth potential using comparative advantages for further economic differentiation via specialization (Zilberman et al., 2018). Productivity differences in one sector can generate substantial spill-over effects on other sectors and via differentiation in production driven by differences in relative productivity down the supply chain further stimulate economic growth. This has been shown at the theoretical level by Acemoglu and Azar (2020) and supported by empirical evidence from the bioeconomy for the European Union by Cingiz et al. (2021). M'Barek and Wesseler (2023) argue that there is no reason to expect why this should not also be the case for Africa and other parts of the world. The growth potential can be further supported by technological change. Many solutions for increasing the productivity of the bioeconomy are already available, as indicated by yield differences in crop production (Tittonell & Giller, 2013). There is a huge potential for increasing the timber supply for local infrastructure and more (Degnet, 2021). The same applies further down the supply chain when assessing food production (Cingiz et al., 2021). Technological change will not only be needed for generating economic growth on its own but also for responding to the challenges posed by climate change.

3 Contribution of bioeconomy in mitigating climate change effects

Estimates show that, globally, food systems are responsible for approximately 60% of terrestrial biodiversity loss, 24% of the greenhouse gas emissions, 33% of the soil degradation, and 61% of the depletion of commercial food stocks (UNEP, 2016). Development of a sustainable bioeconomy to accelerate mitigation and adaptation to climate is particularly relevant for Africa which is warming faster than the global average (Oguntuase & Adu, 2021). Circular bioeconomy, through enhancing and sustainable management of renewable natural resource capital, creates an opportunity to combat climate change and improve environmental health and agricultural productivity while also contributing to economic growth and job creation (Hetemäki & Kangas, 2022; IICA, 2019). For instance, practices such as cultivating on degraded land, regenerative land use, natural pest control, appropriate multi-cropping, and soil and water conservation measures increase agricultural productivity while also contributing to climate change mitigation (IPCC, 2019). Sustainable silviculture, afforestation, and reforestation are crucial in sustaining ecosystems, air quality, and soil carbon sequestration in addition to reducing vulnerability to disease and extreme weather events. By transferring carbon into soils through falling leaves and branches, forests can improve soil organic matter. Generally, forests act as natural carbon sinks and some of their products are substitutes for emissions-intensive materials and forest-based bioenergy involving the utilization of post-consumer wood and forest residues not suitable for production of other materials contributing to emission-free energy production, thereby reducing emissions (Schmid et al., 2022). Bang et al. (2009) projected that industrial biotechnology, biofuels, and bioenergy can potentially reduce global greenhouse gas emissions by 1.0–2.5 billion tons of carbon dioxide (CO_2) per year by 2030. More recent assessments suggest that this is very much at the lower end and expected emission reductions are more than twice as large (Zahed et al., 2021).

4 The importance of biotechnology

Many experts expect that climate change will change pest and disease pressures in agriculture, forestry, and fishery requiring solutions different from those currently available. New developments, such as biotechnology can help to address these challenges and contribute to sustainable development and food security (Wesseler & von Braun, 2017). The use of new plant and animal breeding technologies such as Clustered Regularly Interspaced Palindromic Repeats-CRISPR associated protein(CRISPR-Cas) allows to breed plants and animals with higher temperature tolerance or disease resistance (FAO, 2022). These are considered relatively low-hanging fruits.

The report by FAO (2022) and other publications (Pixley et al., 2022; Falck-Zepeda et al., 2022) provide recent reviews. Some examples are of importance for Africa (Table 18.1). Several Africa-based breeders and their partners (including the Consultative Group for International Agricultural Research) are employing CRISPR-Cas9 technologies to breed for disease and pest resistance, climate resilience, and nutrition enrichment.

The African Agriculture Technology Foundation (AATF) has developed a transgenic pod borer-resistant cowpea (*Vigna unguiculata* [L] Walp.). The first cultivation started in Nigeria in 2019. Cowpea is an important staple crop, often referred to as “poor man’s meat”, and insect resistance has the potential to reduce yield losses while also contributing to food security (Wesseler et al., 2017). Several other applications using modern biotechnology in crop breeding for food security are available. The contribution to reducing malnutrition and micronutrient deficiency is often higher in economic terms than any direct yield effects (Wesseler et al., 2017).

The development of pest resistance using transgenic methods is also expected to protect against the invasion of the Latin American variant of the fall armyworm (Van Den Berg et al., 2021). Biological control methods such as low-risk pesticides offer the opportunity to address pest and disease problems with less negative impacts on the environment than commonly used pesticides. New developments in nitrogen fixation of plants and nitrogen supply via microbes

Table 18.1 Examples of CRISPR/Cas9-based crops for Africa

Species	*Trait targeted*
Banana	Fungus protection
	Protection against bacterial wilt
	Protection against banana streak
Cassava	Reduced cyanide levels
	Virus resistance
	Stress resistance and high yielding
Sorghum	Increased protein content
	Striga resistance
Rice	Resistance to fungal diseases
	Stress resistance and increased yields
Maize	Drought resistance
	Lethal necrosis disease resistance
Beans, maize, and cassava	Nutrition improvement

Sources: FAO (2022), Pixley et al. (2022), and Falck-Zepeda et al. (2022).

hold high promises. They may allow to produce higher yields without the high level of greenhouse gas emissions related to producing nitrogen fertilizer (ammonia) via the Haber-Bosch process which contributes more than 1% to the total emission of CO_2. If the nitrogen rates used in the European Union (EU) are applied to agricultural land in Africa, ammonium fertilizer use worldwide will increase by about 60% implying an increase in CO_2 emission from fertilizer production by about the same magnitude. These are averages but illustrate the challenge and the benefits to be expected from nitrogen supply via microbes. Still, many of the promises made at a very early stage of development may take several years if not decades before reaching farmers in the field.

New developments in the bioeconomy allow not only to improve crop production, but also to increase the efficiency in the use of biological resources more generally. The recycling of biological resources, the use of biological resources for the development of biopolymers, and precision fermentation for food production are only a few of the recent developments. Many of these technical developments are happening outside Africa and transferring those technologies to Africa can help to generate the needed sustainable economic growth. Argentina and Brazil serve as examples where modern biotechnology has generated economic growth of the bioeconomy in the global south (Deciancio & Mac Clay, 2023; Maximo et al., 2022; Follador et al., 2019).

5 Institutional challenges

The development of the bioeconomy in Africa holds promise as described above. The realization of these promises depends on the successful adoption of the technical changes needed, which is largely an institutional problem. Institutions in this context are understood as the norms and values a society holds that are translated into policies generating economic incentives and disincentives (Acemoglu & Robinson, 2012). The institutional environment has to be supportive for the investments needed at farm level and beyond and changes are needed to generate a supportive investment environment. New developments in the bioeconomy are often delayed and blocked by policies. A pest-resistant transgenic event for corn, Bt corn, was submitted for approval for cultivation in Kenya in 1998 by Monsanto (now Bayer Crop Science). In the last year progress has been made by allowing large-scale field trials as part of the approval process, yet still, after 25 years, no Bt corn variety has been approved for cultivation in Kenya. A number of issues related to the cultivation of Bt corn remain, such as the varieties being introduced are hybrid yellow corn varieties for animal feed and not for food. The effectiveness of the protection against insect pests has also been put into question. Still, this is seen as an unreasonable argument for blocking the introduction by many scientists from Africa (Paarlberg et al., 2024). Similarly, trials with a bacterial wilt-resistant banana in Uganda started in 2006 (Kikulwe, 2010). Yet until today, the bacterial wilt-resistant banana has not received approval for cultivation (Falck-Zepeda et al., 2022). Such long delays in approval have a strong negative effect on private sector incentives to invest in developing solutions for farmers (Wesseler et al., 2023). This does not only hold for solutions that are related to gene editing but also applies to the approval of biological control options in agriculture production (Fredericks & Wesseler, 2019). Examples of biological control include management of the red spider mite (*Tetranychus urticae*), a pest in vegetable production, using the naturally occurring fungus *Beauveria bassiana* strain R444 or the naturally occurring virus *Spodoptera littoralis nucleopolyhedrovirus* (SpliNPV) as control against the fall armyworm (*Spodoptera frugiperda*) or the African cotton leafworm (*Spodoptera littoralis*). There are many more solutions under development (see, e.g., Lorsbach et al., 2019). Biological control options are often more expensive as their markets are smaller making them less attractive to be developed for African markets. Regional harmonization of policies can reduce approval costs and increase the market size stimulating private sector investment and increasing the return to public sector investment (Purnhagen & Wesseler, 2019). Yet approval of a new technology for agriculture production will not automatically imply economic success for the technology provider. If Bt corn is not successful economically, the provider of the technology will have to bear the costs of the failure, which provides strong incentives for the provider of the technology to ensure economic success.

The creation of World Trade Organization (WTO) and the subsequent formulation of TRIPS (Trade-Related Aspects of Intellectual Property Rights) created new challenges for intellectual property rights (IPR) policy and law in SSA. Many countries did not have the necessary institutions in place to implement the TRIPS agreement. They were also facing the trade-off between protecting IPR according to the TRIPS agreement and the possibilities of free-riding on intellectual property (Lele et al., 1999; Smith et al., 2021). This necessitated the understanding of the strategic importance of IPR towards social economic development to help in designing policies that address national development goals in SSA (Sikoyo et al., 2006).

Mechanisms for public and private governance will be important as African states embrace the bioeconomy. In so doing, proper Intellectual Property (IP) laws need to be formulated and enforced in Africa to boost its investment in the bioeconomy. This can be achieved by working closely with the global IP community and supported by the regional bodies (LawTeacher, 2013) such as African Union (AU), East African Community (EAC), and Economic Community of West African States (ECOWAS). Bößner et al. (2021) recently reviewed and discussed the different forms of governance that may govern the bioeconomy, ranging from international to national dimensions. The authors argue that although the global bioeconomy governance may play a fundamental role, it may experience several flaws and challenges. They recommended that the regional bodies may be the most promising ways to addressing challenges and opportunities that may arise with governance of bioeconomy pathways. For instance, the global patents and IPR are a relevant issue for bioeconomy pathways as they generate a conducive enabling environment for innovation but may be very costly and stringent for Africa. To break this barrier, proper procedures for IPR, capacity building in IPR management, and building new and strong partnerships are crucial in the governance and regulation of the bioeconomy (Falck-Zepeda et al., 2022).

6 The way forwards

The development of new biotechnology solutions by Africa for Africa and the harmonization of approval processes for new technologies can make new solutions available for the market. The development of Bt cowpea mentioned above serves as an example of how a solution can be developed over a relatively short period of time. Many more opportunities exist as mentioned above with respect to the use of new plant breeding technologies.

Several African countries are currently developing policies for IP. The EAC, for instance, has drafted a Regional Policy for Intellectual Property[2] governing the bioeconomy in East Africa. Article 43 of Protocol on Common Market calls for the EAC partner states to harmonize the IPR. This is aimed to encourage technical innovation and promote industrial and commercial use of such innovations to contribute to the social, economic, industrial, and technological development of Africa. This in turn will simulate research, access to technology, and enterprise growth. A clear and right mix of policies needs to be developed to have the desired bioeconomy future providing the incentives needed for stakeholders to engage (Gatune et al., 2021). The international community can support this process by supporting the institutional change needed (Directorate-General for Research and Innovation of the European Commission et al., 2022).

Notes

1 The term "waste" is somewhat controversial as according to the mass balance nothing is lost, but appears at different places over time and space. At some places we observe accumulation of carbon in the atmosphere or nitrogen in groundwater causing economic damage via climate change-related damages and groundwater-cleaning costs.
2 https://easteco.org/development-of-the-east-african-regional-policy-for-intellectual-property/.

References

Acemoglu, D., & Azar, P.D. (2020). Endogenous production networks. *Econometrica*, *88*(1), 33–82. https://doi.org/10.3982/ECTA15899

Acemoglu, D., & Robinson, J.A. (2012). *Why nations fail: The origins of power, prosperity, and poverty*. Penguin Random House.

Antwi, A. (2013). *Climate change and food security: An overview about the issue*. Friedrich Ebert Stiftung. https://library.fes.de/pdf-files/bueros/ghana/10518.pdf

Bang, J.K., Follér, A, & Buttazzoni, M. (2009). *Industrial biotechnology: More than green fuel in a dirty economy? Exploring the transformational potential of industrial biotechnology on the way to a green economy*. World Wildlife Fund (WWF).

Bouwmeester, H., Blomme, G-, Omondi, A.B., & Ocimati, W. (2023). Banana bunchy top disease in Africa–Predicting continent-wide disease risks by combining survey data and expert knowledge. *Plant Pathology*, *72*(8), 1476–1490. https://doi.org/10.1111/ppa.13764

Bößner, S., Johnson, F.X., & Shawoo, Z. (2021). Governing the bioeconomy: What role for international institutions? *Sustainability*, *13*, 286. https://doi.org/10.3390/su13010286

Caminade, C., McIntyre, K.M., & Jones, A.E. (2019). Impact of recent and future climate change on vector-borne diseases. *Annals of the New York Academy of Sciences*, *1436*(1), 157–173. https://doi.org/10.1111/nyas.13950

Cingiz, K., Gonzalez-Hermoso, H., Heijman, W., & Wesseler., J. (2021). A cross-country measurement of the EU bioeconomy: An input–output approach. *Sustainability*, *13*, 3033. https://doi.org/10.3390/su13063033

Deciancio, M., & Mac Clay, P. (2023). The Bioeconomy in Argentina: Lessons for its development and sustainability (ZEF Policy Brief #43).

Degnet, M. (2021). Determinants of sustainable forest management: The cases of industrial private forest plantations in East Africa and non-industrial private forests in Sweden (PhD thesis). Wageningen University.

Descheemaeker, K., Oosting, S.J., Homann-Kee Tui, S., Masikati, P., Falconnier, G.N., & Giller, K.E. (2016). Climate change adaptation and mitigation in smallholder crop–livestock systems in sub-Saharan Africa: A call for integrated impact assessments. *Regional Environmental Change*, *16*, 2331–2343. https://doi.org/10.1007/s10113-016-0957-8

Diogo, B., Mai, F., Dominique, F., Kemoe, L., Loic, L., Pritha, M., … & Unsal, F. (2022). *Climate change and chronic food insecurity in sub-Saharan Africa*. International Monetary fund. https://www.imf.org/en/Publications/Departmental-Papers-Policy-Papers/Issues/2022/09/13/Climate-Change-and-Chronic-Food-Insecurity-in-Sub-Saharan-Africa-522211

Directorate-General for Research and Innovation of the European Commission, Webb, P., Sonnino, R., Fraser, E., & Arnold T. (2022). Everyone at the table: Transforming food systems by connecting science, policy and society. Publications Office of the European Union, Luxembourg. https://data.europa.eu/doi/10.2777/440690

EAC. (2022). *EAC unveils regional bioeconomy strategy 2021/22–2031/32*. https://www.eac.int/press-releases/2515-eac-unveils-regional-bioeconomy-strategy-2021-22-2031-32

Falck-Zepeda, J.-B., Biermayr-Jenzano, P., Roca, M.M., Fuentes-Campos, E., & Kikulwe, E.M. (2022). Bio-innovations: Genome-edited crops for climate-smart food systems. In *2022 global food policy report: Climate change and food systems* (pp. 90–99). International Food Policy Research Institute (IFPRI). https://doi.org/10.2499/9780896294257_10

FAO. (2015). *Climate change and food security: Risks and responses*. FAO. https://www.fao.org/3/i5188e/I5188E.pdf

FAO. (2022). *Gene editing and agrifood systems*. FAO. https://doi.org/10.4060/cc3579en

Follador, M., Philippidis, G., Davis, J., & Soares-Filho, B. (2019). *Assessing the impacts of the EU bioeconomy on third countries: Potential environmental impacts in Brazil of EU biofuel demand to 2030*. European Commission.

Fredericks, C., & Wesseler, J. (2019). A comparison of the EU and US regulatory frameworks for the active substance registration of microbial bio control agents. *Pest Management Science*, *75*(1), 87–103. https://doi.org/10.1002/ps.5133

Gatune, J., Ozor, N., & Oriama, R. (2021). The futures of bioeconomy in Eastern Africa. *Journal of Futures Studies*, *25*(3), 1–14. https://doi.org/10.6531/JFS.202103_25(3).0001

German Science and Humanities Council. (2023). *Future perspectives for agricultural, food and nutritional sciences. Position paper*. German Science and Humanity Council. https://doi.org/10.57674/a4rt-ke57

Hetemäki, L., & Kangas, J. (2022). Forest Bioeconomy, Climate Change and Managing the Change. Managing forest ecosystems. In L. Hetemäki, J. Kangas, & H. Peltola (Eds.), *Forest bioeconomy and climate change* (pp. 1–12). Springer. https://doi.org/10.1007/978-3-030-99206-4

IFAD. (2011). *Rural poverty report* 2011. *New realities, new challenges: New opportunities for tomorrow's generation*. https://reliefweb.int/report/world/rural-poverty-report-2011-new-realities-new-challenges-new-opportunities-tomorrows

IICA (Inter-American Institute for Cooperation on Agriculture) (2019). *The Bioeconomy: A catalyst for the sustainable development of agriculture and rural territories in LAC*. https://www.iica-ecuador.org/sisbio/doc_informacion/IICA_Cap4_Eng_V4.pdf

IPCC (2019). *Climate change and land*. https://www.ipcc.ch/site/assets/uploads/2019/08/4.-SPM_Approved_Microsite_FINAL.pdf

Kardung, M., Cingiz, K., Costenoble, O., Delahaye, R., Heijman, W., Lovrić, M., … & Xinqi Zhu, B. (2021). Development of the circular bioeconomy: Drivers and indicators. *Sustainability*, *13*, 413. https://doi.org/10.3390/su13010413

Kikulwe, E. (2010). *On the introduction of genetically modified bananas in Uganda: Social benefits, costs, and consumer preferences* (PhD thesis). Wageningen University.

Lacetera, N. (2019). Impact of climate change on animal health and welfare. *Animal Frontiers*, *9*(1), 26–31. https://doi.org/10.1093/af/vfy030

LawTeacher (2013). *Patent and intellectual property issues in Africa*. https://www.lawteacher.net/free-law-essays/international-law/patent-and-intellectual-property-issues-in-africa-international-law-essay.php?vref=1

Lele, U., Lesser, W., & Horstkotte-Wesseler, G. (1999). *Intellectual property rights in agriculture*. The World Bank.

Lorsbach, B.A., Sparks, T.C., Cicchillo, R.M., Garizi, N.V., Hahn, D.R., & Meyer, K.G. (2019). Natural products: A strategic lead generation approach in crop protection discovery. *Pest Management Science*, *75*(9), 2301–2309. https://doi.org/10.1002/ps.5350

M'Barek, R., & Wesseler, J. (2023): The rapid development of bioeconomy policies in the EU and other regions of the world. *EuroChoices*, *22*(3), 5–12. https://doi.org/10.1111/1746-692X.12415

Maximo, Y.I., Hassegawa, M., Verkerk, P.J., & Missio, A.L. (2022). Forest bioeconomy in Brazil: Potential innovative products from the forest sector. *Land*, *11*(8), 1297. https://doi.org/10.3390/land11081297

Oguntuase, O.J., & Adu, O.B. (2021). *Bioeconomy as climate action: How ready are African countries*. African handbook of climate change adaptation. https://doi.org/10.1007/978-3-030-42091-8_82-1.

Paarlberg, R., Bhattacharya, A., Huang, J., Karembu, M., Pray, C., & Wesseler, J. (2024). The uptake of new crop science: Explaining success, and failure. *Food Policy*, *122*, 102572. https://doi.org/10.1016/j.foodpol.2023.102572

Pixley, K.V., Falck-Zepeda, J.B., Paarlberg, R.L., Phillips, P.W.B., Slamet-Loedin, I.H., Dhugga, K.S., Campos, H., & Gutterson, N. (2022). Genome-edited crops for improved food security of smallholder farmers. *Nature Genetics*, *54*, 364–367. https://doi.org/10.1038/s41588-022-01046-7

Poku, A.-G., Birner, R., & Gupta, S. (2018). Is Africa ready to develop a competitive bioeconomy? The case of the cassava value web in Ghana. *Journal of Cleaner Production*, *200*, 134–147. https://doi.org/10.1016/j.jclepro.2018.07.290

Porter, J.R., Xie, L., Challinor, A.J., Cochrane, K., Howden, S.M., Iqbal, M.M., Lobell, D.B., & Travasso, M.I. (2014). Food security and food production systems. In C.B. Field, V.R. Barros, D.J. Dokken, K.J. Mach, M.D. Mastrandrea, T.E. Bilir, … & L.L. White (Eds.), *Climate change 2014: Impacts, adaptation, and vulnerability. Part A: Global and sectoral aspects* (pp. 485–533). Cambridge University Press.

Purnhagen, K., & Wesseler, J. (2019). Maximum vs. minimum harmonization: What to expect from the institutional and legal battles in the EU on gene editing technologies? *Pest Management Science*, *75*, 2310–2315. https://doi.org/10.1002/ps.5367

Schmid, B., Mosley, F., Hassegawa, M., Leskinen, P., & Verkerk, P.J. (2022). *Forest-based Bioeconomy and climate change mitigation*. European Forest Institute.

Schultz, T.W. (1980). Nobel lecture: The economics of being poor. *Journal of Political Economy*, *88*(4), 639–651.

Sen, A. (1983). *Poverty and famines: An essay on entitlement and deprivation*. Oxford University Press.

Sikoyo, G.M., Nyukuri, E., & Wakhungu, J.W. (2006). *Intellectual property protection in Africa status of laws, research and policy analysis in Ghana, Kenya, Nigeria, South Africa and Uganda*. Acts Press.

Smith, V., Wesseler, J., & Zilberman, D. (2021). New plant breeding technologies: An assessment of the political economy of the regulatory environment and implications for sustainability. *Sustainability*, *13*(7), 3687. https://doi.org/10.3390/su13073687

Tang, Q., Feng, J., Zong, D., Zhou, J., Hu, X., Wang, B., & Wang, T. (2023). Potential spread of desert locust Schistocerca gregagia (Orthoptera: Acrididae) under climate change scenarios. *Diversity*, *15*(10), 1038. https://doi.org/10.3390/d15101038

Tittonell, P., & Giller, K.E. (2013). When yield gaps are poverty traps: The paradigm of ecological intensification in African smallholder agriculture. *Field Crops Research*, *146*, 76–90.

Trigo, E., Chavarria, H., Pray, C., Smyth, S.J., Torroba, A., Wesseler, J., Zilberman, D., & Martinez, J.F. (2023). The bioeconomy and food systems transformation. In J. von Braun, K. Afsana, L.O. Fresco, & M.H.A. Hassan (Eds.), *Science and innovations for food systems transformation* (pp. 849–868). Springer. https://doi.org/10.1007/978-3-031-15703-5_45

UNEP. (2016). *Food systems and natural resources. A report of the working group on food systems of the international resource panel*. UNESCO.

Van Den Berg, J., Prasanna, B.M., Midega, C.A.O., Ronald, P.C., Carrière, Y., & Tabashnik, B.E. (2021). Managing fall armyworm in Africa: Can Bt Maize sustainably improve control? *Journal of Economic Entomology*, *114*(5), 1934–1949. https://doi.org/10.1093/jee/toab161

Von Braun, J., Afsana, K., Fresco, L.O., & Hassan, M.H.A. (Eds.) (2023). *Science and innovations for food systems transformation*. Springer.

Wesseler, J., Kleter, G., Meulenbroek, M., & Purnhagen, K. (2023). EU regulation of genetically modified microorganisms in light of new policy developments: Possible implications for the EU bioeconomy. *Applied Economic Perspectives and Policy*, *45*(2), 839–859. https://doi.org/10.1002/aepp.13259

Wesseler, J., & von Braun, J. (2017). Measuring the bioeconomy: Economics and policies. *Annual Review of Resource Economics*, *9*, 17.1–17.24. https://doi.org/10.1146/annurev-resource-100516-053701

Wesseler, J., Smart, R., Thomson, J., & Zilberman, D. (2017). Foregone benefits of important food crop improvements in sub-Saharan Africa. *PLoS One*, *12*(7): e0181353. https://doi.org/10.1371/journal.pone.0181353

Zahed, M.A., Movahed, E., Khodayari, A., Zanganeh, S., & Badamaki, M. (2021). Biotechnology for carbon capture and fixation: Critical review and future directions. *Journal of Environmental Management*, *293*, 112830. https://doi.org/10.1016/j.jenvman.2021.112830

Zilberman, D., Gordon, B., Hochman, G., & Wesseler, J. (2018). Economics of sustainable development and the bioeconomy. *Applied Economic Perspectives and Policy*, *40*(1), 22–37. https://doi.org/10.1093/aepp/ppx051

19 The multifunctional role of livestock in East African food systems: the case for dairy

Simon J. Oosting, Todd A. Crane, An Notenbaert, Ademola Braimoh, Asaah Ndambi, Augustine Ayantunde, Esther Kihoro, Corina van Middelaar, and Jan van der Lee

1 Introduction: food systems, livestock, and sustainability

Food systems comprise all activities and actors associated with food, and the economic, social, and natural environments they are embedded in (Van Berkum et al., 2018). Food systems are complex systems with many objectives. A food system is sustainable when objectives are met to extents that are desired by its stakeholders. Food system sustainability, therefore, is not primarily a hard end point, but rather a political process. Livestock plays many roles in food system sustainability. The present paper reflects on such roles of livestock in East Africa. Dairy cattle and farms with dairy are used as a case study.

One concept that has been developed to address the multiple objectives of food systems is Climate-Smart Agriculture (CSA; Lipper et al., 2014). CSA is an approach for transforming agricultural systems to support food security under the realities of climate change. It comprises three objectives: (1) sustainably increasing agricultural productivity to support food security, (2) adaptation to climate change, and (3) mitigating greenhouse gas (GHG) emissions from agriculture (Lipper et al., 2014).

When multiple objectives are at stake, trade-offs may exist among them: not all objectives may be achieved maximally at the same time and at the same place (Germer et al., 2023; Berlin, 1990). We present here the example of dairy intensification: many policies, programmes, and projects aim at intensification of dairy production by increased use of inputs for feeding, housing, health care, and breeding of livestock (Kihoro, 2022; Oosting et al., 2014). Intensification increases milk output per cow and per hectare of land, and it reduces the GHG emission intensity (i.e., the GHG emissions expressed in kg CO_2-equivalents per unit of milk produced). Moreover, intensification of dairy farming is one of the few strategies that can provide an adequate income for smallholder farmers with insufficient land availability for food self-sufficiency (Giller et al., 2021), and for peri-urban farmers farming on expensive land. Hence, dairy intensification is beneficial for food output, climate change mitigation, and farmers' livelihood. It, however, is not necessarily synergistic with climate change adaptation. Dairy intensification policies often focus on the development of a specialised dairy farm similar to the dairy farm type that developed in temperate regions. Such intensification and specialisation, however, are at the expense of diversity, system modularity (i.e., having system components that can be combined, but that can also function independently from each other), and specific adaptations (e.g., use of crossbred or local cows with high resilience to heat stress and diseases), which are important characteristics of many smallholder farms (Oosting et al., 2014). These characteristics facilitate flexibility in these farming systems, which make them resilient to perturbations including climate change (Ten Napel et al., 2011; Descheemaker et al., this volume). East Africa is a region

DOI: 10.4324/9781032649696-24

where livestock is exposed to climate change, which may articulate in higher temperatures and more frequent incidence of extreme events, such as droughts and floods. Heat stress and reduced availability and quality of feeds due to climate extremes (Rahimi et al., 2021) may affect the feasibility of specialised dairy farms, since they rely on breeds with low thermotolerance and on a constant and abundant supply of relatively high-quality feed to maintain productivity throughout a cow's lactation. Nutritional shortages that last for a few days will give a non-reversible drop in milk yield throughout the rest of the lactation. Hence, climate change resilience of such specialised farms is limited and, since they are specialised, they have no alternative livelihood source. Moreover, the objective of climate change mitigation primarily strives to reduce absolute emissions of a food system. An increase in animal numbers or production volume could increase emissions of the whole food system, even when emission intensity decreases (Wang, 2024). This illustrates the importance of a food system approach to gain insights into trade-offs between sustainability objectives, leaving room for multiple combinations of system components to design sustainable food futures. This chapter aims to contribute to such insights. Therefore, we first give some background information about dairy in East Africa, second present an inventory of food system objectives relevant for dairy, third explore the variety of roles that (dairy) cattle and different dairy farming systems have in East African food systems, and fourth evaluate how such roles contribute to sustainability objectives of these food systems.

2 Dairy farming systems in the East African food system

In Kenya, Tanzania, Ethiopia, Rwanda, and Uganda, and to a lesser extent in Burundi, dairy products are an important source of micronutrients and of high-quality proteins in the human diet. Sufficient supply of dairy products adds to the dietary diversity of low- and middle-income groups. Dairy is produced in three different farming system types: grazing systems, mixed crop-livestock systems, and (semi-)specialised farming systems (Oosting et al., 2014).

2.1 Grazing systems

In dryland regions, mobile grazing systems in which pastoralists herd ruminants on communal lands are dominant, but rangeland systems with private ownership of grasslands do exist. Large portions of dryland regions are too dry for significant crop production and herding is the main agricultural activity supporting livelihoods. Because of the harsh conditions in dryland regions, human and livestock population densities are relatively low. Often, grazing systems exist in symbiosis with crop systems in terms of exchange of food, but also when pastoralists have their livestock grazing on crop residues during the dry season, and crop farmers benefit from manure deposited during this grazing (Ayantunde et al., 2011; Tamou et al., 2018). At a food system level, pastoralists and their livestock provide various roles: cultural (e.g., maintaining rare animal breeds and traditional knowledge, which both can be seen as a cultural heritage), ecological (e.g., contribution to the dynamics of natural grasslands), and agricultural (provision of manure to crop farmers) (Udo et al., 2016). Dairy production is for self-consumption, but there may be some selling of products to traders or town dwellers (Kihoro, 2022). Animals are a form of capital accumulation, contributing not only to wealth and status but also to resilience during droughts (Dahl & Hjort, 1976). The problems of present pastoralist systems are the restrictions to mobility and the reduced availability of communal lands by the conversion of grazing lands to agriculture. Both are related to population growth, climate change, and collapse of traditional institutions (Ayantunde et al., 2011; Tamou et al., 2018). Conflicts between crop farmer communities and pastoralists are common (van Dijk & Syntyche Djindil, this volume; Van Dijk et al., 2022:). In East Africa, (agro-)pastoralist systems combine herding and mixed crop-livestock production. These are often farming systems of pastoralists who have partly or fully become sedentary, a process that has been underway for many decades (Ayantunde et al., 2011; Tamou et al., 2018). In such (agro-)pastoralist farming systems, livestock husbandry is the major activity. In this chapter, we combine pastoralist, (agro-)pastoralist, and rangeland systems, and refer to them jointly as (agro-)pastoralist systems.

2.2 Mixed crop-livestock systems

Mixed crop-livestock systems are found in regions with favourable conditions for both crop and livestock production. Such regions have become densely populated and farms tend to be small. High degrees of integration between crop and livestock activities are observed; various species of livestock are kept to feed on crop residues, household wastes, collected forage, in addition to grazing on communal or private lands. Facilitation of crop production (manure, traction, and provision of cash for crop inputs) is a major function of livestock (Udo et al., 2016). Milk is a valuable by-product, important for home consumption, while surpluses may be traded. The farming family's livelihood is supported by livestock since livestock is a store of wealth and contributes to their status (Moll et al., 2007; Udo et al., 2016). One of the key problems of mixed crop-livestock farms is the fragmentation of farms, as land is divided between multiple children through inheritance, which may lead to farms becoming too small to be economically viable. When a farm is too small to support livestock, farmers can just *hang-in*, simply surviving in a poor way, or partially or completely getting out of farming and depending largely on off-farm income (Dorward, 2009; Giller et al., 2021). Farmers can also *step-up*, transforming their farming system towards relatively high market integration. Increase and intensification of the dairy farming component of their farms is often an activity that is economically feasible for mixed crop-livestock holders. Moreover, livestock holders, specifically dairy farmers, are often better off than farmers having no livestock (Udo et al., 2011; Rao et al., 2021). Having livestock and land facilitates greater investment and input use and consequently higher crop and livestock productivity.

So, within mixed crop-livestock systems, farms may range from crop-oriented farms with livestock supporting crop production and livelihoods, to farms with strong orientation on commercial dairy production with only a small role for crop production. If the latter farms largely depend on external inputs, we classify them under (semi-)specialised dairy farming systems (see Section 2.3).

2.3 Specialised systems

Specialised dairy farming systems are high input-high output systems with a high land use intensity, specialised to dairy production in response to robust markets, which may be informal or formal (Migose, 2020; Nyokabi, 2023). They are typically found in densely populated regions with nearby markets and good infrastructure, allowing farms to source feed externally and market their produce with limited transaction costs (Van der Lee et al., 2018). Animal numbers per farm are relatively low (two to ten adult cows) because of feed limitation; farms are land-limited, and feeds must be imported into the farms, often from small and volatile markets (Migose, 2020). Productivity tends to be relatively high. To achieve high productivity, specialised systems often purchase high-quality feeds (e.g., maize and forages for dairy) cultivated on crop lands. Consequently, land and water use for such systems competes with crop production for human food. The primary role of dairy at specialised farms is income generation through marketed foods. On land-limited farms, manure is considered a waste product, though some farms may market manure or use bio-digesters to produce biogas. Problems of the specialised systems are price stagnation, price volatility caused by seasonal scarcity of input and products, competition with internationally traded imports, and manure accumulation and its disposal (Oosting et al., 2014; Migose, 2020). Fully specialised dairy farms are relatively scarce in East Africa since many farmers combine intensive dairy production with some crop production and/or off-farm jobs. Here we therefore refer to these farming systems as (semi-)specialised.

2.4 Importance of dairy farming system types in societies

In East Africa, a considerable proportion of the human population obtains a livelihood from (agro-)pastoralist and mixed crop-livestock systems (40–60% in Kenya and Ethiopia, Table 19.1). The social and livelihood functions of (semi-)specialised farms on a national scale are more limited. Statistics about such population numbers are scarce, often indirectly estimated, and the delineation of systems is not always clear from the literature sources; hence the information presented in Table 19.1 is approximate at best. The contribution of (semi-)specialised farms to national milk production is estimated to be 30% of production in Kenya (Nyokabi, 2023). Since many of such (semi-)specialised dairy farms are in peri-urban regions, it is likely that this relatively small number of (semi-)specialised producers provides milk for a considerable part of urban dwellers.

Most cattle in both countries are not defined as dairy cattle (Table 19.1). Non-dairy cattle are kept for beef production and functions such as manure, draught, status, and cash and wealth accumulation, whereas milk may be regarded as by-product.

Human demographic dynamics in East Africa drive adaptation of these dairy farming systems: population growth results in shrinking farm sizes and reduces availability of grazing (Giller et al., 2021), urbanisation increases the length and complexity of supply chains (Oosting et al., 2014; Van der Lee et al., 2020), while the growth of middle-income groups boosts demand for dairy product volume and diversity (Oosting et al., 2021). This growth in demand has boosted national dairy production by 4–9% annually for periods of 5–10 years in many East African countries (Hemme, 2021). However, Kenya has struggled to produce sufficient milk to

Table 19.1 Human populations living in farming systems and cattle populations (data from years 2017–2021)

	Ethiopia	*Kenya*
Human population (million)		
• Total	102–120	47–58
• Mixed crop-livestock smallholders[1]	35–50	20–30
• (Agro-)pastoralists	7.5–12	5–9
• (Semi-)specialised farmers	<0.07	<0.02
Cattle population (million)		
• Total	65–70	20–25
• Dairy	8–13	3–8

Sources: ASL2050 FAO (2017); Shapiro et al. (2017); FAO (2019a, 2019b); Management Entity (2021); FAOstat (data retrieved in 2023; data for 2021).

[1] Including farms keeping small ruminants, pigs, and/or poultry.

meet local demand. Between 2014 and 2018, its milk imports from Uganda grew significantly (up to 20% of processed volume in Kenya), benefiting from low Ugandan farmgate prices and the East African Community free market regulations.

3 Food system objectives and dairy farming

To gain insights into the farmers' and value chain stakeholders' perspectives of sustainability of dairy development, the Netherlands East Africa Dairy Partnership (NEADAP) conducted a series of workshops (van der Lee et al., 2022) in Kenya, Uganda, Rwanda, and Tanzania. Stakeholders from farmer cooperatives, input suppliers, farm advisors, dairy processors, scientists, and civil servants expressed concerns regarding dairy development. In the economic domain, important concerns were future land availability and potential profitability of dairy production. Moreover, the ability of dairy production to meet consumer demand was perceived as the greatest challenge. In the social domain, an important concern of farmers was about being included, having a voice in the dairy value chain, in which processors are often most powerful (Nyokabi, 2023), and maintaining some level of autonomy. In the natural resource domain, stakeholders expressed concerns about biodiversity, water management (quality and quantity), and soil health. We translated these stakeholder concerns into food system objectives in the CSA, natural resource, and socio-economic domains (see Table 19.2).

One important additional food system objective is to limit feed-food competition. If dairy production is based on high-quality feeds and forages produced on crop lands, the implication is that these crop lands cannot be used for production of human food crops, and consequently dairy will compete for land (De Boer & van Ittersum, 2018). Crop land is scarce in Africa: Giller and Andersson (this volume) and Schut and van Ittersum (this volume) paint a concerning future regarding the production of sufficient staple crops to feed the East African population in 2050. Land availability is limited, which implies that additional crop production cannot come from expansion of crop land. Schut and van Ittersum (this volume) propose increasing crop productivity on current crop lands as the most promising pathway to meet future staple crop needs and

Table 19.2 Food system objectives

Domain	*Objectives*	*Indicators*
Climate-smart agriculture	• Food and nutrition security • Climate change mitigation • Climate change adaptation	• Quantity and quality (energy, protein, micronutrients) • Land use (including feed-food competition) • Productivity (GHG emission intensities) • Carbon sequestration • Resilience • Diversity • Modularity • Adaptive technologies
Natural resources	• Biodiversity and natural resources (nature, water, soil, nutrients)	• Quantity (land, water, species, nutrients) • Quality • Pollution • Degradation
Socio-economic	• Livelihood and income • Employment and economic development • Social equity and liveability	• Economy • Self-reliance and autonomy • Voice/inclusion • Rural liveability

Notes: The background information in Sections 1, 2, and 3 of this chapter addresses specific objectives within sustainability domains. We have added some examples of indicators of importance for each domain.

Table 19.3 Roles of (dairy) cattle in farming systems and of dairy farming systems in food systems

Farming system	*(Dairy) cattle in farm*	*(Dairy) farm in food system*
(Agro-)pastoralist	Livelihood • Food and nutrition • Resilience • Savings • Sell milk to buy cereals and pulses	• Cultural heritage • Landscape and biodiversity • Animal-sourced food
Mixed crop-livestock	Crop production • Manure • Draught power Livelihood • Food and nutrition • Resilience • Savings	• Plant-sourced food • Animal-sourced food • Employment • Social inclusivity • Economic development • Social exchange
(Semi-)specialised	Income	• Animal-sourced food • Employment in value chain • Economic development

food security. However, fragmentation of farms and limited market development for inputs and outputs currently hamper crop productivity increase (Giller et al., 2021).

We conclude from this that future dairy production, if based on forage production on crop lands, would negatively affect future food crop availability. Therefore, in a sustainable East African food system that is to feed the human population of 2050, dairy and other livestock may have to be fed on by-products and/or from lands unsuitable for crop production (Oosting et al., 2021). In addition, in such a sustainable East African food system of 2050, avoiding feed-food competition on the production side should be complemented by a responsible per capita consumption of animal-sourced food on the consumer side (De Boer & Van Ittersum, 2018). Van Hal et al. (2023) estimated that on a global level, a production of 35 g of animal-sourced protein per capita per day is possible without using crop land for the direct production of animal feeds.

Food system objectives relevant to dairy production are summarised in Table 19.2. Such objectives thus cover those of individual farmers, of regions and nations, and of the planet at large.

Table 19.3 presents an overview of roles of (dairy) cattle in farming systems and of dairy farming systems in food systems.

Table 19.4 presents the contributions of the broad farming systems in meeting the East African food system objectives.

3.1 Climate-Smart Agriculture objectives

Sufficient supply of dairy products adds to the dietary diversity of low- and middle-income groups. All farming systems produce milk as food or tradable commodity for a considerable proportion of the population, but productivity differs widely. (Semi-)specialised systems are the most productive and have potential for additional production to meet urban demand (Oosting et al., 2014). The mixed crop-livestock systems are important for food supply because of the large numbers of these farms.

The high potential productivity of (semi-)specialised farms is associated with a high feed-food competition. (Agro-)pastoral and mixed crop-livestock systems use feeds from lands unsuited for food crop production, crop residues, and waste streams for dairy cattle feeding. Consequently, they have a relatively low feed-food competition.

Table 19.4 Performance of farming systems with (dairy) livestock on food system objectives based on an evaluation done by the authors

	(Agro-)pastoral	*Mixed crop-livestock*	*(Semi-)specialised*
Climate-Smart Agriculture objectives			
Food and nutrition	Low productivity. Limited marketing. Sustains households' nutrition.	Medium productivity. Important dairy provider due to the substantial number of farms.	Potentially high productivity and important for urban milk supply.
Feed-food competition	Low. Production is from land unsuitable for crop production.	Low to medium. Feeding on crop residues and wastes. Livestock's first function is to support crop production and livelihoods.	High. Relatively high-quality feeds (forages and concentrates) to achieve high productivity.
Climate change adaptation	Resilient because of mobility and adaptation to harsh conditions.	Resilient because of diversity and modularity.	Technological and organisational control mechanisms (climate control, irrigation, insurances).
Climate change mitigation	Carbon sequestration in the ecosystem. High emission intensity.	Carbon sequestration in agroforestry. Medium emission intensity, partly attributable to non-production functions.	Low emission intensity because of high productivity.
Natural resources objectives			
Biodiversity and natural resources	Part of dryland ecosystem.	Diversity, but risk of soil mining.	Decoupling of livestock and land causes nutrient accumulation and air and water pollution.
Socio-economic objectives			
Income and livelihood	Livelihood.	Livelihood.	Farm economy.
Employment and economic development	No absorptive capacity for growing labour force.	Limited absorptive capacity for growing labour force because of small farm sizes.	Employment opportunities, improved market access in growing formal value chains.
Social equity and liveability	Heavily affected by conflict and restrained mobility.	Important for many people but impacted by farm fragmentation.	Positive as an economic pillar of communities. Negative on nuisance (odour, pollution).

Climate change adaptation is best in systems that are adapted to harsh climatic conditions ((agro-)pastoral systems) or that have a high diversity and modularity (mixed crop-livestock systems). A (semi-)specialised system requires a different adaptation approach to maintain its function under climate change. Such systems can be robust through high degrees of control, e.g., climate regulation technology, purchase and storage of feed stocks, early warning systems, irrigation, and institutional arrangements such as insurances (Ten Napel et al., 2011).

Climate change mitigation is often expressed as the contribution to reduction of the GHG emission intensity. This emission intensity is often lowest for high-productive animals in (semi-)specialised systems and a co-benefit of productivity-enhancing interventions (Gerber et al., 2011). However, Udo et al. (2016) advocated that GHG emission intensity calculations should account for the multifunctional values generated by livestock rather than assuming that milk or meat is the only value function, which is the standard method. Fair metrics would imply that part of the emissions attributable to other functions than dairy production are not allocated to the milk output. Doing so, emission intensities attributed to dairy production on mixed crop-livestock farming systems are comparable to those of (semi-)specialised farming systems (Udo et al., 2016). Moreover, this implies that part of the GHG emissions should be attributed to non-food-functions of cattle. Such functions can potentially be replaced by other activities: development of banking and insurance systems may reduce the need to keep a high number of low-productive animals as a store of wealth (provided inflation and associated currency devaluations are at reasonable rates), and introduction of mechanisation could reduce the number of animals kept for traction. A reduction of the herd size of non- and low-productive cattle would reduce the absolute emission of GHGs but would also put less pressure on the available high-quality dairy cattle feeds. In addition to reducing emissions, maintaining, or increasing the amount of carbon sequestered in soils is a further way of mitigating climate change, though the impact of carbon sequestration in grasslands as a global mitigation option is limited (Wang et al., 2023). Due to the importance of soil organic carbon for nutrient provisioning and water holding capacity, increasing soil carbon stocks also serves adaptation and food security objectives. Grassland management aimed at preserving soil carbon stocks or carbon sequestration, biomass restoration, and reforestation through agroforestry or silvopastoralism could be an approach by which (agro-)pastoralist and mixed crop-livestock systems contribute to local mitigation and adaptation objectives.

3.2 Biodiversity and natural resource objectives

Soil and water quality, degradation of grasslands, and land use expansion into nature areas are major biodiversity challenges related to dairy production and development. In a well-functioning (agro-)pastoralist system, livestock is an integral element of the dryland ecology and (agro-) pastoralists have traditional ecological knowledge to use their environment sustainably (Tamou et al., 2018), but the functioning of these systems is strongly impacted by constrained access to grazing lands and to corridors for movement through crop regions. Mixed crop-livestock systems are diverse, but soil mining because of under-fertilisation negatively affects soil quality. Grazing pressure on communal grazing lands in mixed crop-livestock regions may lead to severe degradation of these lands. (Semi-)specialised systems are often landless systems and manure disposal is a threat to water and air quality, and overfertilisation of small plots may occur.

3.3 Socio-economic objectives

Since a considerable proportion of the human population of East Africa obtains a (part of their) livelihood from (agro-)pastoralist and mixed crop-livestock systems, these systems are important to absorb rural labour and involve smallholder farmers, in particular the poor, women, and youth in dairy-related activities. Because farmers in these systems rely only marginally on milk marketing, they have a relatively high degree of self-reliance. (Semi-)specialised systems operate in modern dairy value chains that may contribute to development of the formal economy. It is noteworthy, however, that even in Kenya, which has an established formal dairy sector, the informal dairy economy is still larger than the formal one and the lines between formal and informal economies remain blurred (Vernooij et al., 2023). During this coexistence of formal and informal sector, the roles and influence of cooperatives, traders, processors, and sector regulators vary widely, as does the farmers' voice in the formal sector organisation (Nyokabi, 2023). The labour input of (semi-)specialised farming systems and associated value chains requires relatively high skilled, formal labour and may exclude the poor, women, and youth (Tavenner & Crane, 2018; Tavenner et al., 2021).

With the reducing land availability for (agro-)pastoralist systems and the need for farm enlargement for mixed crop-livestock systems (Giller et al., 2021), it is likely that the absorption of labour force in these systems will become negative. With an increase in the number of (semi-)specialised systems and the associated value chains, the contribution of these systems to employment and economic development will grow and be considerable. When (agro-)pastoralists and mixed crop-livestock farmers are forced to step out of farming, marginalisation and exclusion may be unwanted consequences. (Semi-)specialised systems may be economic pillars for communities, but odour and pollution nuisance because of livestock operations in densely populated regions should be minimised.

4 Conclusion

We conclude that East African dairy cattle and farming systems producing dairy play an important role in meeting the food system objectives. These roles are not limited to CSA objectives (i.e., food and nutrition, climate change mitigation, and climate change adaptation) but extend to natural resource and socio-economic objectives. Not all objectives can be met maximally at the same time. Therefore, development efforts should not be overly focused on the development of (semi-)specialised farming systems, since (agro-)pastoral and mixed crop-livestock systems are also essential for meeting part of the food system objectives. So, programmes and projects by international institutions and governmental and non-governmental organisations should aim to maintain and enforce the strong and positive contributions of farming system types (i.e., the ecological and cultural functions of (agro-)pastoralist systems, the diversity and crop-livestock integration of mixed crop-livestock systems, and the economic function and high milk supply of (semi-)specialised systems). Such programmes and projects should, at the same time, aim to overcome the weak and negative aspects of farming system types, i.e., conflict and mobility restrictions in the (agro-)pastoralist systems, farm fragmentation and soil mining in the mixed crop-livestock systems, and pollution and feed-food competition in the (semi-)specialised systems. Across all these livestock farming systems, improving economic opportunities for the poor, women, and youth continue to need attention.

At present, the degree of circularity, i.e., prevention and recycling of losses and wastes, is high in dairy production in East Africa. (Agro-)pastoralist systems source feeds from lands unsuited for crop production, and mixed crop-livestock systems use crop residues and recycle nutrients by using manure as fertilizer. (Semi-)specialised dairy farming systems tend to be less circular since they have higher degrees of feed imports and limited use of manure, because of land limitation. For 2050, we foresee that circularity may be an even more important principle for livestock production, including dairy, than today. Crop land scarcity will force land use to be directed towards food crop production, leaving crop residues, food losses, and wastes as the major feeds for livestock. This will put limits on the number of livestock that can be maintained, on their productivity, and consequently on the human consumption of animal-sourced foods. In such a circular food system, (agro-)pastoral and mixed-crop livestock systems will remain important since they have a high degree of circularity, with limited feed-food competition and a relatively strong resilience to climate change.

Acknowledgements

This chapter is based on a presentation given at the 27th session of the Conference of Parties of the United Nations Framework Convention on Climate Change, November 2022, in Sharm-el-Sheikh, Egypt. This presentation and the elaboration into this chapter were facilitated by the Netherlands East African Dairy Partnership (NEADAP), who we acknowledge with gratitude.

References

ASL2050 FAO. (2017). Africa sustainable livestock (ASL) 2050 livestock production systems spotlight cattle and poultry sectors in Kenya. FAO. Nairobi.

Ayantunde, A.A., De Leeuw, J., Turner, M.D., & Said, M. (2011). Challenges of assessing the sustainability of (agro)-pastoral systems. *Livestock Science*, *139*, 30–43.

Berlin, I. (1990). *The crooked timber of humanity: Chapters in the history of ideas*. Vintage Books.

Dahl, G., & Hjort, A. (1976). Having herds: Pastoral herd growth and household economy. Department of Social Anthropology, University of Stockholm, 335 pp.

De Boer, I.J.M., & Van Ittersum, M.K. (2018). Circularity in agricultural production. Wageningen University & Research, p. 71.

Dorward, A., &erson, S., Nava Bernal, Y., Vera, E.S., Pattison, J., & Paz, R. (2009). Hanging in, stepping up and stepping out: Livelihood aspirations and strategies of the poor. *Development in Practice*, *19*, 240–247.

FAO. (2019a). The future of livestock in Ethiopia. Opportunities and challenges in the face of uncertainty. Rome. 48 pp.

FAO. (2019b). The future of livestock in Kenya. Opportunities and challenges in the face of uncertainty. Rome. 56 pp.

Gerber, P., Vellinga, T., & Opio, C., et al. (2011). Productivity gains and greenhouse gas emissions intensity in dairy systems. *Livestock Science*, *139*, 100–108.

Germer, L.A., van Middelaar, C.E., Oosting, S.J., & Gerber, P.J. (2023). When and where are livestock climate-smart? A spatial-temporal framework for comparing the climate change and food security synergies and trade-offs of Sub-Saharan African livestock systems. *Agricultural Systems*, *210*, 103717.

Giller, K.E., Delaune, T., Silva, J.V., Descheemaeker, K., van de Ven, G., Schut, A.G.T., … & Van Ittersum, M.K. (2021). The future of farming: Who will produce our food? *Food Security*, *13*, 1073–1099.

Hemme, T. (Ed.) (2021). IFCN dairy report 2021. IFCN Dairy Research Center. Kiel, Germany.

Kihoro, E.M. (2022). Achieving low emission development. Anticipating alignment between global strategies and local realities in the Tanzanian dairy sector (Doctoral thesis). Wageningen University, 206 pp.

Lipper, L., Thornton, P., Campbell, B.M., Baedeker, T., Braimoh, A., Bwalya, M., … & Torquebiau, E.F. (2014). Climate-smart agriculture for food security. *Nature Climate Change*, *4*, 1068–1072.

Management Entity. (2021). *Ethiopia's livestock systems: Overview and areas of inquiry*. Feed the Future Innovation Lab for Livestock Systems.

Migose, S.A. (2020). Addressing variation in smallholder farming systems to improve dairy development in Kenya (Doctoral thesis). Wageningen University, 144 pp.

Moll, H.A.J., Staal, S.J., & Ibrahim, M.N.M. (2007). Smallholder dairy production and markets: A comparison of production systems in Zambia, Kenya and Sri Lanka. *Agricultural Systems*, *94*, 593–603.

Nyokabi, S.N. (2023). Bridging the gap: Improving milk quality on smallholder dairy systems in Kenya (Doctoral thesis). Wageningen University, 136 pp.

Oosting, S.J., Udo, H.M.J., & Viets, T.C. (2014). Development of livestock production in the tropics: Farm and farmers' perspectives. *Animal*, *8*, 1238–1248.

Oosting, S.J., van der Lee, J., Verdegem, M., de Vries, M., Vernooij, A., Bonilla-Cedrez, C., & Kabir, K. (2021). Farmed animal production in tropical circular food systems. *Food Security*, *14*, 273–292.

Rahimi, J., Mutua, J.Y., Notenbaert, A.M.O., Marshall, K., & Butterbach-Bahl, K. (2021). Heat stress will detrimentally impact future livestock production in East Africa. *Nature Food*, *2*, 88–96.

Rao, B.K., De Boer, I.J.M., Ripoll Bosch, R., & Oosting, S.J. (2021). Understanding transitions in farming systems and their effects on livestock rearing and smallholder livelihoods in Telangana, India. *Ambio*, *50*, 1809–1823.

Shapiro, B.I., Gebru, G., Desta, S., Negassa, A., Nigussie, K., Aboset, G., & Mechale, H. (2017). Ethiopia livestock sector analysis. Ethiopia Ministry of Livestock and Fisheries and the International Livestock Research Institute.

Tamou, C., Ripoll-Bosch, R., De Boer, I.J.M., & Oosting, S.J. (2018). Pastoralists in a changing environment: The competition for grazing land in and around the W Biosphere Reserve, Benin Republic. *Ambio*, *47*, 340–354.

Tavenner, K., & Crane, T.A. (2018). Gender power in Kenyan dairy: Cows, commodities, and commercialization. *Agriculture and Human Values*, *35*(3), 701–715.

Tavenner, K., Saxena, T., & Crane, T.A. (2021). 'Breaking even' under intensification? Gendered trade-offs for women milk marketers in Kenya. *Rural Sociology*, *86*, 110–138.

Ten Napel, J., van der Veen, A.A., Oosting, S.J., & Groot Koerkamp, P.W.G. (2011). A conceptual approach to design livestock production systems for robustness to enhance sustainability. *Livestock Science*, *139*, 150–160.

Udo, H.M.J., Aklilu, H.A., Phong, L.T., Bosma, R.H., Budisatria, I.G.S., Patil, B.R., Samdup, T., & Bebe, B.O. (2011). Impact of intensification of different types of livestock production in smallholder crop-livestock systems. *Livestock Science*, *139*, 22–29.

Udo, H., Weiler, V., Modupeore, O., Viets, T.C., & Oosting, S.J. (2016). Intensification to reduce the carbon footprint of smallholder milk production: Fact or fiction? *Outlook on Agriculture*, *45*, 33–38.

Van Berkum, S., Dengerink, J., & Ruben, R. (2018). *The food systems approach: Sustainable solutions for a sufficient supply of healthy food*. Wageningen Economic Research.

Van der Lee, J., Klerkx, L., Bebe, B.O., Mengistu, A., & Oosting, S.J. (2018). Intensification and upgrading dynamics in emerging dairy clusters in the East African highlands. *Sustainability*, *10*, 4324.

Van der Lee, J., Oosting, S.J., Klerkx, L., Opinya, F., & Bebe, B.O. (2020). Effects of proximity to markets on dairy farming intensity and market participation in Kenya and Ethiopia. *Agricultural Systems*, *184*, 102891.

van der Lee, J., Alvarez Aranguiz, A., Pishgar Komleh, H., & Ndambi, A. (2022). *Dairy sustainability assessment tool: User manual*. Wageningen Livestock Research, Wageningen University & Research. 17 pp.

Van Dijk, H., Samimi, C., & Zandler, H. (2022). Climate variability and institutional flexibility: Resource governance at the intersection between ecological instability and mobility in drylands. In Kronenburg Garcia A, et al. (Eds.), *Drylands facing change* (pp. 15–31). Routledge.

Van Dijk, H., & Djindil, S.N. (this volume). Political instability and food security: The long view towards 2050.

Van Hal, O., van Zanten, H.H.E., Ziegler, F., Schrama, J.W., Kuiper, K., & de Boer, I.J.M. (2023). The role of fisheries and fish farming in a circular food system in the European Union. *Sustainable Production and Consumption*, *43*, 113–123. https://doi.org/10.1016/j.spc.2023.10.017

Vernooij, V., Vellema, S.R., & Crane, T.A. (2023). Beyond the formal-informal dichotomy: Towards accommodating diverse milk-collection practices in the economic middle of Kenya's dairy sector. *The Journal of Development Studies*, *59*, 1–17.

Wang, Y. (2024). What about the ruminants? Methodological advances to address environmental issues. (Doctoral thesis). Wageningen University, 174 pp.

Wang, Y., de Boer, I.J.M., Martin Persson, U., Ripoll-Bosch, R., Cederberg, C., Gerber, P.J., Smith, P., & van Middelaar, C.E. (2023). Risk to rely on soil carbon sequestration to offset global ruminant emissions. *Nature Communications*, *14*, 7625.

20 Extending the smallholder value-creation frontier: a business perspective on food security

Paul T.M. Ingenbleek, Olawale Rotimi Opeyemi, Dorothy Kanorio Murugu, Cobus Oberholster, and Lucas E. Urbano

1 Introduction

Agribusiness is perhaps the most neglected field in academic research relevant to food security. Because of its direct importance to people's well-being, food production has a central interest for policy-makers. The availability of food at affordable prices provides a stable basis for many government regimes (e.g., Russell, 2022). Perhaps for this reason the role of business in achieving food security is often seen, at best, of secondary interest in agricultural development projects. New technologies to improve food security are developed by publicly funded research institutes or publicly funded scientific projects and then shared with farmers through government-led extension organizations. This picture hardly changed when "the market" became the main philosophy for organizing food systems in the 1990s (Poole, 2010). When business does play a role, it is often referred to as the "private sector", thus again emphasizing the distinction between public and private investments.

Notwithstanding that many businesses are privately owned and have a profit objective, there is a dearth of knowledge on what business is, and what it can contribute to food security. Such knowledge is now more important than ever. Given the complexity and urgency of Africa's "food security conundrum," which is "how to provide cheap, nutritious food to feed the growing urban and rural populations while creating incentives to stimulate increased agricultural production" (Giller, 2020, p. 1, see also Giller & Andersson, this volume), the investments and adaptive actions of companies are now needed more than ever. This chapter aims to shed light on the question how current social, economic, and cultural developments in Africa create opportunities for agribusiness to create value, and whether or not that value will contribute to food security. Central to our approach is that we see businesses as value-creating entities (e.g., Porter, 1985; Hunt, 2000). This approach brings us to the concept of the *inclusion frontier* of smallholders, indicating the point after which value creation gets the upper hand in a farm, thus making it de facto a *farming business* (Teklehaimanot et al., 2017a).

In the following, we first explain some basic elements of businesses as creators of value. We then describe the emerging market opportunities to create value in urban and peri-urban areas. This is followed by a section on the smallholder value-creation inclusion frontier and a section that discusses the opportunities to extend the frontier. This chapter finishes with some concluding comments.

2 Going beyond "the private sector": businesses as value-creators

From a classic business theory perspective, a business is an organizational unit that includes all business functions (such as procurement, inbound logistics, production/processing, selling, and

DOI: 10.4324/9781032649696-25

outbound logistics), thus enabling it to create value for a particular target market (Porter, 1985). Companies can consist of multiple businesses, or even of divisions that each include multiple businesses. The idea that they create value means that healthy businesses create products and services that their customers value to an extent that exceeds the costs for developing, producing, and delivering the products and services. Thus, by creating value for its customers, the business also creates value for itself and can make a profit and/or achieve other (social) goals.

In this regard, it may be surprising that *agribusiness* is typically defined as "all those business and management activities performed by firms that provide inputs to the farm sector, produce farm products, and/or process, transport, finance, handle or market farm products" (Downey & Erickson, 1987, p. 6). This definition emphasizes production rather than the creation of customer value, which also requires more creative tasks like business development, the discovery of customer preferences, new ways of defining markets, and new product development. Shultz and Edwards (2005, p. 57) therefore offer an alternative definition which is more in line with mainstream business theory. They define agribusiness as "the market-oriented sustainable orchestration of food, fibre, and renewable resources".

The term "market-oriented" indicates a business culture that puts the creation of customer value first. A market-oriented business typically acquires and shares insights about customers and uses these in strategic and operational decisions. Because these businesses better understand what customers value, they perform better in their target market then competitors (Kirca et al., 2005). Competition is thus not necessarily a matter of price competition, but rather a continuous struggle for competitive advantage. Businesses that experience disappointing financial returns from their market start to question whether their competitors may offer customers more value than they do, or perhaps the same value but at less cost. They subsequently start to make investments in the resources from which they can improve their market position. If their attempt is successful, their market position and performance will increase. As a result, competitors will likely experience a decrease in the financial returns from their market, leading them in turn to make investments and improvements. This continuous process increases the living standards of consumers, fosters innovation, and generates economic growth (Hunt, 2000).

3 Before the frontier: the vibrant urban food market

The process described above, by which businesses struggle to obtain a position of competitive advantage on the market leading to increased living standards for consumers, is increasingly recognizable in urban areas on the continent. Demographic growth is the main driver of the growing demand for food (see van Wissen, this volume). Increasing urbanization concentrates this demand in relatively smaller areas that attract businesses in different parts of the food system. As compared to the rural areas, the distribution and marketing costs in cities are significantly lower, simply because a truck or an advertisement will reach more people.

Rising incomes, for at least some part of the population, create emerging and floating middle classes that change their lifestyles and eating habits (e.g., Chikweche & Fletcher, 2014). The response of businesses is that they jointly create a larger assortment of foods for the urban consumers, including for example convenient pre-packed and ready-to-eat foods, social status-increasing food from international brands, or foods highlighting cultural heritage and tradition (cf. Baba Daouda et al., 2020). Typical product categories that grow are dairy, meat, fresh foods, and vegetables (especially when provided with consistent quality and with lower safety risks). Western, Chinese, Indian, and Japanese foods are gaining traction in restaurant chains across Africa, and gradually also more imported African specialty foods can be found. In a country like Ghana, consumers become more aware of supporting local business. Food safety is also becoming a competitive element because most African countries lack the institutions to secure food safety for the entire food system. A recent study among consumers in and around Nairobi showed that food safety is also a factor for which consumers are willing to pay higher prices (Chege et al., 2021).

Marketing channels are increasingly becoming a mix of traditional open-air or otherwise informal markets and street vendors, new and upcoming retailing formats like supermarkets and corner shops, out of home like fast food, restaurant, and catering businesses, as well as online and offline takeaways. E-commerce revenue in Kenya in the food and personal care markets, for example, increased from 9 million US$ in 2017 to 79 million US$ in 2020 and is expected to increase even further to over 177 million US$ in 2025.

Accessibility also depends on perishability, with more perishable products requiring better access than products that can be stored for a longer time, as well as whether products are fresh or processed/packaged (with processing plants located at strategic locations to reach cities and be in reach for supplies). On the supply side, businesses will probably look for locations close to, or at least well-accessible to, cities where they will apply more intensive farming systems to cover the costs of the increasing prices for land. In general, these developments will lead (depending on the agroecology) to farming models in peri-urban and urban areas that are larger and more intensive, relatively close to the market and with short supply chains. Besides being a source for the local market, such production centres can supply export markets, including other markets in Africa. The 40,000+ ha Del Monte pineapple plantation in Kenya, which is the largest exporter of agricultural produce from Kenya, is one example.

Notably, a process of competition leads to positive outcomes only when institutions are supportive (cf. Acemoglu & Robinson, 2012 and Kikulwe & Wesseler, this volume). With increasing speed, institutional development may be unable to keep pace with value creation by businesses. Competition among businesses may evolve to a process in which competition focuses on access to scarce strategic resources, such as access to distribution networks, transporting companies, top locations for new stores, and scarce supplies of produce (especially during seasons with bad harvests). When institutional support is not widely available, businesses may compete for permits, subsidies, tax exemptions, etc. This favours companies that use bribes and have close relations with policy-makers and politicians to get the upper hand. Under such conditions, the process of competition does not benefit the customer, but other stakeholders (Hunt, 2000).

Taking the warning of the need for institutional support into account, we conclude that business has the potential to contribute to food security. It can potentially contribute to all dimensions of food security, including availability of food (production and distribution), access to affordable food, utilization (nutritional value and food safety), and stability (long-term sustainability of production and processes) (Zamudio, Bizikova & Keller, 2014).

4 The smallholder value-creation inclusion frontier

The distant rural areas in many African countries are a far cry from the vibrant urban markets in the very same countries. This has consequences for food security. The unreached rural consumer is most vulnerable to food and nutrition insecurity, especially due to poor last mile distribution arrangements and the skewed focus on the urban consumer (Fraval et al., 2019).

The more remote, the fewer companies and institutions are present to provide inputs and support services, or to purchase farm produce. These contexts are sometimes referred to as "thin" markets, because few companies are active in the context (e.g., Dorward et al., 2009). Without a sense of pressure and with scant customer feedback, these companies usually experience little incentive to develop a customer focus. With an internally oriented production focus, their products and services may miss easy opportunities to create value. They may refrain from supplying all their agrodealers in a constant way, thus creating high uncertainty for farmers to make decisions on what to plant for which potential market. The inputs available may be a profound mismatch with what the smallholder clientele want or need to buy (for example, think of fertilizers offered in large bags not affordable for a single farmer, while a farmer who buys also on behalf of the rest of the community is suddenly confronted with complex financing and distribution tasks). Such thin markets are usually less attractive in terms of purchasing power and greater transaction costs. It is unlikely that many new businesses will enter such markets in the coming years (consider, for example, markets in the Sahel where political instability makes the conditions to do business more complex; also see van Dijk & Djindil, this volume). As a consequence, smallholders may be reluctant to produce for the market and will focus more on providing food for their own families.

As the differences between these two contexts are sharp, we speak of a 'smallholder value creation frontier', which distinguishes the main focus of the smallholder farm either as a means to create value or as a means for subsistence. Comparable distinctions have been made in the literature distinguishing substantive (driven by profit or social goals) and subsistence entrepreneurs (who do business out of necessity) (e.g., Babah Daouda et al., 2019; Sridharan et al., 2014). In reality, the distinction is less "black and white" as it may seem. The existing literature offers several segmentation studies of smallholders, usually dividing them into groups with different degrees of commercialization. While farmers close to the urban areas are likely to have the best connections to buyers, inputs, and institutions, in the most remote places such connections may be entirely absent. In between, there are different "shades of grey" (e.g., AGRA, 2017).

Smallholders can create value for different target groups. First, they can play a role as outgrowers to larger commercial farms who use smallholders to increase their production capacity, preferably with a long-term focus on training them in providing the appropriate qualities and increasing quantities. Second, smallholders can create value as specialized suppliers, because their region is particularly suitable for and experienced in certain fruits and vegetables that are demanded by specific consumers (Ingenbleek, 2019). One example is Kersting's groundnut (*Macrotyloma geocarpum*), a legume that is consumed in Benin mostly on special occasions such as Christmas (Babah Daouda et al., 2020). Third, they can create value as local food suppliers to farmers that are fully commercialized and no longer use their land to provide food for their own families. Fourth, as member of a farmer group they can contribute to the production of specific staple or other crops, marketed by the group in specific value chains. The cropping and marketing decisions are then mostly made at the group level.

The smallholder value-creation frontier is not stable, but it shifts when selling to markets becomes either more or less attractive to smallholders. There are both pull and push factors that can extend the frontier. Pull factors include a growing demand for food, leading to higher prices and traders exploring new areas to find the produce that they seek. Hounhouigan and colleagues (2020) report for example that traders from Lagos started turning up in Northern Benin seeking soyabean and other products for animal feed because the supplies within Nigeria were insufficient to satisfy the growing demand. Push factors include a growth of supply. An increased supply may lead to lower prices that make a product attractive for traders to purchase and then explore new markets for it. Sixty-four percent of food consumed in Africa is handled by small- and medium-sized traders, food producers, processors, transporters, and marketers (AGRA, 2019). As the search of new markets is an inherent aspect of the traditional African marketing system (Ingenbleek, 2019), these actors also actively engage in the search for new products.

Babah Daouda et al. (2019) report that high yields of Kersting's groundnut in the south of the country led traders to explore markets further north where they promoted the crop as a speciality for religious ceremonies, and consumers started to adopt it for Ramadan.

When the value-creation inclusion frontier can be extended, it can also be contracted. Companies often ask smallholders to supply timely and at a standardized quality level (Neven et al., 2009). A longitudinal study in Kenya showed that sustaining market linkages is not guaranteed once they are established. The study showed that a substantial number of smallholders dropped out after they initially connected (Fischer & Qaim, 2012). Rejections of offered produce may lead to disappointment with smallholders and sometimes even a resentment against the market system (Jaffee et al., 2011). Hence, for food security, it is important to understand the factors that contribute to an extension of the smallholder value-creation frontier.

5 Extending the inclusion frontier

Business can contribute to extending the "inclusion frontier" in several ways. First, it can play a role as a value-chain developer. The Beninese company Sojagnon, for example, works on the professionalization of providers offering inputs for soyabean to smallholders. At the same time it builds capacity among processors in terms of packaging, branding, pricing, and distribution. Because processors communicate their preferences to the smallholders as well, they create an opportunity for smallholders to increase their income (Hounhouigan et al., 2020). The value chain is also important to increase the commitment of companies at the consumer end of the chain, like retailers and brand manufacturers. Because these companies are eye-catching they may easily be held responsible for food safety and social issues in the value chain. In a well-functioning chain, certification and assessment will be easier to apply than in a sector with short-lived transactions at arm's length.

Second, from the point of logistics, value chains can work in two directions: they help to aggregate produce from farmers, but can also bring consumer products and agricultural inputs back to farming communities. Given the adaptability of traditional African marketing systems (Ingenbleek, 2020), it is likely that once more farmers become integrated with markets (and thus also increase their expendable incomes), food providers will also be more likely to find their way to these currently too-often excluded consumer groups.

Third, many value chains lack a logical party to undertake the role of orchestration, a gap that could be addressed through Information and Communication Technology (ICT) platforms. A recent study showed that ICT platforms are departing from their typical roles of providing only price and other information, hoping that the market mechanism will solve all other problems. They increasingly move towards roles in which they coordinate among different chain members and stakeholders, providing matching services for all members so that the functioning of the chain as a whole improves (Ayesiga et al, 2023).

Fourth, to increase yields as well as quality, the adoption of agricultural inputs by smallholders is often essential. New promising evidence from Ethiopia on the uptake of seed, fertilizers, and rhizobium inoculants showed that while the simple availability of each of these products led to an uptake of just above 10%, the bundling of the products in the right quantities within one box already saw an improvement. The main leap in purchasing came however when farmers were also offered an output market contract, so they could be certain that if they invested in inputs they had a buyer for the produce, as well as a microloan that pre-financed their purchases thus bringing relief to immediate cash constraints. The addition of these two services to the bundle led to a purchase rate of nearly 80% (Abetu et al., 2023).

Fifth, innovations in rural financing help to remove capital constraints of smallholders (think of mobile banking which has a strong impact on social inclusion, particularly of women) (Ouma et al., 2017). An important underlying problem is that financial institutions see agricultural land as collateral for the loans they provide, something that they perceive as risky which makes them reluctant to provide the necessary loans. Value chain financing is a transaction and relationship-based assessment based on the health of the entire system, and not just on the individual borrower or agricultural producer (Oberholster, 2018). This includes, for example, contract farming, lease and management contracts, tenant farming, and share cropping (Sudha & Kruijssen, 2011). More market-driven systems with stable relations are thus seen as less risky and rewarded with higher loans to scale-up and lower interest rates (Stone et al., 2012).

Finally, to prevent farmers from becoming disappointed about their participation in the market because they do not understand the actions of their buyers and suppliers, trainings in marketplace literacy may be needed (e.g., Teklehaimanot et al., 2017b). Previous studies have shown positive effects of marketplace literacy trainings on related outcomes in terms of smallholder well-being (Viswanathan et al., 2021).

6 Conclusion and final thoughts

We provide a business perspective on Africa's food security "conundrum". The rapidly growing demand in Africa's cities for quality food, safe food, and more diverse food creates new opportunities for value creation for businesses at different stages of the food system. These opportunities can also benefit the smallholders that currently operate beyond the smallholder value creation-inclusion frontier. By including smallholders in the value-creating networks, they become more attractive customers to food companies, which again may help to improve food security in the rural areas. Seizing the rapidly emerging opportunities requires agribusiness managers that can see beyond the production roles of their companies and that can think and act in the spirit of value creation.

References

Abetu, T.A., Ingenbleek, P.T.M., & Giller, K.E. (2023). Bundling to overcome the gaps in smallholder markets: A field-based choice experiment in Ethiopia. Working paper. Wageningen University.

Acemoglu, D., & Robinson, J.A. (2012). *Why nations fail; the origins of power, prosperity, and poverty.* Crown Publishers.

AGRA. (2017). *Africa agriculture status report: The business of smallholder agriculture in sub-Saharan Africa.* Alliance for a Green Revolution in Africa.

AGRA. (2019). *Africa agriculture status report: The hidden middle.* Alliance for a Green Revolution in Africa.

Ayesiga, C., Ronner, E., & Ingenbleek, P.T.M. (2023). The evolving role of ICT platforms for smallholders as value chain coordinators: A comparative case study in Uganda. Working paper. Wageningen University.

Babah Daouda, F., Barth, P., & Ingenbleek, P.T.M. (2020). Market development for African endogenous products, *Journal of Macromarketing*, *40*(1), 13–30.

Babah Daouda, F., Ingenbleek, P.T.M., & van Trijp, H.C.M. (2019). Living the African dream: How subsistence entrepreneurs move to middle-class consumer markets in developing and emerging countries, *Journal of Public Policy & Marketing*, *38*(1), 42–60.

Chege, J.W., Ingenbleek, P.T.M., & Fischer, A.R.H. (2021, October 25–26). *Is food safety a satisfier or dissatisfier for African consumers? Evidence on green leafy vegetables from Nairobi, Kenya* [conference presentation], Out of (and in to) Africa Conference, Baltimore, United Sates.

Chikweche, T., & Fletcher, R. (2014). 'Rise of the middle of the pyramid in Africa': Theoretical and practical realities for understanding middle class consumer purchase decision making. *Journal of Consumer Marketing*, *31*(1), 27–38.

Dorward, A., Kydd, J., Poulton, C., & Bezemer, D. (2009). Coordination Risk and Cost Impacts in Economic Development in Poor Rural Areas, *Journal of Development Studies*, *45*(7), 1093–1112.

Downey, D.W., & Erickson, S.P. (1987). *Agribusiness management* (2nd ed.). McGraw-Hill.

Fischer, E., & Qaim, M. (2012). Linking smallholders to markets: Determinants and impacts of farmer collective action in Kenya. *World Development*, *40*(6), 1255–1268.

Fraval, S., Hammond, J., Bogard, J.R., Ng'endo, M., van Etten, J., Herrero, M., … & van Wijk, M.T. (2019). Food access deficiencies in sub-Saharan Africa: Prevalence and implications for agricultural interventions. *Frontiers in Sustainable Food Systems*, *3*, 104.

Giller, K.E. (2020). The food security conundrum of sub-Saharan Africa. *Global Food Security*, *26*, 100431.

Hounhouigan, M.H., Kounouewa, K.M.G., Ayesiga, C., & Ingenbleek, P.T.M. (2020). Sojagnon: Shaping the Beninese soy system to meet the challenges of an emerging market. *International Food and Agribusiness Management Review*, *23*(1), 143–156.

Hunt, S.D. (2000). *A general theory of competition: Resources, competences, productivity, economic growth*. Sage.

Ingenbleek, P.T.M. (2019). The endogenous African business: Why and how it's different, why it emerges now and why it matters, *Journal of African Business*, *20*(2), 195–205.

Ingenbleek, P.T.M. (2020). The Biogeographical foundations of African marketing systems. *Journal of Macromarketing*, *40*(1), 73–87.

Jaffee, S., Henson, S., & Rios, L.D. (2011). *Making the grade: Smallholder farmers, emerging standards, and development assistance programs in Africa*. A research program synthesis. World Bank. Report (62324-AFR).

Kirca, A.H., Jayachandran, S., & Bearden, W.O. (2005). Market orientation: A meta-analytic review and assessment of its antecedents and impact on performance. *Journal of Marketing*, *69*(April), 24–41.

Neven, D., Odera, M.M., Reardon, T., & Wang, H. (2009). Kenyan supermarkets, emerging middle-class horticultural farmers, and employment impacts on the rural poor, *World Development*, *37*(11), 1802–1811.

Oberholster, J.H. (2018). *Possible futures for agricultural financing in sub-Saharan Africa towards 2055* (Doctoral dissertation, Nelson Mandela Metropolitan University).

Ouma, S.A., Odongo, T.M., & Were, M. (2017). Mobile financial services and financial inclusion: Is it a boon for savings mobilization? *Review of Development Finance*, *7*(1), 29–35.

Poole, N. (2010). From marketing systems to value chains: What have we learnt? In H. van Trijp & P. Ingenbleek (Eds.), *Market, marketing and developing countries* (pp. 18–23). Wageningen Academic Publishers.

Porter, M.E. (1985). *Competitive advantage*. Free Press.

Russell, R. (2022). *Price wars: How chaotic markets are creating a chaotic world*. Weinfield & Nicholson.

Shultz, C.J., & Edwards, M.R. (2005). Reframing agribusiness: Moving from farm to market centric. *Journal of Agribusiness*, *23*(1), 57–73.

Sridharan, S., Maltz, E., Viswanathan, M., & Gupta, S. (2014). Transformative subsistence entrepreneurship a study in India. *Journal of Macromarketing*, *34*(4), 486–504.

Stone, R., Agar, J., Carpio, A., Cabello, M., & Hayes, J. (2012). *Study of African and international innovations and best practices in increasing access to rural and agricultural finance*. Oxford Policy Management and Kadale Consultants Ltd.

Sudha, M., & Kruijssen, F. (2011). Linking farmers to market through processing: The role of agro-industry clusters with specific reference to mango in India. In C.A. Silva & M. Mhlanga (Eds.), *Innovative policies and institutions to support agro-industries development*. Food and Agriculture Organization (FAO).

Teklehaimanot, M.L., Ingenbleek, P.T.M., & van Trijp, H.C.M. (2017a). The transformation of African smallholders into customer value creating businesses: A conceptual framework. *Journal of African Business*, *18*(3), 299–319.

Teklehaimanot, M.L., Ingenbleek, P.T.M., Tessema, W.K., & van Trijp, H.C.M. (2017b). Moving toward new horizons for marketing education: Designing a marketing training for the poor in developing and emerging markets. *Journal of Marketing Education*, *39*(1), 47–60.

Viswanathan, M., Umashankar, N., Sreekumar, A., & Goreczny, A. (2021). Marketplace literacy as a pathway to a better world: Evidence from field experiments in low-access subsistence marketplaces. *Journal of Marketing*, *85*(3), 113–129.

Zamudio, A.N., Bizikova, L., & Keller, M. (2014). *Measuring local food systems' resilience: Lessons learned from Honduras and Nicaragua*. International Institute for Sustainable Development.

21 Agricultural development in the poorest countries: insights from the poultry sector in Sierra Leone

Erwin Bulte, Niccolò Meriggi, Michael Rozelle, and Maarten Voors

1 Introduction

The majority of the Sierra Leonean population is dependent on agriculture as a primary source of income. Using low-input production techniques, farmers mainly produce food crops for their own consumption and local markets. There are few commercial farms and there is virtually no manufacturing sector (value added in manufacturing as a share of GDP is around just 2%, WDI, 2024). A large share of the population is unable to afford a healthy diet, similar to countries in the region (Figure 21.1). Food insecurity and lack of wider economic development are tightly linked (de Haas et al, this volume), particularly in rural areas. In a recent survey, we interviewed a nationally representative sample of 6,000 rural households and found that 33% of the participants had eaten "less preferred" food in the last week, 26% had reduced portion sizes for adult household members, and 9% of households said to have passed an entire day without eating anything at all. Food insecurity is particularly intense during the "hungry season", the months

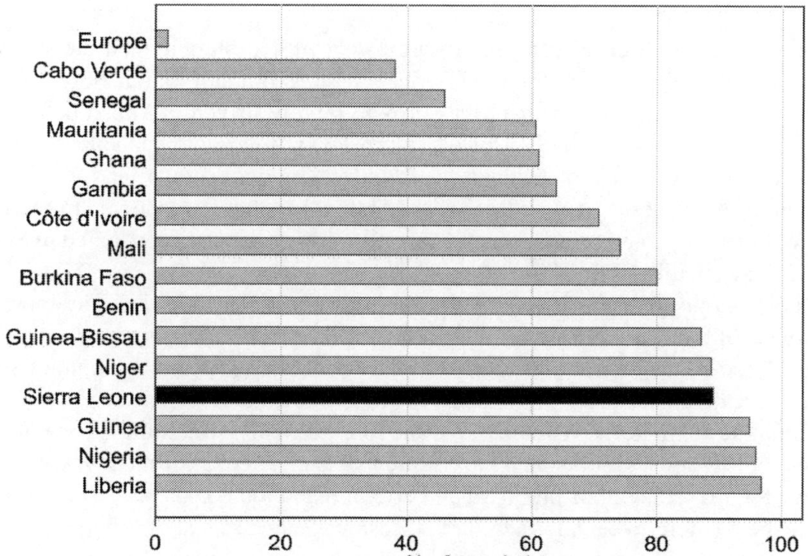

Figure 21.1 Percentage of the population that is "not able to afford a healthy diet".

Note: FAOStat (2024). Figure displays the percentage of the population that "is not able to afford a healthy diet" in 2021 for Sierra Leone and selected countries in West Africa.

DOI: 10.4324/9781032649696-26

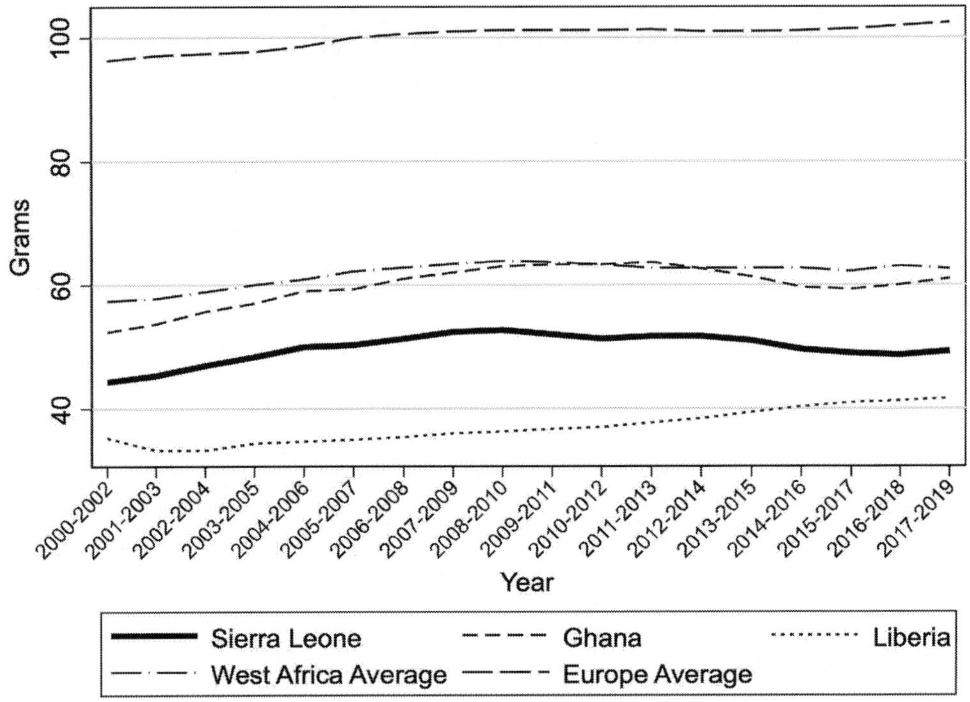

Figure 21.2 Protein supply (2000–2019) for selected countries in West Africa and Europe.

Note: FAOStat (2024). Figure displays the protein supply (in grams per person per day) during 2000–2019 for Sierra Leone and selected countries in West Africa, as well as the average for West Africa and Europe.

before rice is harvested. Sierra Leoneans consume very little protein in their diets (Figure 21.2), and intake sits below the average for West Africa, and is about half that in Northern Europe.

Over the past years, protein demand has grown, largely driven by urban consumers. Even though local production has steadily grown (Figure 21.3), it has not been able to match domestic demand due to low levels of technology adoption (modern inputs) and poorly developed agri-food value chains. As a result, a large share of the food that is consumed in Sierra Leone is imported. With respect to proteins for instance, only 25% of the eggs consumed in Sierra Leone is produced domestically (Soba, 2017).

Imports are mainly paid for by a small basket of export commodities. The most important export products of Sierra Leone are diamonds, rutile and iron ore, fish, and cash crops like cocoa and coffee. These commodities undergo barely any processing within the country so that very little value is created locally. Natural resources are extracted by foreign-owned mining companies and comprise the majority (over 90%) of the foreign currency earnings. The bulk of the fish is harvested by foreign vessel owners. Part of the fish is cleaned domestically, but many fishing vessels have their own facilities on board and do not reach local shores. Sierra Leone lacks the means to inspect catch levels and there are widespread concerns about illegal harvesting and fish stock depletion. The domestic cocoa trade is firmly in the hands of Lebanese traders. While production relies largely on many small-scale producers, a single foreign-owned trading company is responsible for nearly 90% of the raw cocoa exports. While chocolate is a high-value product, most value is added to the cocoa beans in other countries. Despite positive economic growth rates after the devastating 1993–2003 civil war, the agricultural sector remains largely

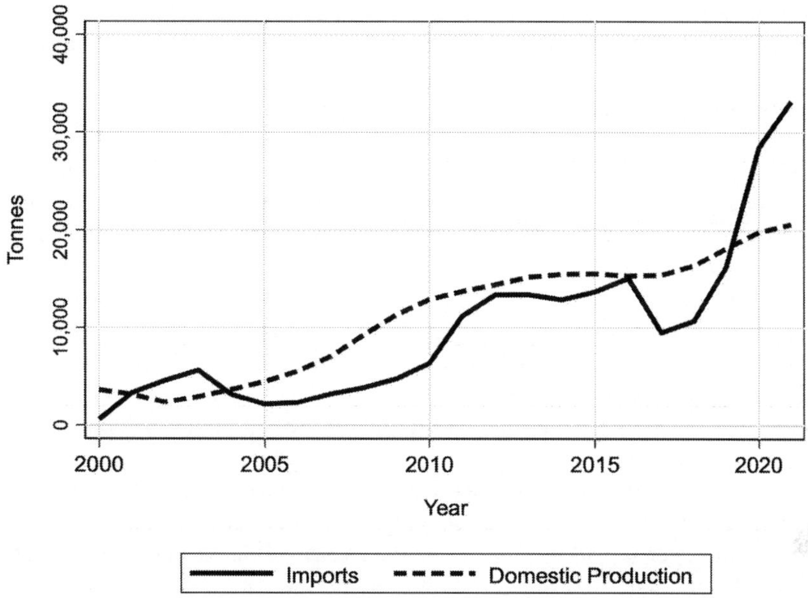

Figure 21.3 Chicken production and trade.

Note: FAOStat (2024). Figure displays the chicken production and trade (in tonnes) for Sierra Leone for the period 2000–2021.

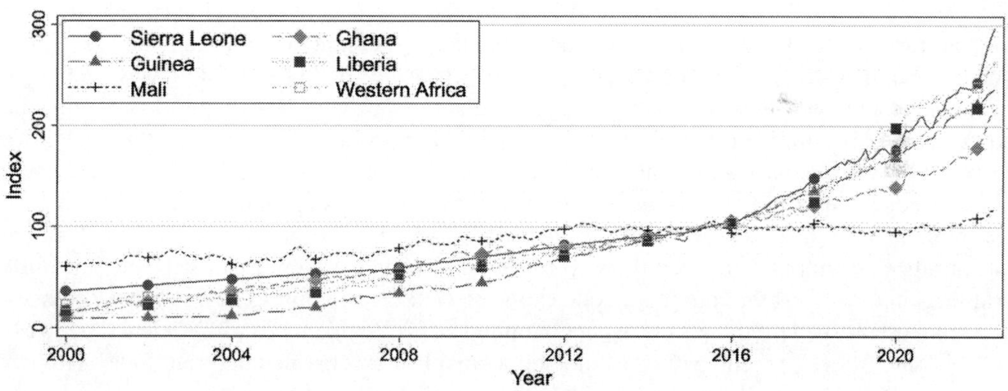

Figure 21.4 Changes in consumer food price index for selected countries in West Africa.

Note: FAOStat (2024). Figure displays the changes in consumer food price index of food for the period 2000–2023 for Sierra Leone and selected countries in West Africa, and index year is 2015.

underdeveloped. Cereal yields are low, estimated at only 0.8 ton per hectare in 2022 (FAOstat). The COVID pandemic and war in Ukraine have impacted trade and the local economy due to disruptions in the supply of food to the country and increases in (farm) input prices—driving up food price inflation (Figure 21.4). A recent survey among cereal farmers suggests that fertilizer, seed, and fuel prices doubled in a year. For poultry farms the price of feed, concentrate, and other inputs sharply increased. Reliance on imports implies vulnerability to trade shocks.

This raises important questions. How can a country like Sierra Leone develop economically if its key export sectors are foreign-owned and controlled, so that most of the profits are expatriated and invested elsewhere? How to pay for imports after rutile and iron ore reserves have been depleted? How can a rapidly growing and urbanizing population be fed if most of the population is employed in low-productive and informal activities?

We argue that there is no apparent reason why Sierra Leone should not be able to feed its own population. Sierra Leone has abundant fertile land and plenty of water. The country is thinly populated and could be exporting rather than importing food. Why does it import rice from Asia and chicken parts from Europe and Brazil when there is a comparative advantage in the production of these foodstuffs?

It is time to rethink development theory popular in economics. The textbook economic development story is based on growth experiences in Western and Asian countries. "Structural Transformation" theory (Kuznets, 1966) describes how land and labour productivity in agricultural areas increases, which frees up labour and makes available cheap food for an expanding industrial sector that exports its manufactured goods to world markets. This is a strategy referred to as the "high road solution" to poverty reduction and growth (De Janvry & Sadoulet, 2016). As the share of the agricultural sector in the economy and overall employment declines, the manufacturing sector and later the service sector grow, absorb labour, and experience ongoing productivity gains that gradually improve the overall living standards. Economists believed that the industry and service sectors are superior to the agricultural sector in terms of long-term growth potential. Due to ongoing technical change and positive spillover effects, industry was considered the most important engine of growth. From this perspective, a process of structural transformation seems necessary for countries like Sierra Leone to experience sustained growth and development. But how appropriate is the structural transformation narrative for these countries?

One weak link in the story is evident. Britain was the first country to develop an industrial sector, and experienced limited competition on expanding global markets for manufactured goods. When Japan industrialized, much later, the ratio of wages in London and Tokyo was about eight to one. In other words, there was a large wage gap that the Japanese could exploit. They could become competitive on international markets even if labour productivity was initially much lower than in England. Countries that subsequently underwent a process of structural transformation likewise benefited from a cheap labour force, relative to incumbent producers, and could specialize in products aligned with this comparative advantage—labour-intensive goods. In almost all industrialized countries, however, the government also played an important strategic role. Governments in most western and Asian countries protected their fledgling industry from competition from established producers elsewhere (Koning, 2017; Frankema & van Waijenburg, 2018).

This same development path may not be available for African countries today (Rodrik, 2018; Frankema & van Waijenburg, 2018; Pieters, this volume). It seems unfeasible for African countries to invest and grow in sophisticated manufacturing industries. On world markets for manufactured goods, they have to compete with producers in countries like China and Vietnam. There, wages are still relatively low, and labour productivity is high. African countries are far removed from a comparative advantage in the production of labour-intensive manufactured goods. While average wages are low they are still high relative to productivity levels. Per-unit production costs are therefore high, and undercutting Asian competitors seems a huge challenge. While investments in the manufacturing sector could change this, it is not clear how much room there is on world markets for labour-intensive manufacturing products from Africa. Ongoing progress in automation and robotization further diminish the scope for new entrants to conquer these markets based on a "cheap labour" strategy. It is also unclear where investments in African manufacturing should come from. Western and Asian investors target the primary sector when investing in Africa—think of natural resource extraction and plantation agriculture. Recent political trends in many African countries do little to change this outlook.

Such traditional and narrowly defined development paths (agriculture first, then manufacturing, onwards to services) do not make much sense for many African countries. The continent's comparative advantage may lie in the primary sector, not only during the earliest stages of the development process but also later. Agriculture provides opportunities for sustained primary product-based processing, value creation, and export—hopefully creating sufficient employment opportunities for the growing population. Recent research based on experiences in other parts of the world suggests that long-run growth of agriculture is possible due to new varieties, the adoption of modern complementary inputs, and mechanization. This scenario is unlikely to unfold spontaneously, without appropriate government intervention. Farm sizes have remained small, and are getting smaller (Giller & Andersson, this volume). In addition, productivity growth among African smallholders has been dismal, and local processing and value creation are only just getting started. We thus need a broader view on the development in Africa that focuses on agriculture-led growth, where the share of agriculture in GDP does not necessarily fall, as predicted by standard Structural Transformation theory. Rather, the continent should use its comparative advantage in the sector, and invest in "agricultural industrialisation", rather than sophisticated manufacturing, or services in the tech sector (De Janvry & Sadoulet, 2016).

How to get African farming going? Africa is a wildly diverse continent and generic statements about "African farming" are typically silly. At the risk of oversimplification, nevertheless, we will make some broad statements.

While there are examples of highly productive agriculture in Africa, many farmers are engaged in semi-subsistence farming. In recent decades, a range of interventions was rolled out to promote productivity—mainly to develop and implement "supply-side" solutions. These include new cultivars and production technologies, farmer extension, digitalization, credit and insurance, land reform, and institutional innovations (such as farmer associations). Several African governments have invested in subsidy programmes for modern inputs such as hybrid seeds or fertilizer. While impact analyses have demonstrated the successes and failures of these interventions often on a small scale, none produced the silver bullet for agricultural development (Suri & Udry, 2022).

While change is possible, the focus on incremental steps to raise agricultural productivity fails to address the complexities that challenge the sector. African smallholders are constrained along multiple dimensions, and a multipronged plan of attack is necessary to transform the agricultural sector. Think of "bundled interventions" where farmers receive technical assistance as well as access to credit and insurance (Deutschmann et al., 2022; Boucher et al., 2024), and are connected to remunerative markets. We believe that especially this latter part is an important and often overlooked aspect.

While past efforts promoting development focused on lifting *supply-side* constraints, it may be more important to address *demand-side* constraints. If farmers cannot sell surplus crops because of prohibitive transaction costs, if they face highly uncertain prices, or if markets do not reward quality, then why would they work hard and expose themselves to all sorts of risks and produce high-quality crops for the market? Giller et al. (2021) points to general lack of incentives for farmers to intensify their production as a key explanation for the underdevelopment of the agricultural sector. A recent study showed that providing farmers access to markets with quality premiums immediately increases productivity and income (Bold et al., 2022). This reveals the importance of demand-side constraints as a limiting factor in raising rural income and productivity growth. It also illustrates that many African farmers quickly respond to incentives and can work their way around perceived constraints.

2 Responding to local demand locally

Let us return to the case of Sierra Leone, where we have been involved in a project that focused on developing the poultry supply chain. Chicken meat and eggs are coveted food items in urban areas, and potentially important components of a nutritious diet. The current poultry production sector consists of backyard farming and a fledgling commercial poultry sector.

First, we turn to backyard farm production. Chickens are kept throughout the country, mainly relying on scavenging. Over 90% of farmers keep some chicken, and flocks are small (less than 20 birds on average). Birds are largely kept for own consumption or as "liquid assets"—sold whenever cash is needed. Few are kept as an additional source of protein. Very few owners sell chickens or eggs on a regular basis. Mortality is high due to predation and diseases. Eggs are often kept just to replenish flocks. While small-scale chicken rearing could increase nutrition and welfare, many challenges need to be overcome. These relate to access to feed, high-quality breeds, veterinary services, and market opportunities.

There are several capital-intensive commercial chicken farms. Many are small firms with typically a few hundred chickens, supplying meat and eggs to local markets. There are only a few larger firms—with thousands of birds. Backward and forward linkages imply these commercial farms are connected with specialized input suppliers and processors, which create economic value and formal employment. We think of these firms as a stepping stone towards primary sector-based "industrialization".

We documented the trials and tribulations of local companies that try to engage in commercial poultry production for domestic markets and implemented a national firm-level survey among poultry businesses. In total, we identified 80 poultry farms owning more than 100 chickens—a low threshold. Most firms rely on Isa Brown layers, imported as day-old chicks from Europe. Important bottlenecks concern access to imported inputs, such as layers, maize, and quality feed concentrate. There is no parent stock for commercial layers in Sierra Leone. Day-old Isa Brown chicks are ordered in bulk (thousands of chicks). Since most small and medium-sized firms cannot afford the cost of acquiring large quantities of inputs, they resort to costly and often unsuccessful coordination efforts with other firms. About half of the companies recently closed down. The dominant reason was supply-side constraints—poultry farming was expensive because of high input costs and high bird mortality. The country has one veterinarian and access to vaccines and disease treatments is severely constrained.

Demand-side constraints also matter. Despite inefficient domestic production and value chains, poultry products are still fairly cheap because they are imported. Chicken prices are just above the regional average (about 9% higher according to regional price data from the International Comparison Program) and about 40% higher compared to the main exporter Brazil (World Bank, 2020). About 85% of the chicken meat and eggs consumed in Sierra Leone are currently imported. In 2021, Sierra Leone imported poultry meat and eggs to the value of US$47M (OEC, 2024). Frozen eggs and chickens (whole or in parts) are shipped in from Europe and South America, at considerable environmental costs and depleting scarce reserves of foreign exchange. The presence of these imported chickens and eggs on local markets reduces the price of animal-sourced proteins. This does not imply that "food is cheap"—poultry products in urban areas are still rather expensive compared to price levels in other parts of the developing world when we control for income levels (also see Frankema, this volume). Lower prices benefit the growing urban population. But it is detrimental for potential local suppliers, who face formidable competition and struggle to even recover their production costs.

Traditional economic textbooks provide a simple framework to organize our thinking about these issues. Below we provide insights into the core ideas that have become popular in economic theories of development. Of course, modernization theories have been critiqued over the past decades. They have however remained stubbornly present in textbook theories and still underpin policy debates about development and international development donors (such as the World Bank, International Monetary Fund (IMF)). The core model is sketched in Panel A of Figure 21.5, with the quantity of chicken products (Q) displayed on the horizontal axis, and the price of chicken products (P) on the vertical axis. D provides the demand curve for chicken products (capturing the preferences and budgets of domestic consumers), and S_0 provides the supply curve (capturing the per-unit production costs of poultry products by domestic farmers). In the absence of trade, the market moves towards an equilibrium characterized by quantity Q_0 and associated price P_0. The contribution of the poultry sector to the overall welfare would be the sum of consumer surplus, or the area encapsulated by the demand curve and price (*a*), and producer surplus, or the area encapsulated by the price and the supply curve (*b*+*e*). Hence, the overall welfare without trade is (*a*+*b*+*e*)

Enter trade. Assume that chicken parts and eggs can be produced at lower costs in very large quantities elsewhere in the world. Specifically, assume that a virtually endless supply of chicken products is available at a given world market price P_w. What happens to welfare? Consumers benefit from the reduced price, and consumer surplus increases (*a*+*b*+*c*+*d*), i.e. the entire area between the demand curve and the new price level. Farmers lose out after the inflow of imports. Production is reduced as several domestic producers are driven out of business, and producer surplus shrinks to only area (*e*). The overall welfare increases due to trade, as the gain for consumers is greater than the loss for producers. The new welfare level is (*a*+*b*+*c*+*d*+*e*), so theoretically opening up to trade has increased welfare in this country by the area (*c*+*d*).

These static welfare gains from trade are an important foundation for the economist's conviction that "trade is good" and underlie the efforts of organizations like the World Trade Organization to reduce trade barriers globally. The winners could compensate the losers of open trade, and still be better off. A simulation exercise for the poultry sector in Ghana using micro data confirms the negative overall impacts for the local economy of trade restrictions. Local production is low so the gains from trade tariffs do not outweigh the costs to consumers (Knößlsdorfer & Qaim, 2023).

However, such static analysis omits an important aspect. We became convinced of the importance of technology adoption (more suitable chicken breeds, see below), as well as "learning" (about how to produce chicken products efficiently) and "coordination" (about linking the different actors in the poultry value chain, see also Wilson et al., 2022). We elaborate on these points below. For now, assume that through adoption, learning, and improved coordination, farmers can learn-by-doing and gradually reduce their production costs over time. However, learning-by-doing takes time, and domestic producers run the risk of being driven out of business by cheap imports in the meantime.

Assume that the government shields the domestic market from cheap imports by engaging in some form of trade regulation (tariffs, quotas, or minimum product requirements). Prices on local markets remain sufficiently high for local farmers to produce, learn, and become more efficient *over time*. This is captured, in Figure 21.5, by shifting the supply curve to the right. Assume that after several years of production the new supply curve is S_1, provided in Panel B. The overall welfare level associated with the new market equilibrium (Q_1,P_1) is given by the area (*a*+*b*+*c*+*f*+*g*). Adopting a dynamic perspective may therefore reverse the results of a static welfare ranking. Regulating trade and enabling local producers to reduce their production costs over time generate greater overall welfare than free trade if area (*g*) is greater than area (*d*+*e*)—as drawn in Panel B. In this new equilibrium, consumers still pay a bit more for their meat and eggs than before ($P_1 > P_w$), but the increase in welfare for farmers is greater than the loss in welfare for consumers.

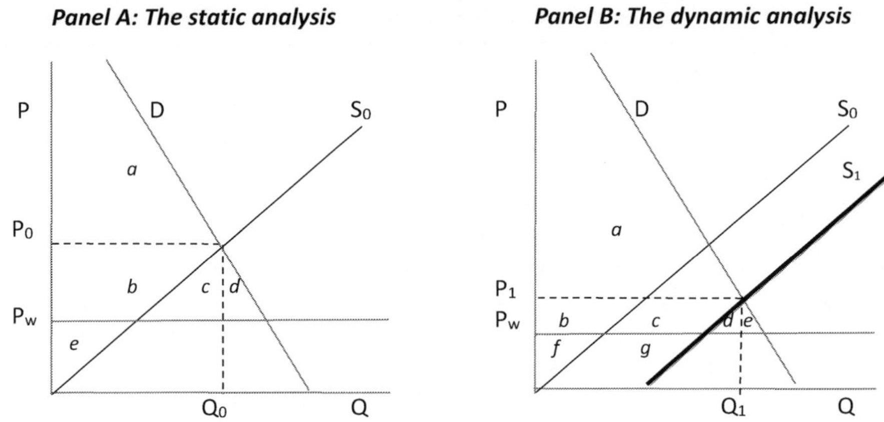

Figure 21.5 Static and dynamic analysis of trade measures.

Notes: According to the static analysis, restricting trade will always cause a welfare loss (of *c*+*d* in Panel A). According to the dynamic analysis, restricting trade may increase welfare if it causes efficiency improvements.

As mentioned, there are other considerations. Countries with higher domestic production are likely less vulnerable to disruptions in global supply chains and can use their foreign exchange reserves to purchase items that cannot be produced locally—typically from the manufacturing sector. There may be positive environmental side-effects. Sierra Leone has extensively managed land. Thus, expanding and intensifying local production may not immediately erode nature conservation.

Importantly, Panel B is a partial analysis that only focuses on the poultry sector. But other parties in the economy may have a stake in the outcomes as well. An increase in domestic production may result in a higher employment beyond poultry production, such as in the feed sector. In total, 70% of egg production costs is feed (maize, oyster shell, rice, or wheat bran). Downstream firms process birds and eggs. This generates additional employment for workers and profits for capital owners—part of which will be consumed or invested domestically, resulting in additional business opportunities for other entrepreneurs.

The type of trade regulation that we envisage is not new but rooted in a much older paradigm from the 1960s and 1970s (see also De Janvry & Sadoulet, 2016). During these years, many low-income countries tried out a development strategy based on "import substitution", protecting specific "infant industries" to transition from agriculture to manufacturing. The idea was that, through learning-by-doing, these protected sectors would become competitive players on international markets. Most countries experimenting with import substitution had to abandon this strategy during the structural adjustment years, in the 1980s and 1990s, after accumulating unsustainable debts. Corruption also mattered, and it became clear that some infant industries were better at lobbying for continued trade protection than at innovating to reduce production costs.

The development process we propose deviates from this conventional import substitution strategy in key ways. First, we do not propose to target the industrial sector for protection but instead propose to focus on agricultural activities that are much closer to current production modes. Second, we do not envisage that domestic producers will eventually become global players and conquer world markets, but instead believe that there is scope for them to cater for domestic and regional markets. Third, while we imagine that the government plays an active and coordinating role in the development process, as elaborated below, we do not believe that large-scale borrowing is necessary to finance this development strategy.

This strategy hinges on the crucial yet untested assumption that (temporary) protection of domestic agricultural producers will enable these producers to reduce their production costs "over time". How realistic is that story?

3 Technology, learning, and coordination

The current state of affairs in the poultry sector in Sierra Leone is bleak. Without investments and appropriate policies change is unlikely. The challenges are large and many, mainly relating to breeds, feed, management, and trade policy.

3.1 Breeds

What is the growth potential of backyard farms? Most farms rely on small, local hens, and households rarely consume eggs. There are potential, yet largely untested, benefits from new chicken breeds. In some African countries, new more robust breeds have been introduced that are both productive and adapted to the heat and poor feed quality. SAPPSA (Sustainable Access to Poultry Parent Stock for Africa) is a recent initiative of dual-purpose chicken breeds. These breeds, known as SASSO, maximize both on layer and on broiler capacity (compared to a standard model where broiler and layer production lines are separated). Within SASSO, there are breeds more geared to egg production, and others to meat production. In several African countries, SASSO chickens have been introduced using an outgrower model, where the so-called mother units sell day-old chicks to brooder or mother units that raise these chicks for 45 days and make sure they are properly fed and vaccinated. Afterwards, chicks are sold to village communities where they will scavenge their own feed and at the same time providing eggs and meat. As part of our project in Sierra Leone we are learning about the potential economic returns to keeping such different breeds. As these birds largely rely on scavenging, feed intake costs are low and the laying rate, at least theoretically, is decent (50–60% of the days). While the economic benefits to households are not yet known, it appears that backyard poultry production is a low-cost activity that many farm households can incorporate—selling eggs and meat in local markets. However, the scope for scaling is unclear. As flocks grow, they need to be housed and fed, which increases costs (see also Wilson et al., 2022).

Commercial poultry firms in Sierra Leone fall far behind productivity levels observed in Europe or the USA, despite using high-quality genetic breeds that, under perfect conditions, reach a laying percentage of over 95%. Firms in Sierra Leone typically reach just 50%. Part of the problem is that the breed, Isa Brown, is not well suited for the climate, due to heat stress, sensitivity to varying feed quality, and the disease environment. Temperatures are between 25°C and 30°C year-round, with varying humidity, ranging from 70% in the dry season to 100% in the rainy season. Low feed quality and diseases further suppress performance. These problems are not unique to Sierra Leone. It is unclear what gains can be made in the type of breeds utilized on farms. Developing more appropriate breeds needs to be a priority for breeders. The adoption of dual-purpose birds (discussed above) is slowly advancing, but it is an open question if these are a cost-effective alternative for larger firms. In scavenging systems, they may perform well, but as flock sizes grow, they would need to be housed and fed—eroding profitability. In addition, farmers cannot readily access these birds as these need to be flown in batches, which hampers an optimal production cycle. To create an independent chicken production system in Sierra Leone, parent stocks are necessary. Fertilized eggs can be hatched, and day-old chicks kept under good conditions in so called "mother units". Rearing day-old chicks to become grown laying hens involves high levels of biosecurity, management practices, and access to vaccines.

The combination of a parent stock, hatchery, mother units including good feed, and vaccines could result in a sustainable supply of young production birds for poultry farmers in Sierra Leone. Given the current prices, however, it is hard to imagine how investments in the expansion of this sector are financially attractive. Even socially oriented investors are reluctant to enter.

3.2 Feed

About 75% of the egg production costs is feed, the majority of which is imported. Feed composition typically contains 20% concentrate (a mix of minerals, vitamins, protein), 50% corn, 10% oyster shell, and 7% rice or wheat bran. Concentrate is provided by only few companies and mostly imported. Due to the limited availability of local sources of protein, the share of concentrate in the feed mix is large. It is also the most expensive unit item. Maize is often imported from Mali, as local production has been stagnant. In a recent survey we found that whilst 40% of farmers grow some maize, very few produce a significant marketable surplus. Maize is typically consumed at home or grilled and sold at the roadside. Poultry firm owners too mentioned a preference for imported maize due to lower maize quality (including toxins and insects). Maize farm yields have been very low (700 kg per hectare), even below the regional standards of 1 ton per hectare, Freduah et al., 2019). Farmers use very few inputs—just 13% use any fertilizer, herbicides, insecticides, or fungicides—and most never use any form of mechanization.

However, we have learned that the potential gains in local maize production are large. To stimulate a shift towards more market-oriented maize production, interventions would need to not only promote the adoption of productivity-enhancing technologies, but also address structural barriers such as access to credit, markets, and quality inputs. We recently engaged in a project with a local bank that provided loans to farmers to grow maize and facilitated connections to a poultry farm to purchase maize for their chickens. Importantly, maize is not considered a main staple, implying that there is limited risk of crowding out food consumption due to increased feed production. The enthusiasm for the project in participating communities was overwhelming. Farmers planted several hectares of maize, outperformed control group farmers in terms of yield, and managed to produce high-quality maize suitable for chicken feed. The maize was largely free from mycotoxins which would pose a risk to animal and human health. The quality was much better than what is typically available in markets or imported from neighbouring countries. In the pilot study, all maize was purchased by a local poultry farm.

Increased local maize production, as well as other feed like soyabean, is one step to design the feed mix for local layers, making it more affordable. A more radical rethinking is necessary, and the design of the feed mix should take locally available inputs as a starting point and include oyster shells, rice bran, cassava, or by-products such as palm kernel cake.

3.3 Management

Around farm management core constraints include veterinary services, and vaccines and drugs are in limited supply. Many chickens suffer from preventable diseases such as Newcastle Disease, Avian Influenza, and Infectious Bursal Disease (Bertran et al., 2020). In addition, most farms have low levels of biosecurity, leaving ample scope to reduce the incidence of infections.

3.4 Coordination and policy

In addition to raising productivity at each step in the poultry value chain, more coordination among key stakeholders is necessary. This requires re-modelled incentive schemes and coordination mechanisms (contracts). We learned from conversation with farmers that developments in the poultry sector are, somewhat ironically, constrained by the well-known "chicken-and-egg problem". Lack of coordination creates uncertainty and stifles investments. Potential maize growers do not supply feed because they are uncertain whether there will be sufficient demand for their crop. At the same time, potential poultry producers do not invest because they are uncertain whether they can source feed at an affordable cost in local markets. Improved coordination could result in a large push across multiple nodes of the chain simultaneously. It is evident that the government could (and should) play an important role in this process.

What else can the government do to help this process forward? One area is to refocus attention away from global to regional trade networks and policies. Progress to date has been minimal, and many African nations continue to trade more globally than within the continent (UNESC, 2024). There is an obvious need to facilitate technology development and transfer. In addition, trade regulation must be rethought—shielding domestic producers from cheap imports and providing them with a market environment characterized by relatively high and stable prices during the learning stage. This is a plea for raising import tariffs on drumsticks, chicken wings, and eggs, and perhaps similar food items that can be produced domestically. To keep welfare losses due to taxation manageable, we recommend that trade regulation should be targeted and temporary and goes hand in hand with market policies. Governments should be careful not to create local monopolies, because then "shielded firms" can abuse their market power to charge excessively high prices to consumers, for excessively long periods. We need multiple firms competing for market shares, where the most successful firms can grow and expand. This should be doable because the economies of scale in poultry production are not massive.

4 Discussion

As is the case for other poor countries in the region, investing in strategies to improve food security and address growth challenges in Sierra Leone should be a top priority. The country faces a population growth of 2.2% per year. The prospects for the country to progress through the stages of the traditional structural transformation process, as advocated by economists, are limited. Most of the hard-earned foreign currency is currently spent on inputs and food that the country could produce itself. The prospects of further growth and development along a path of extensive agricultural expansion seem limited. As reserves of exhaustible resources are slowly being depleted, it becomes increasingly important to ask how future imports can be financed.

We argue for a growth strategy based on agriculture-based industrialization for domestic and regional markets. The emphasis needs to be on technology transfer, learning by doing, and overcoming coordination challenges. The government should step in to provide temporary incentives for local production and value chains to develop. This implies a focus on both assisting smallholders to reduce food insecurity in the hinterland and helping larger commercial farms to serve a growing urban population. Supply-side interventions to help farmers become more productive are necessary, but governments should also consider supporting domestic producers by regulating international trade. Temporary trade regulation enables domestic producers to gain a competitive edge in both the short and long term (after trade regulation has been abandoned). We focused on the poultry sector as an example, but other sectors would benefit from some degree of trade protection as well. This includes rice and other primary food stuffs that can easily be grown in the country. The overarching idea is to prioritize those food items that currently deplete scarce foreign exchange reserves, and where low-hanging fruit productivity gains are within reach and coordination failures are evident. Policy support should also be guided by the question whether there is scope to promote domestic processing and value addition—upstream and downstream linkages should matter as well. Most countries that experienced a process of structural transformation supported their infant industries, and strategic trade regulation has always been a component of successful policy strategies (Chang, 2002).

It is evident that government intervention in markets entails a trade-off. There are obvious costs to getting it wrong. Trade regulation, such as through tariffs on imports, implies that domestic prices will rise, which hurts consumers. Urban consumers form a powerful constituency. It may also rub multilateral donors the wrong way. African governments have been uneasy about the hands-off approach prescribed by the Washington Consensus for a long time, but they depended on institutions like the World Bank, IMF, and country governments for budget support. In addition, depreciation of the local currency helps to make imports less attractive, strengthening the competitiveness of local producers. It is however evident that urban elites benefit from overvalued exchange rates, and the absence of import tariffs, as it makes imports of luxury goods cheaper, so there will be resistance to change from those politically well-connected. But it is hard to imagine alternative development paths for a country that one day will run out of minerals and fish, and that is unlikely to gain a competitive edge on international markets for manufactures. Growth in agriculture is achievable with local resources (i.e. land is still abundant and labour is readily available) and likely creates more broad-based prosperity than investing in other sectors that are more susceptible to rent capture (by foreign businesses or local elites). A policy focus on sustainable intensification of farming seems to be the best approach for many African countries.

References

Bertran, K., Cortey, M., & Díaz, I. (2020). The use of H-index to assess research priorities in poultry diseases. *Poultry Science*, *99*, 6503–6512. https://doi.org/10.1016/j.psj.2020.09.017

Bold, T., Ghisolfi, S., Nsonzi, F., & Svensson, J. (2022). Market access and quality upgrading: Evidence from four field experiments. *American Economic Review*, *112*(8), 2518–52. https://doi.org/10.1257/aer.20210122

Boucher, S., Carter, M., Flatnes, J., Lybbert, T., Malacarne, J., Mareyna, P., & Paul L. (2024). Bundling genetic and financial technologies for more resilient and productive small-scale farmers in Africa. *The Economic Journal*, ueae012. https://doi.org/10.1093/ej/ueae012

Chang, H.-J., (2002). Kicking away the ladder: An unofficial history of capitalism, especially in Britain and the United States. *Challenge*, *45*(5), 63–97. https://doi.org/10.1080/05775132.2002.11034173

De Janvry, A., & Sadoulet E., (2016). *Development economics: Theory and practice*. Routledge.

Deutschmann, J., Duru, M., Siegal, K., & Tjernstrom, E. (2022). Relaxing multiple agricultural productivity constraints at scale. SSRN working paper. https://doi.org/10.2139/ssrn.4479905

FAOstat. (2024). Food and agriculture organization of the United Nations. FAOSTAT Statistical Database. FAO.

Frankema, E., & M. van Waijenburg (2018). Africa rising? A historical perspective. *African Affairs*, *117*(469), 543–568. https://doi.org/10.1093/afraf/ady022

Freduah, B., MacCarthy, D., Adam, M., Ly, M., Ruane, A., Timpong-Jones, E., … & Adiku, S. (2019). Sensitivity of maize yield in smallholder systems to climate scenarios in semi-arid regions of West Africa: Accounting for variability in farm management practices. *Agronomy*, *9*, 639. https://doi.org/10.3390/agronomy9100639

Giller, K.E., Delaune, T., Silva, J.V., Descheemaeker, K., van de Ven, G., Schut, A.G., … & van Ittersum, M.K. (2021). The future of farming: Who will produce our food? *Food Security*, *13*(5), 1073–1099.

Koning, N. (2017). *Food security, agricultural policies and economic growth: Long-term dynamics in the past, present and future*. Taylor & Francis

Knößlsdorfer, I., & Qaim, M. (2023). Cheap chicken in Africa: Would import restrictions be pro-poor? *Food Security*, *15*, 791–804. https://doi.org/10.1007/s12571-022-01341-5

Kuznets, S. (1966). *Modern economic growth*. Yale University Press.

OEC (2024). *Poultry meat in Sierra Leone*. https://oec.world/en/profile/bilateral-product/poultry-meat/reporter/sle accessed: 1 February 2024.

Rodrik, D. (2018). An African growth miracle? *Journal of African Economies*, *27*(1), 10–27, https://doi.org/10.1093/jae/ejw027

Soba. (2017). Sierra Leone opportunities for business action, results presentation.In Poultry & Maize, unpublished report, https://investsalone.com/wp-content/uploads/2020/07/SOBA-Poultry-Business-Opportunity-Vet-Services.pdf, accessed: 1 February 2024.

Suri, T., & Udry C. (2022). Agricultural technology in Africa. *Journal of Economic Perspectives*, *36*(1), 33–56. https://doi.org/10.1257/jep.36.1.33

UNESC. (2024). Overview of recent economic and social developments in Africa, brief as part of UNESC forty-second meeting Victoria Falls, Zimbabwe, 28 February–1 March 2024. https://www.uneca.org/eca-events/sites/default/files/resources/documents/com/2024/e_eca_coe_42_4_rev.1_e.pdf accessed: 1 February 2024.

WDI. (2024). *World development indicators*. The World Bank.

Wilson, W., Slingerland, M., Oosting, S., Baijukya, F., Smits, A.-J., & Giller, K. (2022). The diversity of smallholder chicken farming in the Southern Highlands of Tanzania reveals a range of underlying production constraints, *Poultry Science*, *101*, 102062, 1–14. https://doi.org/10.1016/j.psj.2022.102062

World Bank. (2020). *Purchasing power parities and the size of world economies: Results from the 2017 international comparison program*. World Bank.

Conclusion

22 Growing towards a food-secure Africa in 2050: reflections and pathways

Michiel de Haas, Ewout Frankema and Ken E. Giller

1 Introduction

Achieving food and nutrition security in Africa by 2050 is one of the most challenging and important 'wicked' problems facing humanity. Without food security, it is hard to imagine inclusive and equitable distribution of opportunity and wealth on a global scale. Meanwhile, the impacts of climate change are increasingly evident, with rapid increases in temperatures and the incidence of extreme weather events. Widespread civil unrest, insurgencies and political instability, especially in the Sahel and the Horn of Africa, put progress at severe risk. The lack of attention to addressing poverty and food security from politicians prompts concerns about leadership and commitment. Whilst the future impacts of climate change and political conflict in Africa and the world at large are unpredictable, demographic trends and associated challenges are much better understood. By 2050, approximately an extra billion people will be added to Africa's population.

The population of Africa has grown from 0.28 billion in 1960 to 1.46 billion in 2023 – almost a sixfold increase. This rapid growth has happened without a long-term worsening of food insecurity. Rather, the incidence and severity of famine has declined since the 1980s. Moreover, the *rate* of population growth has been declining for decades already. So why should we be particularly concerned now about African food security? First, in the coming decades, Africa will see the addition of an unprecedented absolute number of people. This matters both for Africa and for the world. Overall, Africa has long been thinly populated (Austin, 2008). Yet in the coming decades, the continent will shift to become comparatively populous, with a population density greater than the world average by 2050 (van Wissen, this volume). Moreover, this demographic ascendancy takes place in a context where concerns about planetary boundaries are urgent. In other words, although the African population has doubled several times in the past century, the next doubling is a much larger absolute increase, which will take place in a very different setting.

Second, creating sufficient jobs to absorb the large numbers of young labour market entrants is a massive challenge. Most of Africa's population growth will be urban: 77% (878 million) of population growth between 2020 and 2050, compared to 54% (387 million) between 2020 and 2050 (UN-DESA, 2018, 2022). While the population employed primarily in agriculture is declining (McMillan & Headey, 2014; de Haas et al., this volume), the rate of structural change is slow, and non-agricultural jobs are primarily informal and poorly paid (Pieters, this volume). The prospect of jobless growth presents a key concern about food affordability and nutrition.

Third, gains in agricultural productivity have been small and slow, and what gains we observe are limited to a small set of countries. Increases in food production marginally keep pace with population growth and reliance on food imports increases steadily. They have resulted

DOI: 10.4324/9781032649696-28

mainly from the expansion of the cropland area rather than the intensification of agriculture. In many regions of Africa the most suitable land for crop production, with reliable rainfall and favourable soils, was settled long ago (Giller & Andersson, this volume). Further agricultural expansion to feed Africa's growing population must either take place in areas with lower rainfall and on more marginal soils with less potential for production, or compete directly with other crucial land uses, including forest preservation and urban expansion. The question if Africa is heading towards a Malthusian precipice or instead finds itself on the cusp of a Boserupian revolution, triggered by increasing population pressure, is hardly new, but has only gained in salience (e.g. Jayne et al., 2014; Koning, 2017).

As this book illustrates, Africa is a hugely diverse continent and the issue of food and nutrition security is multifaceted, requiring an interdisciplinary analysis. In this concluding chapter, we draw on the contributions to this volume. We present a simple framework to identify four interrelated focus areas for food systems change and explore four potential drivers of change that could serve as pathways towards a food-secure Africa by 2050.

2 Accelerating food systems change in Africa

2.1 An integrated framework

Food systems need to change at an accelerating pace to achieve food security by 2050 for Africa's growing population. Figure 22.1 synthesizes the dimensions of challenge in an overarching framework. It embodies the interdisciplinary nature of this book, drawing attention to the importance of both the *production* and *consumption* side of food security, and considering both the

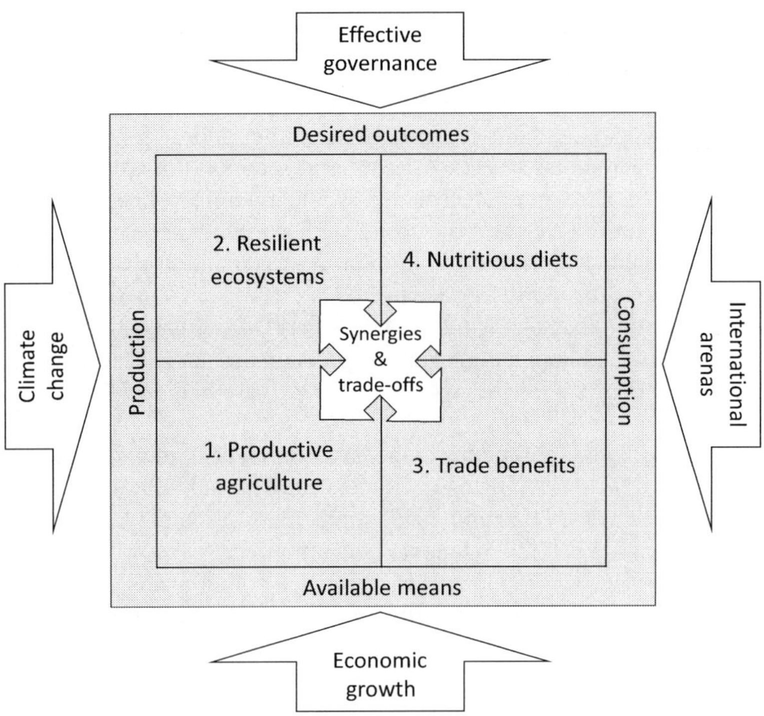

Figure 22.1 Focus areas for food systems change.

desired outcomes and the *available means*. By intersecting these two dimensions, we identify four interrelated focus areas, all of which are crucial when food systems are considered: *agriculture*, *ecosystems*, *trade* and *nutrition*. Each comes with its own set of considerations, which we approach through the lens of challenges, threats and opportunities that have run as a core thread throughout this volume. The four focus areas have their own dynamics, but they also interact, creating a range of potential synergies and trade-offs. An example of a potential trade-off is that too much focus on increasing the yields of a few cereal crops may result in the marginalization of more nutritious crops in diets and production systems. An example of a synergy would be that trade (through food imports) can relieve pressure on domestic ecosystems.

The framework also identifies four external forces that affect the functioning of food systems. *Economic growth* crucially augments the available means for food systems change but it certainly does not automatically translate into nutritious and healthy diets for all, or into flourishing natural ecosystems. Inversely, *effective governance* is necessary to steer change in desired directions, but it can only do so if the available means are present. In other words, both economic growth and food systems governance are necessary conditions for food systems change, but neither is sufficient on its own. The only way to work effectively towards food systems transformation is to generate the necessary means, while also putting these means to the desired use. *Climate change* is a key threat to food systems, directly affecting agriculture and ecosystems (the production side). Contestation in *international arenas* (from regional collaborations to global governing bodies) shapes the extent to which trade, investments or migration (remittances) can be harnessed to improve food and nutrition security. Of course, international negotiations also impact local agriculture and ecosystems, for example through their impact on markets and prices. Next, we discuss each of the focus areas in more detail, using examples from the volume's chapters.

2.2 *Agriculture: support smallholders to boost domestic food supply and support rural incomes*

Africa's farmers are predominantly smallholders, often with diversified and precarious livelihoods and farming on ever smaller plots (Giller and Andersson, this volume). Given the continued population growth in rural areas in most countries (van Wissen, this volume) and the slow growth of high-quality non-agricultural jobs (Pieters, this volume), this situation is likely to persist. Agricultural growth will involve larger farms and plantations, but small-scale farms will continue to be vital for food security, poverty alleviation and sustainability in the coming decades. Increasing smallholders' productivity is key not only to supplying growing numbers of urban consumers (van Wissen, this volume) but also to sustainably improve rural incomes and food security. The primary challenge is to work with the great variety of smallholders across a broad range of contexts to increase yields and curb further land expansion. Agronomists have long addressed the 'supply side' of this challenge (increase yields), while economists have stressed the 'demand side' (create markets for smallholders' produce). Yet gains have been modest, and it is clear that both approaches must be addressed simultaneously by policy makers and other relevant actors for smallholders to thrive.

Large gaps exist between potential and realized yields for all the major crops (Schut & van Ittersum, this volume). Despite a clear understanding of the causes of yield gaps, progress to close them is slow and variable. Overall, African smallholders have yet to take advantage of the yield gains unlocked by plant genetic improvement (Tittonell & Giller, 2013). This highlights the need to advance the debate on agricultural development beyond a focus on crop yield gaps at the field level towards a broader consideration of the reasons why the required increase in productivity is often not taking place. Only in this way can critical barriers be identified and addressed. Ethiopia (Sida et al., this volume) and Senegal (Hathie, this volume) stand out as examples where, through political commitment to agricultural transformation, meaningful progress in average crop yields has been booked over the past decades. By contrast, after experiencing something of an agricultural revolution in smallholder agriculture in the 1980s, Zimbabwe's agricultural sector has experienced sustained decline, which shows that past gains can be reversed when governance and economic performance are poor (Nyandoro & Andersson, this volume).

Across the continent, smallholders face a multitude of intricate poverty traps. As the majority of farmers have less than a hectare of land it is difficult to make a living from farming (Giller & Andersson, this volume). The need for rural households to seek income opportunities outside the farm in turn leads to labour shortages at critical times in the crop production cycle. Moreover, the institutional context within which smallholder farming takes place is often unfavourable (Frankema, this volume). Transaction costs are high due to small volumes and scattered production. Weak institutional regulations and corruption lead to rent-seeking across the value chain, and limit access to finance and inputs. The bargaining power of unorganized farmers is weak, and little of the value transmitted back to the producer. On top of this, transport, storage and processing costs are high due to poor infrastructure. Inefficiencies in supply chains cause large post-harvest losses, which not only reduce the profitability of farming, but also raise food prices in African cities (Frankema, this volume). All this results in what we have coined as the African "food security conundrum" (Giller, 2020): despite increasing demand for nutritious food from a burgeoning population, many farmers still lack the economic incentives to invest in food production.

Climate change and degrading soils will further aggravate challenges faced by African farmers. Although climate change is likely to cause increased rainfall and enhance crop yield potential in some areas, such as the highlands of East Africa, the impacts of climate change on food production are generally negative (Descheemaeker et al., this volume). Optimistically, crop yields might be reduced by only 10% if temperature increases are restricted to 1.5°C, but yields of many crops could halve if temperatures increase by 4°C. Thus climate change will exert substantial extra pressure on agricultural systems that are already highly stressed. Of particular concern is that increasing incidence of extreme heat stress, which could render large areas of Africa unliveable for both livestock and people (Nelson et al., 2024; Thornton et al., 2024).

The existence of large yield gaps is not only a sign of lagging productivity but also a testament to the success of plant breeding over the past decades. Various chapters in this book identify opportunities to reap this potential and bring African smallholders' plots closer to the yield frontier. Greater application of mineral fertilizer will be absolutely necessary to achieve this. Nutrient limitations are the primary factor constraining agricultural productivity, which in turn reduces the amount of organic matter returned to the soil, creating a negative spiral. The contributions from nitrogen fixation are important they are insufficient to boost and sustain agricultural productivity (Falconnier et al., 2023). Although still very much at the development stage, the prospect of locally produced 'green' nitrogen fertilizers, produced using renewable energy, represents a potentially revolutionary technology for Africa (Schut & van Ittersum, this volume). The question has only gained in salience. Maintaining and where possible improving moisture supply is also crucial. While only 3–6% of the agricultural land is currently irrigated, it contributes ca. 25% of agricultural value. The extent of and possibilities for irrigation are larger than thought, with scope for improving the functioning of current irrigation practices on the river-basin level, and expanding irrigation access with a focus on inclusivity (Veldwisch et al., this volume). Senegal provides an interesting example where investment in irrigation led to impressive increases in rice productivity and production over the past 15 years (Hathie, this volume).

In Egypt, food production is being increased through planned agricultural expansion by irrigating dry land outside the Nile Delta not currently used for crop production (Ramadan & Castet, this volume). Trees can play a surprisingly large role in retaining moisture, and regreening efforts will likely boost agricultural output in many African regions (Sheil, this volume). Controlling yield-suppressing weeds and pests is also key, especially in a changing climate. New advances in gene editing hold great promise to increase the speed with which new pest- and disease-resistant cultivars of crops can be developed. To bring these innovations to the field hinges on political choices around intellectual property rights, which remain highly contested and require local involvement (Kikulwe & Wesseler, this volume).

Intensification may also be induced by land pressures resulting from population growth. Most African farmers long had few incentives to intensify their production, given a relative abundance of land. Now that many African regions are becoming population-dense, incentives are changing. Indeed, experience shows that when markets are functioning well, farmers respond rapidly to new opportunities (Giller et al., 2011; Verkaart et al., 2018). This means that if Boserupian pressures begin to mount, increasing numbers of farmers may seek to intensify (see also Jayne et al., 2014). However, land pressure alone is not sufficient for such intensification to occur. In pockets of the East African highlands, populations have been as dense as 300 people per km^2 for more than a century (Crowley & Carter, 2000). In extreme cases, populations now exceed 1,000 people per km^2 – equivalent to the densities found in many cities – with concomitant decreases in farm size (Giller & Andersson, this volume). Yet yield gaps remain huge and the tendency has been to diversify incomes off-farm. In Central Ethiopia, households have even scaled down their livestock from cattle to small ruminants due to land pressures (Giller & Andersson, this volume).

Thus, while rural population densification provides an important – perhaps necessary – incentive for smallholder intensification, more needs to happen. In particular, favourable market conditions and supportive policies are crucial. Dispersed patterns of urbanization might catalyse a breakthrough in this respect. Until about 1990, population growth and urban sprawl in Africa concentrated around large conurbations – the national or regional capitals such as Accra, Dakar, Dar es Salaam, Kampala, Nairobi and Lagos. Recent decades have seen a shift towards growth in smaller, rural towns (Meier zu Selhausen, 2022). Closer proximity of urban markets for produce and employment provides opportunities for the integration of rural and urban livelihoods. It also creates increased demand for safe, diverse and high-quality food. Businesses, supported by enabling institutions and policies, can – and in many cases already do – play a crucial role in linking smallholders to markets not only for their produce, but also for inputs and capital (Ingenbleek et al., this volume). Of course, the benefits of a thriving business sector go beyond supplying farmers. Businesses can generate substantial value and jobs across entire agri-food value chains, and especially in agro-processing industries (Bulte et al., this volume; Hathie, this volume).

2.3 *Ecosystems: foster sustainable synergies between agriculture and natural ecosystems*

Over the past decades, increases in food production in Africa have been largely driven by expansion of the area of farmland (de Haas et al., this volume). Given that most of Africa was comparatively thinly populated, the environmental consequences of this expansion were, in general, not disastrous. However, pressure on land will continue to intensify given the continued growth of rural population, food demand from urban areas and large-scale commercial agriculture. Such pressures will happen along the 'intensive margin' (in already densely settled areas) as well as along the 'extensive margin' (in currently lowly populated areas, including forests) (Sassen, this volume). The environmental pressures that agricultural expansion involves, whether by smallholders or large, commercialized plantations, will vary across the continent but will be substantial and manifold. Trade-offs between agriculture and the environment will be compounded by climate change, which results in greater scarcity of key resources, water most notably, through increasingly erratic precipitation and the spread of new pests and diseases (Descheemaeker et al, this volume). Thus, climate change will reduce the agricultural potential of some regions which could drive further agricultural land expansion. The challenge, then, is to limit trade-offs and where possible foster synergies between ecosystems and agriculture. This is not a simple matter of choosing between 'land sparing' by intensifying production and 'land sharing' by using more biodiverse farming methods; both have a role to play depending on the local circumstances and opportunities (Baudron & Giller, 2014; Sassen, this volume).

Forests are key not only for biodiversity and carbon storage but also for moisture regimes across the continent which support agricultural productivity. Thus, agricultural expansion into currently forested land represents a major threat – not only to ecosystems but also to agriculture. The Congo basin harbours one of the largest remaining rainforests, which is of global importance for several reasons – for biodiversity, as a carbon store, and for its role in the continental water cycle (Sheil, this volume; Sassen, this volume). The latter role is much less known but critically important. A large proportion of the rainfall in much of East Africa is derived from evapotranspiration from the Congo rainforests (Sheil, this volume). Loss of this forest could have catastrophic impacts on food production. At a more local level, trees result in a greater proportion of rainfall being captured into the soil leading to greater water availability at the catchment level, and at the same time reducing run-off and associated problems of soil erosion. Forests also support other important ecosystem services, including pollination, natural pest control and soil fertility that benefit the livelihoods of many (Sassen, this volume). The same is true of savanna woodlands, which attract less attention although they potentially face greater pressure of agricultural expansion.

Further degradation of agricultural soils and the continuing spiral of declining yields are also key threats. However, planting trees on farmland through agroforestry provides many direct and indirect benefits. Fruits and nuts from trees are important in the diversification of diets in addition to the provision of wood for construction and fuel. Indirect benefits include rainfall capture through infiltration that enhances recharge of aquifers, and enhanced nutrient cycling to maintain soil fertility (Sheil, this volume; Sassen, this volume). The potential to sustainably expand irrigation for food production in Africa in the face of climate change is often considered to be limited. However, Veldwisch et al. (this volume) argue that the area currently under irrigation is strongly underestimated, and that inefficient use of water at the farm level is not a major problem if the water is not lost to the system as a whole. A farmer-led approach to support innovations in small-scale irrigation has great potential to expand the total area under irrigation, and to bring important benefits in crop diversification for healthy diets (Veldwisch et al., this volume).

Livestock is often seen as a major threat to natural ecosystems. Yet the extensive livestock of pastoralists found in semi-arid and arid regions is resource-efficient given the lack of opportunity for crop production where rainfall is marginal (Oosting et al., this volume). Assessing the impact of meat and milk production requires a consideration of broader set of roles that livestock play in the livelihoods of both pastoralists and sedentary farmers (Oosting et al., this volume). In particular, livestock can be an important contributor to the resilience of livelihoods in the face of climate variability and climate change through their role in accumulating and protecting wealth. To avoid competition over scarce cropland, livestock intensification should focus on farming systems where the animals can feed on by-products or graze on lands unsuitable for crop production. The productivity of livestock can be strongly increased. In the case of dairy, this brings co-benefits such as a reduced greenhouse gas emission intensity which can be seen as a climate change mitigation measure (Oosting et al., this volume).

2.4 Trade: enhance and diversify domestic and regional markets to reduce food imports

African consumers have become increasingly reliant on lengthening supply chains. This includes imports from outside the continent, especially of wheat, rice, oil and animal products. High and rising levels of food import dependency have multiple causes, which are product- and context-dependent, and are linked to both consumption and production (van Berkum & de Steenhuijsen Piters, this volume). Several country cases illustrate the breadth of experiences across the continent. Senegal has long focused on groundnut exports and in turn has been dependent on rice and wheat imports. Self-sufficiency in rice might be attainable, but this is not the case for wheat (Hathie, this volume). Egypt, having a highly productive agricultural sector but with limited fertile and irrigated land, has long been both a major producer and importer of wheat and will continue to be so in the coming decades (Ramadan & Castet, this volume). In Zimbabwe, wheat imports have increased markedly after the collapse of its agricultural sector, and the future of its food import dependency will depend to a large extent on whether this sector can be put back on track (Nyandoro & Andersson, this volume). Sierra Leone relies on imports of eggs and poultry meat to satisfy demand, even though domestic production potential is substantial (Bulte et al., this volume).

While touted by the 'Washington consensus' in the 1980s and 1990s, dependency on food imports has since come to be seen as undesirable, for various reasons. Food imports eat into the trade balance, outcompete fledgling domestic value chains and expose consumers to volatile prices. Peaks in global wheat prices in the past two decades have each time renewed interest in national food self-sufficiency in many African countries, often framed in a broader agenda of food sovereignty. Bending the trend of rising food import dependency will not be easy. In the long term, food has become cheaper globally (Frankema, this volume). This is a double-edged sword for food-importing developing countries. Consumers in sprawling cities and beyond can easily be supplied with cheap imports. Rural producers, meanwhile, face a more challenging competitive environment in open markets (Bulte et al., this volume), although the ability to import cheap staples also opens up scope for specialization in more lucrative crops for local consumption or export. Future population growth and fast urbanization are likely to further stimulate food imports. Consumers' growing preference for imported food items is likely to continue, while opportunities to achieve greater self-sufficiency for these essential staples are limited. Rice can be produced efficiently in many parts of Africa, although it is unclear whether it can be economically competitive in open markets with heavily subsidized production in rice-exporting countries (e.g. India, Thailand, the USA). By contrast, few African countries produce wheat in substantial quantities and opportunities to expand production are limited, as agroecological conditions provide little or no comparative advantage.

For reasons stipulated above, in the coming decades, international food trade will continue to be an inevitable, and sometimes desirable, component of (urban) food security in many African countries. Thus, reducing import dependency for its own sake might not be the most helpful policy objective. It is more important that consumers have sufficient purchasing power, and that export revenues from agricultural commodities and minerals not only serve elites but drive inclusive prosperity. That said, there are instances where food import dependency is a symptom of underlying problems. Where food imports are symptomatic of poorly developed domestic value chains, enhancing such value chains can be an effective way to increase rural purchasing power and foster food self-sufficiency. In such cases, governments might consider putting in place carefully targeted and temporary tariffs or other protective measures to help domestic sectors (for example maize, rice, poultry or milk) overcome major inefficiencies in production and supply chains and be able to compete with imports (Bulte et al., this volume). Such measures hold promise to reduce import dependency and to generate opportunities for employment and structural change in the value chain. Given that job creation through manufacturing is extremely hard to achieve for African countries (Pieters, this volume), import substitution in agro-food value chains might be a promising avenue (Bulte et al., this volume). Protection might hurt consumers in the short run. However, it is important to note that, even though food imports are 'cheap' in much of Africa today compared to what they would have cost half a century ago, urban consumers still pay more for their food than they would have to if their food was supplied through efficient domestic value chains. Ultimately, therefore, consumers will also benefit (Frankema, this volume).

There are also examples of food crops, such as millets, sorghum, specific rice varieties, highland banana, cassava, fonio and teff, in which African producers do have a comparative advantage, but which have often been neglected in research, extension and marketing. Although achieving major productivity and marketability breakthroughs for most of these crops will be costly, investing in specifically African crops may reduce their relative prices and increase their attractiveness to consumers (van Berkum & de Steenhuijsen Piters, this volume; Hathie, this volume). Targeted policies, for example mixing local flours in with bread flour, or using food subsidies, such as school feeding programmes, to increase the consumption of domestically produced crops might further contribute to a reduction of food import dependency and stimulate domestic agricultural development (Hathie, this volume).

Finally, whether this happens under some form of trade protection or not, and whether it involves 'local' crops or globally traded staples, food security would benefit from more regional trade. Removing barriers to regional trade will enlarge markets, allow for greater specialization and thus increase opportunities to benefit from intra-regional comparative advantages. The development of an internal market under the African Continental Free Trade Area (AfCFTA) agreements contains the promise that such benefits can be reaped, but only if free intra-African trade is also stimulated through reductions in overland transportation and transaction costs.

2.5 Nutrition: prioritize improved nutritional outcomes

Enhancing food security goes beyond increasing the supply of calories: diverse diets are critical to address the problems of mal-, over- and undernutrition (Melse-Boonstra et al., this volume). van Berkum and de Steenhuijsen Piters (this volume) highlight how Africa is following a global trend of homogenization of the food system, with increasing reliance on a small number of staple grains and oil crops. In fast-growing urban areas, diets are shifting rapidly, with processed fast foods proliferating, such that even if food *availability* is addressed, the *quality* of diets is not assured. A good example is Egypt, where government subsidies on bread, rice and sugar have kept particular foods affordable, but do not necessarily contribute to healthy diets (Ramadan & Castet, this volume).

In a global food market where flours, fats and sugar are cheap, there is often only one generation between hunger and obesity, and a transition towards more nutritious food must be actively pursued. However, in light of more acute, short-term challenges, nutrition is often not prioritized. Short-term and production-oriented measures may even be counterproductive for nutrition, if they focus on a few staples. In contexts of conflict, short-term concerns about availability and affordability are urgent, and issues of dietary quality and diversity are not prioritized. Lack of access to nutritious food jeopardizes the development of young generations, as shown by the effect of protracted food insecurity on stunting in conflict-exposed areas of Chad (van Dijk & Djindil, this volume).

Melse-Boonstra et al. (this volume) argue for a radical refocus of production systems to a more nutrition-sensitive agriculture. Diverse diets need a range of different food stuffs. Grain legumes – such as common beans, cowpea, groundnut and soyabeans – are important for healthy diets, providing some essential amino acids, minerals and vitamins (de Jager et al., 2017). Vegetables and fruits are also important, and provide a market for smallholders, particularly where (farmer-led) irrigation is available (Veldwisch et al., this volume; Ramadan & Castet, this volume). At a global scale, meat consumption is a contentious issue in debates relating to food systems transformation due to the environmental footprint of livestock. In Africa, an increase in the consumption of animal-sourced foods in moderation would go a long way to addressing the problems of undernutrition (Melse-Boonstra et al., this volume). The rapid growth of an urban middle class is driving the demand for fish and poultry (both meat and eggs) and leads to a burgeoning aquaculture and poultry production sector, often within or on the peripheries of the major conurbations. This provides a double opportunity of creating a market for smallholders to diversify their farms to produce soyabean as feed while also reducing the dependence on imports of feed and meat.

There are many potential synergies between a focus on nutrition sensitivity and other focus areas outlined in Figure 22.1. Nutritious diets should therefore be a central target for all initiatives on food systems transformation, whether they focus on production, technology, trade or policies.

3 Four drivers of food systems transformation in Africa

As noted throughout this book, the pace of food system change needs to shift several gears compared with the past decades. Given Africa's huge diversity, it is highly unlikely that all countries will do this at the same speed or follow an identical trajectory. Here we sketch four possible drivers that may set change in motion and leverage transformation.

3.1 Driver 1: crises breed change

The first driver of change will play out when the main obstacles to sustainable intensification and equitable access to affordable and healthy diets remain in place. If this is the case, especially in densely populated countries and regions, competition over land increases, while farm sizes continue to decline. Climate change compounds these challenges. In urban areas, food prices remain high and volatile, while rapidly increasing numbers of urban consumers generate continuous upward pressure, which domestic agricultural sectors cannot alleviate. The reliance on a limited range of imported food items will grow. In a context of poor governance and lagging economic performance, food insecurity becomes an ever-pressing problem, which contributes to social unrest or even violent conflict. Emerging food crises undermine wellbeing and economic activity and cause necessity-driven migration. International intervention will be focused on providing food aid necessary for immediate survival, while investment in the health and nutrition of future generations will be scant. Only after conditions first deteriorate, do new windows to improve food security open up. As systems perform poorly, some actors will rally around improvement and be ready to implement positive change when the right conditions arrive. Crisis breeding change is not without historical precedent. Europe's common agricultural policy emerged from the ashes of a war-torn and famished Europe (Tracy, 1982). In Africa, the post-famine and post-genocide eras in Ethiopia and Rwanda respectively have seen major improvements in agricultural productivity growth and poverty reduction. However, even if crisis drives change and doomsday will be averted in the medium to long run, the human costs in the coming decades would be enormous.

3.2 *Driver 2: institutional ripple effect*

The second driver of change is less bleak and may be referred to as an institutional ripple effect. The logic would be as follows. Some African countries establish an impressive record of effective governance of their food systems, possibly including trade protection of domestic producers and targeted subsidy schemes for producers or consumers, focused on diverse and nutritious food. The welfare gains of these policies are tangible, effectively reducing transaction costs and eliminating too powerful intermediaries in the value chain. Food prices stabilize, and may in the longer run even decline without affecting farmers' incomes adversely. These outcomes are most likely in countries that have seen relatively strong agricultural productivity growth in the past decades, such as Côte d'Ivoire, Ethiopia and Senegal. Past gains remain fragile when exposed to conflict, poor governance or major economic shocks. However, with 54 countries, Africa is the world's most 'decentralized' continent. This increases the chances that at least some countries 'get their policies right' for an extended period of time. Such policies, when proven successful, can become trendsetting, being hailed as best practices. Policies, and the institutions that underpin them, may then spread to other places through learning effects or through vocal segments of African electorates, business or consumer groups who call for their adoption. If institutional ripple effects prevail as a driver of change in the coming decades, this would result in large intra-African inequalities, with notable winners and losers.

3.3 *Driver 3: regional or pan-African initiatives*

Over the past decades, African states have developed a patchwork of free trade zones and other forms of regional collaboration. In recent years, important steps have been made towards the establishment of an internal AfCFTA, signalling that African politicians recognize the welfare gains of increasing intra-African trade and competition, and take a more assertive stance in global arenas. The AfCFTA agreements also sent powerful signals to entrepreneurs that politicians are aware of the importance of a reliable business environment without red tape, and a healthy investor climate. Meanwhile, the African Union has expressed strong commitment to transform food systems and achieve food security, and translated this into concrete and periodically monitored targets, such as the allocation of 10% of government budgets to agriculture. Should these multilateral initiatives indeed live up to expectations and lead to increased food production and trade within Africa, food systems can be strengthened substantially. If successful, regional collaboration could reduce reliance on food imports from outside Africa, or at least avoid further increases. Food producers will see their markets grow and opportunities for specialization increase. Revenues from food production and trade can be re-invested within the region, for example to improve infrastructure, and expand agricultural research and extension services. Urban consumers will become better off as a result of enhanced market integration and new investments in cross-border infrastructures. Targeted policies can be put in place to steer the food system towards diverse and healthy diets. Ecosystem preservation can be managed more effectively through cross-border agreements under continental supervision. Given the very large coordination challenges that have to be overcome before such deep regional or even pan-African collaboration is realized, they will likely not materialize in the next few years. By 2050 (or 2063, the target year of the African Union) notable progress is feasible, if ongoing developments are further harnessed through supra-national governance structures.

3.4 Driver 4: global responsibility

The fourth driver of enhanced food security in Africa would emerge from global collaboration. In such a situation, a new world order coalesces around efforts to arrest global warming and to ensure planetary boundaries are not breached. This shift of mindset would involve acknowledging that the largest (historical) emitters have a global responsibility, and that African food security is inextricably linked to global environmental and economic progress. Protecting African forests as reservoirs of biodiversity, regulators of rainfall and stores of carbon is given high priority by the international community. The dependence of past global economic development on the human and natural resources of Africa is recognized and, potentially as part of a reparations agenda, Europe and North America radically step-up commitment to African food systems transformation. Meanwhile, African governments become more accountable to their citizens and invest heavily in creating conducive conditions for investment in (smallholder) farming, which leads to major gains in crop and livestock productivity. Through firm policy making and successful moral appeals, the interests of powerful international corporations and research institutes are aligned with fair value addition along the supply chains and healthy food for consumers. Increased population density is met with investment in infrastructure and institutions that reduce transaction costs and create positive feedbacks to further drive investment in food systems. Attitudes to migration within and outside Africa change, enabling much more mobility towards countries facing critical labour shortages. Circular migration of Africans to fill the jobs created leads to increased remittance flows. The youthful population of Africa thus becomes the catalyst to addressing both ends of the major demographic challenges of the coming decades. This final driver of change is certainly the most ambitious and, sadly, the least plausible, as it requires the highest degree of coordination and alignment, both of ideas and of interests – but it would also be the most effective pathway towards a food-secure Africa in 2050, with the least human and environmental costs.

4 Concluding thoughts

We began this chapter by discussing the burgeoning population in Africa in a context of sluggish agricultural productivity growth. These are challenges, but they also present major opportunities to move towards a more food-secure continent. Rapid population growth means there will be many young people looking for jobs and there will be many more mouths to feed. Large yield gaps threaten food security, yet they also imply major opportunities for farmers to increase their productivity. The large body of research conducted over the past decades shows that we have the knowledge and technology to increase crop yields two- to threefold immediately in nearly all parts of the continent. Moreover, such gains can be made without adversely impacting nutrition or ecosystems. How do we turn challenges into real opportunities for the future?

We have presented four drivers of change, from least to most ambitious in terms of the extent of transformation and coordination involved. The threat of climate change, the difficulty of creating high-quality jobs and the complex set of constraints that hamper smallholder agriculture in the context of fast population expansion and urbanization are daunting. This means that transformations are complex but that ambitious efforts are all the more necessary. The main factor that will decide where and when which chain of events will play out is the ability of private and public actors, on a national, continental and global scale, including governments, researchers, multinational organizations and corporations in Africa and beyond, to step out of the shadow of private and partisan interests, accept responsibility for the common good and coordinate effectively. This process will not be without political contestation, winners and losers.

The politics and power struggles that arise from clashes of interest – both domestically and internationally – are inevitable and potentially fruitful in raising awareness and mobilizing people and resources. However, abstract and decontextualized ideologies about the 'proper' model for African farming or the 'right' characteristics of African food systems or diets are unhelpful. In particular, Global North actors have often tended to unduly transpose challenges faced by the wealthiest nations – overconsumption of animal proteins, overuse of fertilizers and herbicides, animal welfare and disease concerns arising from intensive livestock farming – onto an African context. Moreover, appeals to merely invest in Africa out of an 'enlightened self-interest' might yield some positive results, but they eventually will fall flat. For example, while poverty-induced migration is often invoked as an argument to invest in African development, ample research shows that as economies develop, migration rates will not decline. To the contrary, migratory aspirations tend to increase faster than local opportunities until rather late in the development process (De Haas 2010; De Haas & Frankema, 2022). Ultimately, a world in which the most vulnerable disproportionately bear the burden of climate change and persistent poverty and food insecurity while others live in wealth is perfectly imaginable yet morally unacceptable. How the path towards food security in 2050 should be governed is far beyond the scope of this book. Instead, it should be the outcome of endogenous democratic processes and committed leadership, which will be required to harness available means and achieve desired outcomes.

Acknowledgements

We thank Jeroen Candel and Leo van Wissen for their useful comments on the chapter.

References

Austin, G. (2008). Resources, techniques, and strategies south of the Sahara: Revising the factor endowments perspective on African economic development, 1500–2001. *The Economic History Review*, *61*(3), 587–624. https://doi.org/10.1111/j.1468-0289.2007.00409.x

Baudron, F., & Giller, K.E. (2014). Agriculture and nature: Trouble and Strife? *Biological Conservation*, *170*, 232–245.

Crowley, E.L., & Carter, S.E. (2000). Agrarian change and the changing relationships between toil and soil in Maragoli, western Kenya (1900–1994). *Human Ecology*, *28*, 383–414.

De Haas. H. (2010). Migration and development: a theoretical perspective. *International Migration Review, 44*(1), 227-264.

De Haas, M. & Frankema, E. (2022). Migration and development: lessons from Africa's long-run experience. In M. de Haas & E. Frankema (Eds.), *Migration in Africa: Shifting patterns of mobility from the 19th to the 21st century* (1st ed., pp. 379–389). Routledge. https://doi.org/10.4324/9781003225027

de Jager, I., Abizari, A.-R., Douma, J.C., Giller, K.E., & Brouwer, I.D. (2017). Grain legume cultivation and children's dietary diversity in smallholder farming households in rural Ghana and Kenya. *Food Security*, *9*, 1053–1071. https://doi.org/10.1007/s12571-017-0720-0

Falconnier, G.N., Cardinael, R., Corbeels, M., Baudron, F., Chivenge, P., Couëdel, A., … & Giller, K.E. (2023). The input reduction principle of agroecology is wrong when it comes to mineral fertilizer use in sub-Saharan Africa. *Outlook on Agriculture*, *52*(3), 311–326. https://doi.org/10.1177/00307270231199795

Giller, K.E. (2020). The food security conundrum of sub-Saharan Africa. *Global Food Security*, *26*, 100431. https://doi.org/10.1016/j.gfs.2020.100431

Giller, K.E., Murwira, M.S., Dhliwayo, D.K.C., Mafongoya, P.L., & Mpepereki, S. (2011). Soyabeans and sustainable agriculture in southern Africa. *International Journal of Agricultural Sustainability*, *9*, 50–58. https://doi.org/10.3763/ijas.2010.0548

Jayne, T.S., Chamberlin, J., & Headey, D.D. (2014). Land pressures, the evolution of farming systems, and development strategies in Africa: A synthesis. *Food Policy*, *48*(C), 1–17. https://doi.org/10.1016/j.foodpol.2014.05.014

Koning, N. (2017). *Food security, agricultural policies and economic growth. Long-term dynamics in the past, present and future*. Routledge.

McMillan, M., & Headey, D. (2014). Introduction - understanding structural transformation in Africa. *World Development*, *63*(C), 1–10. https://doi.org/10.1016/j.worlddev.2014.02.007

Meier zu Selhausen, F. (2022). Urban migration in East and West Africa since 1950. Contrasts and transformations. In M. de Haas & E. Frankema (Eds.), *Migration in Africa: Shifting patterns of mobility from the 19th to the 21st century* (1st ed., pp. 281–307). Routledge. https://doi.org/10.4324/9781003225027

Nelson, G.C., Vanos, J., Havenith, G., Jay, O., Ebi, K.L., & Hijmans, R.J. (2024). Global reductions in manual agricultural work capacity due to climate change. *Global Change Biology*, *30*(1), e17142. https://doi.org/10.1111/gcb.17142

Thornton, P., Herrero, M., Nelson, G., & Mayberry, D. (2024). Resilient livelihoods in Africa's pastoral–agropastoral transition zones will increasingly depend on heat stress adaptation and systemic change *Nature Food*, 5, 967–981.

Tittonell, P., & Giller, K.E. (2013). When yield gaps are poverty traps: The paradigm of ecological intensification in African smallholder agriculture. *Field Crops Research*, *143*, 76–90. https://doi.org/10.1016/j.fcr.2012.10.007

Tracy, M. (1982). *Agriculture in Western Europe. Challenge and response, 1880–1980* (2nd ed.). Granada.

UN-DESA. (2018). *World urbanization prospects 2018 revision*.

UN-DESA. (2022). *World population prospects 2022 revision*.

Verkaart, S., Mausch, K., Claessens, L., & Giller, K.E. (2018). A recipe for success? Learning from the rapid adoption of improved chickpea varieties in Ethiopia. *International Journal of Agricultural Sustainability*, *17*(1), 34–48. https://doi.org/10.1080/14735903.2018.1559007

Index

African Continental Free Trade Area (AfCFTA) 31, 106, 182, 274, 276
African Union 11, 217–218, 228, 276
agricultural credit 26–27, 42, 62, 92–94, 106–107, 153, 175, 177, 255, 260
agricultural extension 37, 62, 92, 95, 107, 153, 243, 255, 274, 276
agricultural growth 10, 12, 15, 36–37, 128, 192, 269
agricultural inputs: access 66, 118, 234–235, 246, 256, 260, 270; supply 8, 27, 43, 93, 95, 106, 117, 175, 189, 237, 244–245, 247, 253, 255–256, 271; use 154–155, 163, 190, 251–252, 260
agricultural intensification 14–15, 36, 40, 81, 93–95, 116, 192, 217–219
agricultural investments 11, 60, 95, 127
agricultural policies 26, 32, 66, 168, 275
agricultural processing 252, 256
agricultural stagnation 12, 15, 86, 126, 128
agricultural technologies 5, 40, 44, 51, 175, 187, 190, 195
agriculture: capital-intensive 51, 178, 218, 256; degrees of input use 65, 106, 152–155, 163, 168, 190, 234, 247, 251–252, 260; energy-intensity 36; environmental impact 8, 165, 226, 272; yields 36, 65, 92–93, 151, 153–154, 168, 190, 237, 271 (*see also* yields, cereal)
agro-climatic zones, arid 67
agro-ecological zones 9, 24, 28, 37, 64, 66, 68, 116, 118, 187, 189, 208; arid 152, 201, 272; desert 51, 55; dryland 140, 155, 203, 209, 233, 238–239; Guinea savannah 90; humid 223; rangelands 142, 233–234; Sahel 89; savanna 89, 160–161, 163–166, 202; semi-arid 23, 67, 150, 152, 155, 189, 217, 272; temperate 232; tropical rainforest 90; upland 203
agro-ecological zones, tropical highlands 189
agroforestry 95, 153, 161, 167–169, 202–203, 210, 238–239, 272
aid: donor 35, 60, 64, 67–68, 94, 107, 142, 156, 257, 262; emergency 60, 139, 142, 201; food 12, 38, 40, 59–61, 102
Algeria 13, 57, 82, 84–85, 91, 103, 130, 141
amino acids 124, 275
anaemia 53, 125–126, 128
Angola 13, 82–83, 85, 103, 112–114, 130, 163
animal proteins 26, 125, 217, 278
Argentina 100, 227
Asia 12, 26, 75–78, 84, 100, 102, 112–114, 121, 125, 177, 179, 181, 183, 189, 191, 194, 223, 254
atmosphere 205, 207–208, 229

bananas 92, 160, 226–227
Bangladesh 12–14, 175, 178–179, 181
banking 64, 111, 118, 239, 260
barter 62, 66, 176
beans 26–27, 275
Benin 82, 85, 111, 131, 246–247
biodiversity 4, 89, 94–95, 130, 152, 166, 168, 193, 225, 236–239, 272, 277
bioeconomy 223–229, 235
biomass 151, 153, 224, 239
biotechnology 223, 225–227, 229
birth rate 55, 77–81, 145
blood pressure, high 124, 143
body mass index (BMI) 39, 128
Boserup 11, 78, 163, 268, 271
Botswana 13, 82–85, 91
Brazil 112, 126, 161, 227, 254
bread *see* wheat consumption
breeding: animal 226, 232, 259; plant 95, 107, 153, 226, 228, 232, 270
Bretton Woods institutions 26, 30
broken rice 23–24, 26
Burkina Faso 82, 85, 139, 177, 183, 202, 204, 209
Burundi 82, 84–85, 111–113, 128, 144, 163, 233
business 6, 8, 31, 66, 106, 176–180, 182–183, 187, 243–248, 256–259, 261–262, 271, 276

Cabo Verde 82, 85, 114
calorie-dense foods 8, 53, 108, 125, 127, 129, 132, 152, 274
Cameroon 82, 85, 124, 139–140, 162–165, 177, 179

Canada 100–102
capacity-building programmes 43, 218, 228
carbohydrates 105, 108
carbon 225, 229, 239, 272, 277; organic 204, 239; stocks 94, 166, 239; storage 236, 238–239, 272
carbon dioxide (CO_2) 151, 201, 226–227, 232
cash crops 11, 51, 132, 167, 190, 192, 252; *see also* export, agricultural
cassava 8, 26, 92, 103–104, 113, 127, 226, 260, 274
Central Africa 9, 95, 150, 161–165, 169, 201, 206, 223
Central African Republic 13–14, 82, 85, 113, 139, 163
cereals 8–9, 11, 23, 26–31, 35, 38–40, 45, 49, 61, 66, 68, 91–92, 94, 100, 102, 107–108, 114, 116, 124–125, 128, 130, 132, 151, 156, 164, 193, 202, 269, 274
cereal, yields *see* yields, cereal
Chad 13–14, 78, 82–83, 85, 105, 139–140, 144, 146, 275
chicken *see* poultry
children: development of 39, 124, 140, 143, 146; undernourishment of 46, 141
China 78, 92, 119, 182, 244, 254
chocolate 113, 252
cities 4, 7, 26, 31, 55, 86, 111–115, 117–119, 121, 244–245, 248, 270–271, 275; *see also* urban, urbanization
climate change 4–5, 10, 12, 15, 23–24, 150–157, 271–272, 277–278; adaptation 6, 91, 130, 157, 169, 225, 232, 236, 238, 240; effects of 4, 35, 49, 54–56, 95, 118, 150–151, 155–157, 161, 169, 201, 214, 216, 223, 225–226, 233, 267, 269–270, 275; future projections 24, 80, 150, 166, 205, 213; induced risks 119, 152–154, 157, 214–215, 218, 223; maladaptation 150, 154; mitigation 6, 56, 94, 155, 157, 160, 225, 232–233, 236, 238–240, 273; resilience 150–151, 156–157, 226, 233, 240; vulnerability to 68, 155, 193, 223
Climate-Resilient Development (CRD) 154–155
Climate-Smart Agriculture (CSA) 232, 236–238
climate variability 24, 27, 60, 63, 68, 95, 150–154, 272
climatic shocks *see* shocks
cluster farming 41, 43–44, 46
cocoa 12, 104–105, 160, 163–164, 167, 169, 190, 252
coffee 12, 104–105, 160, 163–164, 188, 252
colonialism 10, 25–26, 59–62, 64–65, 68, 145, 152, 161, 164, 190–191, 193
commodities, agricultural 101, 104–105, 162, 164, 181, 190, 273
Comoros 82, 85, 114
comparative advantage 100, 182, 254–255, 273
competitive advantage 45, 244
conflict 4, 6, 13, 16, 23, 39, 76, 82, 85, 91, 101, 112–113, 118–119, 124, 139–147, 161, 163, 168, 206, 223, 233, 238, 240, 274–276, 278; cattle rustling 142; farmer-herder 144, 152; insurgencies 62, 140–141, 144, 267; rebels 139, 142, 144–146; *see also* war
Congo Basin 160, 163, 169, 205, 208, 272
Congo Democratic Republic *see* Democratic Republic of the Congo
Congo Republic *see* Republic of the Congo
consumer-centric food systems 129–132
consumer groups 247, 276
consumer preferences 5, 8, 11, 29–31, 49, 54, 106–107, 113, 125–127, 129–131, 224, 236, 244–247, 273
consumer prices *see* price
consumer subsidies *see* subsidies
consumption: animal-sourced food 126, 217, 237, 275; dairy *see* dairy consumption
contract farming 26, 30, 66, 107, 248
convenience foods 125–126, 130
corn *see* maize
corruption 66, 120, 245, 258, 270
Côte d'Ivoire 13–14, 82, 85, 103–104, 121, 163, 169, 179, 276
cotton 45, 51, 60, 164
Covid-19 pandemic 5, 23, 39, 49, 52–57, 79, 100–101, 111, 119, 124, 126, 253
cowpea 26, 226, 275; Bt 226, 228
crime, organized 118, 139
crisis: food 23, 26, 124, 147; protracted 139–143, 146, 275
crop failure 167, 214, 219
cropland expansion 8–9, 50–51, 65, 89, 92, 94–95, 144, 152, 160–162, 164–166, 168–170, 193, 236, 239, 261, 268–269, 271–272
crop residues 155, 190, 233–234, 237–238, 240
cultivars 94–95, 153, 255, 271
cultural heritage 233, 237, 244
culture 31, 78, 127, 143, 187
currency devaluation *see* devaluation, currency

dairy 31, 38, 116, 232–237, 239–240, 272
dams 51, 60
deforestation 25, 94, 154, 160–164, 169, 203, 206, 208
Democratic Republic of the Congo (DRC) 4, 13–14, 27, 82, 85, 140, 161, 163, 165, 168
demographic dividend 4, 80, 195
demographic transition 4, 75–79, 81, 86, 194
demography *see* population
devaluation, currency 49, 54, 64, 165
diabetes 124, 128
dietary: change 4, 8, 24–26, 31–32, 53–55, 75, 91, 94, 102, 107, 113, 126–127, 233, 272, 274–275; guidelines 128–129; preferences *see* consumer preferences; quality 5, 8, 24, 32,

53–54, 102, 105, 107–108, 111, 124–129, 132, 139, 156, 167, 195, 219, 248, 251–252, 256, 269, 271–272, 274–276
diets, sustainable healthy diets 5, 125–126, 129–132, 213
disease: animal 151, 223, 232, 256, 259–260, 278; human 5, 105, 124, 152; plant 94, 151–152, 167, 220, 223, 225–226, 271–272
Djibouti 82, 85, 114
drought, resistant crops 8, 152, 156, 169, 226
droughts 32, 59, 61, 66, 95, 106, 140, 144, 151, 155, 157, 160, 193, 201, 209, 214, 217, 219, 233
dry season 125, 183, 205, 208–209, 233, 259

early warning systems 104, 238
East Africa 4, 9, 11, 44, 83, 92, 95, 115, 125, 150, 160–161, 163, 165, 169, 193, 201, 203, 223, 228, 232–237, 239–241, 270–272
Economic Community of West African States (ECOWAS) 30, 106, 228
economic growth 12, 15, 36–37, 44, 53–54, 61–62, 81–83, 86, 108, 126–127, 140, 156, 165, 179, 183, 188, 192, 223, 225, 227, 244, 254–255, 261, 269, 276
ecosystem services 8, 160, 162, 168–169, 272
education 45–46, 53, 57, 62, 75, 78–81, 86, 115, 131, 139–141, 143–144, 146, 180–181, 183, 192, 218
eggs 49, 113–114, 125, 128, 252, 256–259, 261, 273, 275
Egypt 7, 12–14, 49–57, 81–82, 84–85, 91, 103, 112, 120, 216, 240, 271, 273–274
elites 102, 218, 262, 273
employment: agricultural 7, 12–15, 76, 89, 151–152, 155, 163, 175–178, 180–181, 267, 269; informal 68, 195, 254; International Labour Organization definition 183; job-seeking 61, 64, 182; non-agricultural 7, 11, 44, 60, 64, 68, 86, 105, 108, 152, 165, 167, 176–179, 181–183, 192–195, 235, 239, 256, 270–271; self 44, 178–183, 195; underemployment 176, 190; unemployment 4, 63–64, 68, 178, 193; youth 81, 108, 179, 183, 195, 267
energy, fuel 54, 64, 160–162, 164, 167–168, 224–225, 253, 272
Enset ventricosum 193
entrepreneurship 44, 118, 180, 191, 246, 258, 276
Equatorial Guinea 82–83, 85, 114, 163
equity, social 132, 236, 238
Eritrea 82, 85, 91
erosion 25, 95, 205
estates 188, 218
Eswatini 82, 85
Ethiopia 4, 12–14, 35–46, 55, 82, 85, 91–92, 103, 112–114, 128, 130, 163, 176–177, 182–183, 189, 192–193, 202, 217, 233, 235, 247, 270–271, 275–276
Europe 59, 76–77, 84, 92, 100, 160–161, 182, 194, 252, 254, 256, 259, 275, 277
European Union 64, 162, 189, 223, 225, 227–228
evaporation 204, 207
evapotranspiration 216, 272
exchange rates 65–66, 100, 113, 117–118, 262
export: agricultural 11–12, 24, 26–27, 37, 45, 51, 56, 59–60, 63, 65, 68, 101, 104–106, 114, 121, 190, 217, 245, 252, 273; non-agricultural 117, 181, 252, 254
export diversification 31, 62, 105–106
export restrictions 11, 30, 101
export revenue volatility 12, 30, 273
extensification 193
extensive agricultural growth *see* agriculture, cropland expansion
extremist movements 140–141

fall armyworm 226–227
fallow 168, 190
family planning 78, 80, 86; contraceptives 78, 80, 181
famine 35, 38–39, 61, 79, 139, 142, 267, 275
farmer 6–8, 31, 106, 118, 129, 132, 235, 257–259, 261, 270, 274, 276; *see also* smallholder
farm fragmentation 4, 8, 35, 40, 51, 177, 192, 234, 237–238, 240
farming: commercial 51, 60, 64, 176, 217, 234, 246, 256, 261; subsistence 51, 255
farming systems 23, 44, 78, 107, 130, 155, 187, 190, 193, 214, 232–235, 237–240, 245, 272
farms: cluster 41–44; large 51, 165
farm size 10–11, 15, 40–42, 51, 64–66, 68, 84, 94, 106–107, 120, 154, 156, 161, 167, 176–177, 187–195, 203, 234–235, 238, 240, 255, 271, 275
fats 108, 113, 116, 124–126, 128, 274
feed, oyster shells 258, 260
fertility (human) 4, 53, 55, 75, 77–81, 86, 95, 151, 155, 180–181, 183
fertilizer *see* soil fertilization
firm *see* business
fishery 23, 38, 226, 252
flood 51, 151, 201, 205, 223, 233
fodder 152, 167, 192, 201, 203
food 4–12, 23–24, 30–32, 49–50, 53–57, 93–95, 100, 104–109, 111–113, 115–121, 124–132, 160, 162–163, 167–170, 223–224, 235–240, 243–246, 251–254, 267–270, 273–277; demand 11, 23–24, 26, 32, 35, 49–51, 55–56, 89, 91, 94, 105, 126, 157, 170, 271; fortification 130; plant-based 217, 237
food crops 10, 12, 29, 60, 63, 67–68, 107, 127, 236–237, 240, 251, 274
food import 11–12, 15, 23, 29–31, 59–60, 87, 89, 95, 102, 104–105, 107–109, 112, 115–118, 120–121, 217, 244, 254, 257, 273–274, 276;

cereals 29, 49–50, 52–54, 100, 104, 107, 112, 117, 254, 273; costs of 100, 102, 106
food import dependency 11, 14, 24, 29, 31, 51, 53–54, 61, 66–68, 100, 102–105, 107–109, 116, 119–120, 273–274; *see also* food self-sufficiency
food industry 31, 56, 129
food insecurity 4–6, 12, 15, 35, 38–39, 60–61, 64, 66–67, 104, 139–144, 146, 251, 275, 278
food markets 24, 104; domestic 8, 108, 164, 252; globalized 100; international 100–102, 104; local 142
food processing 8, 44, 142
food production 9–11, 23, 54–56, 60–62, 68, 89–90, 124–125, 140, 142–143, 154–155, 160, 162–163, 165–168, 209–210, 270–272, 276–277
food quality *see* dietary quality
food riots 54, 64, 67
food safety 5–6, 50, 107, 127, 244–245, 247
food security 4–16, 23–32, 35–36, 38, 40, 42, 44–46, 49–57, 59–69, 76–86, 90, 92, 94–96, 100–102, 104–109, 112, 116, 118, 120, 126, 130, 132, 139–147, 150, 152, 154–158, 162, 164–170, 175–184, 188, 190, 192–196, 201–210, 213–220, 223–224, 226, 228, 234, 236, 238–240, 243–248, 252–262, 267–278; acceptability 131; access 12, 31, 49, 53, 113, 115, 120; affordability 5–6, 50, 108, 124, 126, 132, 267, 274; availability 5–6, 8, 31, 38, 49–51, 53, 102, 108, 124–125, 127, 131–132, 152–153, 165, 213, 223–224, 237, 243, 245, 274; conundrum 195, 243, 270; rural 27, 139, 248, 251; sustainability 5–6, 42, 50, 126, 129, 220; utilization 49, 53
food self-sufficiency 8, 26, 32, 38–40, 45–46, 49, 89, 91–92, 95–96, 100, 109, 154–156, 190, 192, 195, 213–214, 232, 269, 273–274
food sovereignty 23, 30, 196, 273
food supply 4, 8, 12, 23, 25, 27, 31, 44, 49–50, 55–56, 87, 93–94, 101–102, 104, 106, 108, 111, 117–118, 124, 130, 139, 142, 237, 245–247, 252–253, 255–257, 260–261, 274
food supply chain 5, 7–8, 45, 54, 93, 106–107, 130, 225, 235, 245, 258, 273–274, 277
food system 4–7, 11, 13–15, 30, 100–101, 104–106, 125–127, 129–132, 157, 232–233, 236–238, 240, 243–245, 268–269, 274–278; change 4, 23, 59, 127, 129, 131–132, 141, 155, 192, 218, 268–269, 275–277; types 130
food value chains 12, 31, 108, 113, 120, 252, 271, 274
food waste 7–8, 53–54, 89, 217, 224, 229, 234, 238, 240
forage 234, 236, 238
foreign currency 56, 60, 252, 261
foreign direct investment (FDI) 64, 107, 182
foreign exchange 12, 64–65, 68, 256, 258, 262
foreign reserves 105, 121
forest 6, 8, 15, 94, 142, 160–170, 201–210, 224–226, 272, 277; conservation 6, 161, 163, 165, 167–169, 268; loss 8, 160–161, 163–164, 167–168, 202, 205, 208–209; products 160–162, 164, 168, 225; recovery 162, 167, 225
forest-agriculture interactions 160–163, 165, 167, 170, 201, 203, 205, 207, 209
forest transition 162, 164–165, 168–169
France 25, 53, 102
free trade zones 106, 276
fruits 31, 45, 53–54, 105, 107–108, 113, 116, 124–125, 127–130, 132, 167, 272, 275
fuel *see* energy

Gabon 85, 128, 163, 165
Gambia 82–83, 85
GDP *see* Gross Domestic Product
gender 6, 131, 180–181
genetic modification 224, 226–227, 269, 271
Ghana 82, 85, 103–104, 128, 130, 150, 155–157, 163, 169, 178–179, 189, 215, 244, 257
GHI *see* Global Hunger Index
Global Food Security Index (GFSI) 5–7, 9, 50
Global Hunger Index (GHI) 35, 46, 144, 155
governance 13, 16, 27, 37, 108, 120, 132, 140–141, 143–144, 166, 219, 228, 268–270, 275–276; *see also* policymaking
government: expenditure 63, 105, 132, 276; fiscal imbalances 53, 64, 104–105, 108, 113, 120, 125, 128, 252, 262, 275; intervention 26, 45, 60, 62, 68, 108, 255, 262; investments 5, 10, 59–60, 66, 219; military 43, 144; policies 43, 53, 55, 65–66, 107, 156
governments 23, 26–27, 30, 51, 53–54, 59–60, 62, 64, 66, 106–108, 119, 140–141, 220, 254–255, 261–262, 277
grain *see* cereals
grasslands 25, 94, 233, 239
grazing 94, 142, 152–153, 161, 164, 166, 192–193, 203–204, 209, 217, 233–235, 239; communal 64, 188, 239
greenhouse gas emission 89, 91, 94–96, 125, 132, 150, 154, 166, 225–227, 232–233, 236, 238–239, 273
Green Revolution 84, 89, 91, 195
Gross Domestic Product (GDP) 6, 11, 24, 27, 37, 49–50, 60, 68, 78, 104, 111–113, 120, 124, 126, 131, 155, 181, 183, 194, 251, 255
groundnut 23, 25–27, 29, 155, 275; Kersting's 246–247
groundwater 90, 154, 204–205, 209, 214, 229
Guinea 82, 85, 105, 114, 163
Guinea-Bissau 13–14, 82, 85

habitat 160, 164
habitat loss 166

health: human 31, 53, 80, 86, 124–131, 139, 143, 146, 152, 181, 248, 260; planetary 126, 129
healthcare 62, 75, 111, 115, 124, 141, 143, 146, 232
heat stress 151–152, 232–233, 259, 270
height, human 145–146
herbicides 43, 260
herders 144, 152–153
highlands 161, 169, 192, 206, 270
HIV-Aids 77
honey 113
horticulture 23, 26–27, 51, 62
household incomes 105, 180, 188
human capital 37, 56, 75, 79, 105, 124, 143
human rights 86, 141
humidity 152, 189, 206, 208, 259
hunger 5, 35, 38, 46, 59, 61, 111, 120, 124–125, 139, 141, 143–144, 223, 274
hunting 162, 164
hydrology 151, 209–210, 216

imports 11–12, 23–24, 32, 39–40, 49–53, 61, 89, 100, 102–103, 105, 182, 208, 252–254, 261–262, 273–275
import substitution 12, 30, 62, 109, 258, 274
inclusivity 12, 15, 44, 46, 54, 81, 105–106, 108, 127–128, 180, 213, 218, 243, 247–248, 270
income: farm 23, 27, 119, 154, 180, 192, 195, 218, 247, 251; living 176, 190, 192; off-farm 56, 60, 62, 66, 154, 180, 192–194, 218, 234, 270–271; rural 8, 36, 157, 181, 193, 255, 269, 274
income groups 11, 114, 119, 233, 235, 237
India 53, 100, 115, 119, 121, 126, 175, 177, 181, 189, 244, 273
Indonesia 104, 119, 126
industrialization 31, 36–37, 141, 180–181, 183, 195, 256, 261
industrial policies 175, 181–182
industries, agro 31, 44, 51, 118–119, 129, 255, 271
industry 31, 36–37, 44, 60, 64, 114, 116, 155, 181, 183, 192–193, 218, 254, 258, 262; *see also* manufacturing
inequalities 39, 119, 126, 131, 154, 157, 213, 244, 276
infiltration 203–205, 209, 272
inflation 23, 49, 55, 68, 101, 105, 239
informal sector 56, 66, 239
Information and Communication Technology (ICT) 107, 247
infrastructure 6, 31, 37, 42–43, 59, 65, 84, 86, 94, 105–107, 139, 153, 160–161, 182, 195, 213, 219–220, 225, 234, 270, 276–277; harbour 106, 118, 166; railway 118; road 59, 62, 80, 93, 106–107, 118, 142
innovation 31–32, 43, 127, 129, 131, 167, 224, 228, 244, 248, 271; institutional 26, 255; technological 101, 220, 228
institutions 6, 118, 120, 141–142, 144, 146, 153, 157, 227–228, 233, 245, 270–271, 276–277; property rights 64, 228
insurance 77, 92, 111, 118–119, 153, 175, 183, 238, 255
intellectual property rights (IPR) 228, 271
intensification 4, 8, 10–11, 15, 24, 66–67, 89–95, 106–107, 116, 162–164, 166, 168, 170, 193, 214, 217–218, 232, 234, 258, 268, 271–272
Intergovernmental Panel on Climate Change (IPPC) 24, 80, 153–154, 225
intermediaries 44, 118, 276
international arenas 91, 268–269, 276
International Comparison Program (IPC) 113, 116
International Monetary Fund (IMF) 54, 64, 257, 262
investments 10–12, 51, 62–65, 90–91, 95–96, 117–119, 129, 144, 153–154, 165, 177, 181, 187, 192, 227–228, 243–244, 254, 259–261, 269–270, 275–277
iron ore 252, 254
irrigation 10, 27, 45, 51, 62–63, 65, 106, 151–154, 188–189, 205, 213–220, 238, 270–273, 275
Islam, jihadism 139–141, 144

Japan 77, 112, 244, 254

Kenya 78, 82, 85, 91, 94, 103, 107, 178–179, 193, 227, 233, 235–236, 239, 245, 247

labour: child 78; forced 190; markets 64, 175, 178–183, 195, 267; migration 44; reserves 191; saving 44, 163, 177, 182, 188, 195; scarcity 8, 78, 142, 270, 277; skilled 182, 239; unlimited supply of 190–191; working conditions 108
labour-intensive manufacturing 116, 181–182, 254
Lagos 115, 246, 271
land: abundance 95; acquisition 65; arable 65; availability 15, 95, 141, 163–168, 187–188, 190, 192, 224, 232, 234, 236, 240, 262, 270–271, 275; as collateral for loans 248; communal 234; consolidation 43, 177, 191, 194–195; cultural value 194; degradation 4, 23, 25, 60, 95, 157, 161, 209, 225; grabbing 144; marginal 166, 193; ownership 68, 234; policies 191; reclamation 51; reform 43, 60, 62–68, 144, 191, 255; rented 94; sharing 168–169, 272; sparing *see* intensification; tenure security 213
Latin America 76, 78, 81, 84, 124, 225–226
legumes 23, 53, 93, 125, 127, 129, 132, 151, 156, 190, 246, 275
Liberia 82, 85, 112–113, 161
Libya 82–85, 91, 113
life expectancy 77, 79, 81, 145
livelihoods 60, 144–145, 153–155, 164–166, 192, 195, 223, 232–239, 272; *see also* income

livestock 11, 23, 60, 62, 105, 151–153, 155–156, 160, 190, 193, 232–235, 237–241, 270–273, 275; as capital 153, 233–235; cattle 144, 156, 193, 232–233, 235, 237, 239–240, 271; mixed farming 155–156, 192, 233–235, 237–240; multifunctional value 232, 238; (semi-) specialised farming 232–233, 235, 237
living standards 37, 62, 100, 111, 143, 176, 201, 244, 254, 256–258, 261, 275–276
Lord's Resistance Army (LRA) 139–140, 142, 144
Low and Middle Income Countries (LMICs) 12–15, 124, 126, 132, 175, 177–182

Madagascar 13–14, 82, 85, 114, 131
maize 8, 11, 26–28, 55, 63–67, 102–104, 107, 114–118, 125, 127, 151–153, 155–156, 226–227, 260; Bt 227–228; hybrid 60, 227
Malabo Declaration on Accelerated Agricultural Growth and Transformation 11
Malawi 62, 82–85, 95, 103, 114, 130, 176, 183, 187, 189, 202
Maldives 114
Mali 82, 92, 105, 139, 141, 189, 193, 202, 260
malnutrition 5, 35, 39, 53, 56, 61, 124–127, 129, 139, 143, 146, 223, 226
mango 26–27
manufacturing: capital-intensive 178, 182, 195; competitiveness 262; export 181, 254; labour-intensive 254, 274
manufacturing sector 37, 56, 60, 62, 64, 68, 177, 181–183, 251, 254, 258
manure 63, 66, 90, 155, 190, 193, 217, 233–235, 237, 239–240
Maputo Declaration 11
market: actors 44, 106, 244, 246; competition 44, 120, 236, 244, 273; demand 94, 124, 217, 227, 255; diversification 100, 105, 108, 119, 244, 246; domestic 30, 104, 132, 256–257; equilibrium 257; foreign 105, 274; functioning of 101, 142, 175, 224, 227, 234, 243, 245–246, 271; informal 54, 113, 120, 245; infrastructure 105, 118, 245, 261; integration 31, 44, 100, 114–115, 118–120, 228, 234, 274, 276; international 27, 53, 56, 96, 100, 102, 104–105, 111, 115–117, 127, 245, 254, 257–258, 262; intervention 120, 132, 191, 261–262; local 30, 87, 106, 111, 127, 165, 234, 245, 251, 256–257, 261, 274; mechanism 247; places 131, 180, 245, 248; power 119, 261; share 261
marketing boards 27, 119
marketing costs 244
marketing, seed 45, 59
Mauritania 82, 85, 91
Mauritius 82–85, 111, 114, 177
Mayotte 85
meat 53–54, 108, 112–113, 124–125, 128, 132, 162, 164, 217, 257, 259, 272, 275
mechanization 94, 101, 106, 177, 188, 214, 239, 255, 260
Mexico 119, 126
microbes 226–227
micronutrients 5, 53, 108, 160, 233, 236; deficiencies 5, 53, 143, 226
Middle East 76, 216
migration 4, 15, 60, 76, 79, 153, 157, 162, 183, 192–194, 269, 275, 277–278
millets 8, 26–28, 31–32, 66–67, 107, 113, 151–152, 155–156, 274
mining 11, 68, 90, 164, 166, 193, 252
moisture 6, 203, 205–210, 214–216, 271–272
monsoon 208
morbidity 124, 126, 143
Morocco 13–14, 57, 78, 82–85, 91, 103, 113
mortality 75, 77–79, 124, 126, 256; child 46, 75, 77, 124, 128, 143, 145–146
Mozambique 13, 62, 82, 103, 111, 177

Nairobi 113, 115, 245, 271
Namibia 82, 84–85, 103, 128, 177
natural disasters 4, 139
nature 164, 168, 187, 236, 239; conservation 6, 12, 258
New Zealand 77, 112
Niger 78, 82, 84–85, 89, 95, 105, 128, 139–140, 189, 202
Nigeria 4, 7, 82, 85, 91, 95, 103, 107, 112, 115, 117, 124, 128, 130–131, 139–140, 142, 146, 176, 178, 189, 226, 246
Nile 50–51, 55, 89–90, 93, 216, 226–227, 229, 271
nitrogen 90, 229; fixation 27, 190, 226, 270; production 91
non-governmental organizations (NGOs) 144, 240
norms 27, 131, 155, 192, 227
North Africa 9–10, 56, 78, 83, 104, 112–113, 119, 150, 215–216
North America 77, 84, 117, 277
nutrients 5, 89–90, 93, 108, 125, 127, 129, 156, 167, 190, 236; decreased 151; depleted 44; essential 127, 129; micro 90; minerals 108, 124, 260, 262, 275; net offtakes 89; recycle 240
nutrient stress 152
nutrition (human) 5, 7–8, 15, 24, 31, 35, 53, 56, 107–109, 125–130, 132, 143, 160, 167–169, 201, 226, 233, 237–238, 240, 245, 256, 267, 269, 274–275, 277
nutrition security 5, 10, 15, 24, 36, 100, 108–109, 112, 124, 126–127, 160, 162, 168, 170, 193, 224, 236, 245, 267, 269
nuts 125, 127–129, 167, 272

obesity 5, 53, 124, 128, 274
oceans 111, 210, 216

oil: cooking 113, 116; edible 45, 53, 60, 114, 125; petroleum 11–12, 111, 113–114, 116, 120, 144, 161
oil crops 11, 25, 102, 108, 273–274
oil mills 27
oil palm *see* palm oil
onions 26
organizations, international 132, 240, 277
overconsumption 53, 124, 132, 278

packaging 31–32, 44, 247
palm kernel cake 260
palm oil 102–104, 114, 160, 164
pastoralism 142, 152, 217, 233–235, 237–240, 272
pastures 142, 217
peace 94, 139, 141–144, 146
pensions 78, 86, 192, 194–195
pest control: biological 226–227; natural 160, 225, 272; pesticides 43, 217, 226, 260
pests 94, 151–152, 164, 167, 220, 223, 227, 271–272
planetary boundaries 4, 89, 106, 125–126, 129, 267, 277
plantations 169, 255, 269, 272
planting times 153, 216
plants, wild 160, 164
Poland 112–113
policy: dietary 5, 8, 107–108, 124–127, 129, 132, 139, 167, 195, 251, 269, 272, 274–276; price 62, 105, 119–120, 132, 257; taxation 107–108, 111, 118, 120, 142, 146, 190, 193, 245, 261; trade 29, 106, 108, 127, 132, 175, 182, 189, 257–259, 261–262, 274, 276
policymakers 8, 11, 24, 43, 102, 120, 129, 131–132, 209–210, 269
policymaking 26–27, 30–31, 54–55, 62–64, 86, 95, 104–105, 107–109, 120, 132, 156–157, 164–165, 180–183, 191, 217–218, 224, 227–228, 232, 261–262, 273–277; *see also* governance
political instability 139–147, 246, 267
political patronage 60, 64–65
political stability 10, 139, 144, 146
pollination 151, 160, 272
pollution 236, 238, 240
population: density 8, 55, 75, 84, 86, 89, 140, 163–164, 166, 168, 187, 190, 193, 267, 271–272; growth 4–5, 8–9, 12–15, 23–27, 29–32, 35–36, 38–39, 44–45, 49, 52, 54–56, 59, 61, 75–76, 79–86, 89, 91, 95, 100–102, 104, 125–127, 141, 146, 154, 157, 164, 175, 177, 179, 181, 183, 190, 193, 195, 213, 233, 235, 244, 255, 261, 267–271, 273, 277; indigenous 59, 79, 86, 141, 267; projections 16, 23, 35–36, 79–80, 86–87, 213, 267; rural 94, 195, 269; urban 36, 82, 84, 86, 94; working-age 24, 53, 68, 108, 195
post-harvest losses 8, 31, 45–46, 118, 270
potatoes 26, 49, 51, 114, 127; sweet 92, 114
poultry 12, 26, 31, 38, 51, 54, 125, 128, 235, 251, 253–254, 256–261, 273–275; imported 256
poverty: alleviation 4, 6, 10, 54, 56, 108, 116, 154, 157, 180, 187, 191, 254, 269, 276; conflict-induced 146; rates 4, 35, 37, 155–156; rural 63, 141, 146, 155, 157, 191, 195, 209, 223; traps 15, 90, 146, 153, 157, 189–190, 196, 270; urban 15
precipitation 24, 150, 205–208, 214, 216, 272
price: consumer 7, 11, 23, 49, 56, 64, 67, 89, 100–101, 104–109, 111–121, 125–126, 213, 244–246, 253, 256–257, 261, 270, 274–276; gaps 11, 23, 49, 53, 93, 101–102, 104, 107, 111–119, 131, 153, 164–165, 253, 255–257, 269, 274; land 63, 245; levels 11, 30, 53, 100–102, 106–107, 114–115, 117–121, 243, 246, 256–257, 260–262; producer 10, 44, 118, 153, 236; volatility 11–12, 23, 30, 49, 51–52, 93, 100–101, 104, 106, 111, 114, 116, 119, 142, 164, 216, 235, 261, 273
private investments 26, 30, 37, 165, 243
private sector 27, 43, 106, 132, 140, 183, 218, 224, 228, 243
processing, agricultural 30–32, 43–44, 106, 117–118, 152, 236, 239, 245–247, 252, 255–256, 262, 270; *see also* industries, agro
producer prices *see* price
producer subsidies *see* subsidies
production costs 107, 116, 121, 168, 256–260
productivity: agricultural 12–15, 24–25, 31, 35–36, 38, 41–42, 44, 51, 56, 64–65, 68, 81, 84, 86, 89, 93, 100, 105–106, 116–120, 124, 126–128, 130, 132, 146, 151, 153, 156–157, 163, 165, 170, 175–177, 180–181, 183, 187–193, 195, 213, 225, 234, 237–238, 240, 255, 260–261, 267, 269–270, 272, 274–277 (*see also* intensification); labour 40, 89, 93, 105, 116, 146, 151–152, 167, 177, 187–188, 191, 194–195, 254, 259; land 9, 41, 68, 91, 95, 116, 156, 162, 165, 220; livestock 106, 150, 155, 234, 273, 277; non-agricultural 68, 116, 165, 177–179, 181–183, 188, 192, 195, 225
profit 8, 31, 63, 65, 94, 113, 119, 153, 213–214, 218, 243–244, 246, 254, 258–259, 270
proletarianization 190–191
protected areas 163, 165, 167
proteins 5, 101, 151, 226, 236, 252, 256, 260
pulses 53, 108, 128, 130, 237
purchasing power 6, 8, 23, 49, 55, 100, 104, 108, 113, 121, 125, 223–224, 246, 273

racism 59, 68
rainfall 24–25, 27, 59, 65–66, 90, 95, 118–119, 150, 152, 155, 201, 203–210, 214–218, 223, 268, 272

red tape 118–119, 276
reforestation 209, 220, 225, 239
refugees 139, 142, 144
regreening 162, 209, 271
remittances 60, 66, 193–194, 269, 277
rent-seeking 116, 120, 262, 270
Republic of the Congo 13–14
resilience 30, 95, 100, 104–105, 150–151, 153, 156–157, 164, 168, 232–233, 236–238, 272
resource depletion 36, 45, 164
resource efficiency 8, 41, 43, 67, 152, 217, 272
resource pressures 4, 12, 54, 67
resources: non-renewable 261; renewable 225, 244, 270; scarcity 5, 12, 14, 90, 150, 245
restaurants 131, 244–245
retailers 44, 118, 120, 132, 245, 247
rhizobium inoculants 247
rice 8, 11, 23, 26–31, 100, 102–104, 107, 114, 116–117, 125, 127, 260, 273–274; imports 26–27, 29
rivers 25, 89, 209, 214–216, 270
root and tuber crops 8, 51, 93, 102, 108, 116, 124, 128, 130
rubber 160, 164
ruminants 193, 233, 235, 271
runoff 203–204, 209, 215, 272
Russia 100–101, 104
Rwanda 14, 82, 84–85, 103, 112–113, 119, 163, 177, 233, 236, 275; army 140

Sahel 23–24, 83, 86, 140, 144, 151, 204, 206, 209, 246, 267
salinity 25, 45
salt 124–126, 128, 209
sanitation 75, 130
Sao Tome 82, 85
school feeding programmes 30, 131, 274
seasonality 45, 51, 53, 61, 66–67, 90, 93–95, 118, 125, 150–152, 155, 161, 164, 168, 176, 178, 183, 193, 201, 205–206, 208, 216, 227, 235, 245, 251, 259, 270
Second World War 49, 60–61, 116, 160
seeds 26–27, 116–117, 125, 128, 157, 217; improved 42, 66, 94, 106, 255
Senegal 7, 13–14, 23–27, 29–32, 91
service sector 116, 155, 177–178, 182, 188, 192, 255
settler farmers, white 59–65
shocks: climatic 24, 26–27, 105, 118, 150–151, 154–155, 162, 207–208 (*see also* natural disaster); economic 11–12, 23, 26, 30, 49, 54, 64, 66, 68, 104–105, 115, 126, 164, 253, 276
Sierra Leone 12, 14, 82, 146, 161, 251–261, 273
skills 43–44, 46, 183, 219, 224; *see also* human capital; labour
slavery 76
smallholder 36, 40, 42–43, 46, 51, 60–68, 93–94, 106–108, 119, 142, 150, 152–154, 157, 161, 163–164, 177, 183, 187, 190–193, 213–214, 217–218, 220, 232, 246–248, 255, 261, 269–272, 275, 277; capital constraints 38, 167, 187–188, 190, 220, 234–235, 247–248, 256; constraints 61, 154, 163, 176, 188, 255, 277; diversification 153–154, 191, 193, 269, 275; environmental impact 163; flexibility 232; innovation 217, 247; market access 26–27, 31, 42–46, 51, 62, 67, 94, 105–107, 120, 156, 164–166, 183, 187, 193, 234, 238, 243, 245–248, 254–256, 259, 261, 271, 274–275; productivity 41, 45, 65, 68, 94, 106, 187, 189, 195, 255, 269–270; risk aversion 90, 93, 95, 119, 152, 154, 191, 195, 246–247; self-consumption 7, 45, 233, 251, 256; vulnerability 150, 153, 157
smallholders, environmental impact 187
social security 53, 56, 119, 157, 183, 194, 213; *see also* pensions
soil: carbon sequestration 225; conservation 220, 225; degradation 8, 11, 24–25, 30, 32, 44, 93, 95, 190, 203, 205, 225, 270, 272; depletion 25, 93–94, 116, 238–240; fertility 27, 59, 63–66, 89–90, 93, 95, 151–152, 155, 160–161, 163, 166, 189–190, 193, 215, 254, 272–273; fertilization 8, 10–11, 26–27, 41–44, 46, 59, 63, 66, 89–96, 101, 106–107, 116–117, 121, 152–154, 156–157, 190, 214, 217, 227, 246–247, 253, 255, 260, 270, 278; improvement 93, 130, 156, 167, 190, 201, 204, 209, 225; nutrients 11, 44, 89–91, 93, 156, 190, 236, 238–239, 270, 272
soil types: clay 65; dark loam 59; red loam 59; saline 25; sandy 59, 63, 65, 204
solidarity 54, 120, 144
Somalia 82, 85, 142, 144
sorghum 8, 26–28, 51, 66–67, 107, 151–152, 155–156, 169, 226, 274
South Africa 9–10, 12–14, 57, 82–85, 91, 103–104, 126, 128, 130, 177, 182, 188, 225
South Asia 78, 113–114, 125, 181, 195, 216
South Central Africa 161, 163
Southeast Asia 78, 112–114, 124, 195
Southern Africa 9, 11–12, 59, 62, 66, 83, 106, 150–151, 191
Southern Rhodesia *see* Zimbabwe
South Sudan 82–83, 85, 139–140, 142, 144
soyabeans 246–247, 260, 275
staple crops 23, 91, 100, 102–105, 113, 115–116, 125, 127–130, 132, 156, 160, 236, 273–274
starvation 4, 61, 120
storage 7, 31, 44, 53, 107–108, 117–119, 142, 153, 214, 238, 245, 270
structural adjustment programme (SAP) 11, 27, 64, 68, 105, 116, 258
structural change 4, 7, 12–15, 24, 26, 36, 42, 46, 68, 75, 81, 126, 128, 175, 177–179, 181, 183, 188, 190–193, 195, 218, 254–255, 258, 267, 274

structural transformation 7, 37–38, 67–68, 89, 126, 188, 191–192, 194, 254–255, 262
stunting 38, 46, 139, 143–144, 146, 275
subsidies: consumer 49, 53–54, 56–57, 102, 113, 119–120, 132, 220, 274, 276; producer 59, 92–93, 95, 107, 119, 121, 132, 157, 189, 195, 220, 245, 255, 273
subsistence 26, 51, 78, 115, 118, 153, 164, 183, 218, 246, 255
Sudan 13–14, 82, 84–85, 142
sugar 45, 51, 53–54, 113, 124–126, 128, 132, 151, 274
supermarkets 113, 130–131, 245
sustainability 10, 30–32, 43–44, 56, 80, 89, 125, 127, 132, 153–154, 157, 175, 219, 225, 232–233, 236, 269, 271
sustainable and healthy diets (SHDs) 125–126, 129–132
sustainable development 154, 156, 168, 224, 226, 254
Sustainable Development Goals (SDGs) 125, 162
sustainable intensification 8, 10–11, 15, 89, 93, 153, 262, 275
synergies 15, 131, 168, 268–269, 272

Tanzania 13–14, 82, 85, 91, 95, 103, 112–115, 121, 163, 176–177, 180, 182–183, 188–189, 204, 217, 233, 236
tariffs *see* policy, trade
tea 105, 188
technological change 102, 105, 162, 177, 182, 188, 213, 223–225
technology 6, 26, 37, 44, 46, 91, 95, 132, 156, 177, 182, 187–188, 190, 192–195, 213, 216–220, 224, 227–228, 238, 243, 252, 257, 259–261, 270, 275, 277
teff 41, 107, 274
temperature 151, 207, 259; rising 24, 49, 54, 150–152, 156–157, 201, 207, 216, 223, 233, 267, 270
terrain 25, 204, 208–209
textiles 60, 111, 224
Timor-Leste 143
tobacco 45, 60–62, 188
Togo 13–14, 82, 91, 112
town 4, 142, 233, 271
trade 56, 105, 178; agreements 106, 108; balance 29, 273; costs 120; diversification 105; free 6, 106, 257; global 45–46, 164, 182–183, 261, 274; intra-African 61, 89, 105–106, 121, 164, 182, 261, 274, 276; protection 258; terms of trade 105
traders 43, 106, 118, 120, 142, 233, 239, 246–247
traditional knowledge 37, 130, 153, 233, 239
training 43–44, 180–181, 183, 246, 248
transaction costs 7, 106–107, 118–120, 234, 246, 255, 270, 274, 276–277
transpiration 204, 206–209
transportation 105, 114–119, 131, 142, 178, 195, 244, 246, 270, 274
transport costs 91, 94, 96, 101, 108
trees 67, 154, 161–162, 167–170, 201–206, 208–210, 216, 271–272
trust 118, 132, 146
Tunisia 13–14, 57, 82–85, 91, 103, 112

Uganda 13–14, 85, 103, 107, 111, 114, 128, 139, 142, 144, 163, 176, 182–183, 189, 227, 233, 236
Ukraine, Russian invasion of 11–12, 39, 49, 52–53, 56, 101, 252, 254
undernourishment 46, 126, 141, 143, 213
undernutrition 5, 38, 124–125, 274–275
unpaid work 57, 177–179
urban 15, 31–32, 64, 118–119, 157, 183, 192–194, 235, 244, 246, 256, 262, 271, 275; consumers 7, 29–31, 61, 111, 117, 121, 237, 245, 252, 262, 269, 274–276; peri-urban 232, 243, 245; population 4, 23–24, 26, 32, 55, 81, 102, 160, 213, 219, 254
urban growth 4, 55, 82–85, 271
urbanization 4–5, 8, 12, 16, 23–24, 26, 30, 32, 49, 53, 55, 75, 78, 81, 84, 86, 94, 126, 130, 187, 193, 195, 235, 244, 271, 273, 277

value chains 7, 31, 45, 106–108, 119–120, 157, 213, 239, 247–248, 256, 270, 274; domestic 8, 11, 27, 120, 261, 273–274; global 127, 182; grain 45–46
value creation 127, 243, 245, 248, 255
vegetables 26, 31–32, 49, 51, 53, 56, 62, 105, 108, 113, 116, 124–125, 127–132, 160, 162, 167, 219, 244, 246, 275
vegetation 25, 94, 155, 203, 205–206, 208–210
Venezuela 120
veterinary services 256, 260
Vietnam 12–14, 178–179, 181, 254
villages 62, 142, 145–146, 259
violence 66, 139–146, 275; *see also* conflict
vitamins 108, 124, 128, 260, 275
voucher systems 93, 120
vulnerability 23, 30, 54, 104, 141, 152–155, 157, 160, 253

wages 61–62, 68, 89, 111, 115–116, 177–183, 191, 193, 254; *see also* employment
war 56, 61–62, 76, 79, 100–101, 139–140, 142, 144, 146, 160, 252–253, 275; *see also* conflict
Washington Consensus 262, 273
water: availability 35, 49, 54–55, 92, 95, 142, 151–152, 160, 201–205, 208–209, 213–217, 220, 224, 254, 272; cycle 166, 203, 205–209, 214, 272; erosion 25; ground 55, 204; management 37, 51, 214, 220, 236; preservation 205, 210, 216, 239; productivity

213–214, 218, 220; quality 55, 96, 130, 220, 238–239; regulation 160–161; security 214; use 45, 50, 54–55, 60, 62, 91, 116, 125, 151–153, 205, 209, 213–220, 234
water security 201, 210, 217
wealth 78, 162, 164, 217, 233–235, 239, 267, 272, 278
weather 25, 60, 68, 101, 150, 153, 155, 157, 201, 225, 267; *see also* climate
weeds 94, 151, 216
welfare *see* living standards
welfare states 194–195
West Africa 9, 11, 23, 76, 78, 83, 106, 113, 115, 151, 161–164, 167–169, 205–206, 251–253
Western Sahara 85
wheat: consumption 26, 29, 102; import 11–12, 23–24, 29, 36, 51–53, 55, 100–102, 104, 107, 117, 273 (*see also* food import, cereals); irrigated 38, 45, 51; price 11, 53, 101, 111, 114, 117, 125, 132, 153; production 36, 38, 41, 44–45, 51, 53–55, 60, 63, 66, 102, 127, 151, 153, 273
wildlife 162, 164
wind 91, 206, 208, 223
winter 51, 63
women 39, 53, 55–57, 77–81, 86, 128–130, 142, 177–178, 180–181, 213, 239–240, 248
wood 142, 167–168, 201, 272; timber 160, 167
woodfuel, extraction of 161
woodlands 160, 162, 165, 201, 208, 272
work *see* employment, labour
work hours 176, 180, 183
World War II 116, 194

yams 92, 113
yields, cereal 4, 8–14, 28, 41, 49, 51, 66, 91–93, 102, 107–108, 126, 130, 156, 253
youth 4, 24, 75–76, 80, 108, 139–143, 179, 183, 239–240, 275, 277

Zambia 13–14, 62, 82, 85, 91, 95, 103, 128, 179
Zimbabwe 13–15, 59–69, 82, 84–85, 95, 128, 150, 155–157, 179, 270, 273

FSC
www.fsc.org
MIX
Papier | Fördert
gute Waldnutzung
FSC® C083411

Kolibri 360 GmbH
Ferdinand-Jühlke-Straße 7
99095 Erfurt, Deutschland
produktsicherheit@kolibri360.de